Workshop on High Symmetries in Physics 2008

Journal of Physics: Conference Series Volume 175

Madrid, Spain
6-8 November 2008

Editors:

R. Campoamor-Stursberg **J. M. Ancochea Bermúdez**
M. Castrillón López

ISBN: 978-1-61738-308-3

Printed from e-media with permission by:

Curran Associates, Inc.
57 Morehouse Lane
Red Hook, NY 12571
www.proceedings.com

Some format issues inherent in the e-media version may also appear in this print version.

Printed by Curran Associates, Inc. (2010)

Institute of Physics
Dirac House, Temple Back
Bristol BS1 6BE UK

Phone: 44 1 17 929 7481
Fax: 44 1 17 920 0979

techtracking@iop.org

TABLE OF CONTENTS

SHORT COURSES

Cohomology of Filippov Algebras and an Analogue of Whitehead's Lemma 1
J A De Azcárraga, J M Izquierdo
The Homogeneity Conjecture for Supergravity Backgrounds 25
José Miguel Figueroa-O'Farrill
Some Results on Cubic and Higher Order Extensions of the Poincaré Algebra 40
M Rausch De Traubenberg

PAPERS

(Super)integrability from Coalgebra Symmetry: Formalism and Applications 88
A Ballesteros, A Blasco, F J Herranz, F Musso, O Ragnisco
Black Hole Entropy in Loop Quantum Gravity: The Role of Internal Symmetries 114
J Fernando Barbero G
Vafa's Approach: F-Theory and GUTs 122
Luis J Boya
Superintegrable Hamiltonian Systems: An Algebraic Approach 129
J A Calzada, J Negro, M A Del Olmo
Inequivalent Sets of Commuting Missing Label Operators for the Nuclear Surfon Model 145
R Campoamor-Stursberg
A Geometric Approach to Higher-order Riccati Chain: Darboux Polynomials and Constants of the Motion 155
José F Cariñena, Partha Guha, Manuel F Rañada
Functional Description of $S^1 \times S^2$ and S^3 Gowdy Cosmologies 170
Daniel Gómez Vergel, Eduardo J S Villaseñor
Symmetries and Singularities in Hamiltonian Systems 178
Eva Miranda
Hamiltonian Systems Related to Invariant Metrics 194
Gabriela P Ovando
Symmetry Reduction and Superintegrable Hamiltonian Systems 204
M A Rodríguez, P Tempesta, P Winternitz
Quantum Hall Effect on Riemann Surfaces 212
Carlos Tejero Prieto
Tensor Product Decompositions for su(3) of an Irreducible Representation with Itself and with Its Conjugate 226
Maria S M Wesslén
Author Index

PREFACE

This volume of *Journal of Physics: Conference Series contains* the Proceedings of the Workshop on Higher Symmetries in Physics (WHSP), held at the Universidad Complutense of Madrid (UCM) on 6–8 November 2008. This meeting constituted one of the activities of the research group GEODISIM-920920 of the Universidad Complutense, through the research project CCG07/ESP-2922 of the UCM/CAM for the academic year 2008/2009.

The objective of this meeting was to provide a forum to facilitate the opportunity for interaction between specialists working in different fields of physics and mathematics, but who share a common interest in group theoretical, geometrical and symmetry methods applied to physical phenomena. This goal was achieved by means of lectures and technical presentations on different subjects, the only constraint being the current academic interest.

The multidisciplinary character of the meeting allowed an effective exchange of ideas between different topics having a symmetry background, like higher order and n-Lie algebras and their cohomology theories, supergravity backgrounds, the geometric approach to the Quantum Hall effect, integrable and superintegrable systems, loop quantum gravity, master symmetries, constants of motion, Gowdy cosmological models, new methods for the Kronecker product decomposition of multiplets, the internal labelling problem or recent developments concerning Grand Unified Theories.

The workshop consisted of three microcourses of three hours each and some plenary talks of one hour, as well as a small number of short communications. The Proceedings have been divided into two main sections, according to the structure of the meeting. The first one corresponds to the papers of the courses, which in addition to the material presented in the lectures also contain new and original results. The second part is devoted to the papers of the plenary talks and the remaining contributions. In some cases, the corresponding contributions are completely original, and expand or complement the topic presented at the workshop. Unfortunately, some of the speakers were not able to submit their contributions in time, for which reason they do not appear in these Proceedings. In addition to the contributions of the participants, other specialists in the field that could not attend the meeting, as well as some members of the scientific committee, were also invited by the Editors to submit their papers for this volume.

The main motivation for the three courses was to provide a short and updated introduction to to current research topics, as well as to provide an overview for the non-specialists. We briefly describe the principal results of these lectures.

The first course, given by Professor J A de Azcárraga (IFIC and University of Valencia) dealt with generalized Lie algebras and Filippov algebras. These structures, which enlarge naturally the notion of ordinary Lie algebras, have been shown to be of interest for the description of various physical phenomena, like the low energy dynamics of coincident M2-branes. The lecture presented an exhaustive review of the mathematical apparatus of these generalized structures, and recent developments on their cohomological properties were announced. The paper underlying these lectures expands the material covered during the course, and new results concerning the non-existence of central extensions and the cohomological rigidity of Filippov algebras are obtained.

Professor J M Figueroa-O'Farrill (Maxwell Institute, University of Edinburgh) presented recent results concerning the homogeneity conjecture for supergravity backgrounds. The procedure to associate a Lie (super)algebra to a spin manifold with additional geometrical constraints was explained, and the most important examples of supersymmetric supergravity backgrounds commented. The lecture focused on two main results: on one hand, the recovery of the compact forms of the simple algebras B4, F4 and E8 by means of geometrical techniques, and the construction of the Killing superalgebra of 11-dimensional supergravity backgrounds. Criteria ensuring local homogeneity were described in terms of supersymmetry.

Finally, the course of Professor M Rausch de Traubenberg (IPHC, Université de Strasbourg) reviewed the present status of higher order extensions of the Poincaré algebra. In this lecture, basing on some features of the Wess-Zumino model, additional algebraic structures are added in order to obtain a hierarchy of non-trivial extensions of the Poincaré algebra. Two different types, with interesting applications in the corresponding context, are presented. The first type corresponds to finite dimensional cubic extension in D-dimensional space-time. The latter induces a symmetry on generalized gauge fields, and the corresponding invariant Lagrangians are constructed explicitly. The remaining possibility is shown to be an infinite dimensional higher order extension inducing a symmetry that allows to connect relativistic anyons. This procedure presents some analogies with supersymmetry.

All papers published in this volume of *Journal of Physics: Conference Series contains* have been peer reviewed through processes administered by the proceedings Editors. Reviews were conducted by expert referees to the professional and scientific standards expected of a proceedings journal published by IOP Publishing. This meeting was possible thanks to the financial and infrastructural assistance of the following Spanish institutions and projects:

- Universidad Complutense de Madrid (UCM)
- Instituto de Matemática Interdisciplinar (I.M.I.) of the UCM
- The Geometry, Mechanics and Control Network (GMC)
- CCG07/ESP-2922 of the UCM/Comunidad Autónoma de Madrid

- MTM2005-00173 of the Ministerio de Educación y Ciencia (MEC)
- MTM2006-09152 of the Ministerio de Educación y Ciencia (MEC)
- Consolider-Ingenio 2010 "Programa de Investigación Intensiva sobre Mecánica Geométrica y Teoría de Control"

Finally, on behalf of the Organizing Committee, we would like to express our gratitude to the participants and assistants in the WHSP meeting for their presence and contributions, as well as to the members of the Scientific Committee for their help and outstanding efforts, with special mention to E Padrón from the Universidad de La Laguna and the GMC Network.

R Campoamor-Stursberg, M Castrillón López and J M Ancochea Bermúdez
Universidad Complutense de Madrid
Editors of the WHSP Proceedings

Workshop on Higher Symmetries in Physics IOP Publishing
Journal of Physics: Conference Series **175** (2009) 012001 doi:10.1088/1742-6596/175/1/012001

Cohomology of Filippov algebras and an analogue of Whitehead's lemma

J.A. de Azcárraga†and J. M. Izquierdo‡

† Dpto. de Física Teórica and IFIC (CSIC-UVEG), University of Valencia, 46100-Burjassot (Valencia), Spain

‡ Departamento de Física Teórica, Universidad de Valladolid, 47011-Valladolid, Spain

E-mail: j.a.de.azcarraga@ific.uv.es, izquierd@fta.uva.es

Abstract. We show that two cohomological properties of semisimple Lie algebras also hold for Filippov (n-Lie) algebras, namely, that semisimple n-Lie algebras do not admit non-trivial central extensions and that they are rigid *i.e.*, cannot be deformed in Gerstenhaber sense. This result is the analogue of Whitehead's Lemma for Filippov algebras. A few comments about the n-Leibniz algebras case are made at the end.

1. Introduction and outlook

In the last years there has been an increasing interest in the applications of various generalizations of the ordinary Lie algebra structure to theoretical physics problems. In these generalizations, which we shall denote generically as n-ary algebras, the two entries of the standard Lie bracket are replaced by $n > 2$ entries. There are two main ways of achieving this, depending on how the Jacobi identity (JI) of the ordinary Lie algebras is looked at. The JI can be viewed as the statement that (a) a double Lie bracket gives zero when antisymmetrized with respect to its three entries or that (b) the Lie bracket is a derivation of itself. Both (a) and (b) are equivalent for ordinary Lie algebras and (a) is indeed an identity that follows from the associativity of the composition of the Lie algebra generators.

When a n-ary algebra is defined using the *characteristic identity* that extends property (a) to a n-ary bracket, one is led to a generalization denoted *higher order Lie algebras* or *generalized Lie algebras* (GLA) $\mathcal{G}$ [1, 2], and the characteristic identity satisfied by its multibracket is called *generalized Jacobi identity* (GJI). This generalization is natural for n even (for n odd, the *r.h.s.* of the GJI, rather than being zero, is a larger bracket with $(2n-1)$ entries [3]). Similar algebras have also been discussed in [4–8]; GLAs may also be considered as a particular case (when there is no violation of the GJI [3]) of the strongly homotopy algebras of Stasheff [9–12]. When possibility (b) is used as the guiding principle, then one is led to the *Filippov identity* (FI) [13] and correspondingly to n-Lie or *Filippov algebras* $\mathfrak{G}$ [13] (both terms, Filippov and n-Lie, will be used indistinctly), for which the characteristic identity is the FI. For $n = 2$, both algebra structures coincide and determine ordinary Lie algebras $\mathfrak{g}$; when $n \geq 3$, the GJI (n even) and the FI become different characteristic identities and define, respectively, generalized Lie algebras[1]

[1] The GLA were called *Lie n-algebras* in [4], where they were independently considered (see also [5]). But, rather

(GLA) $\mathcal{G}$ and n-Lie or Filippov algebras (FA) $\mathfrak{G}$.

Filippov algebras [13–16] have recently been found useful in the search for an effective action describing the low energy dynamics of coincident M2-branes or, more specifically, in the Bagger-Lambert-Gustavsson model (BLG) [17–23]. The field theory BLG model contains scalar and fermion fields that take values in a 3-Lie algebra plus gauge fields that are valued in the adjoint representation of the Lie algebra of the automorphisms of the 3-Lie algebra; the model has $\mathcal{N}$=8 supersymmetries. The uniqueness of the euclidean 3-Lie algebra A_4 was in fact found in the context of the BLG model, where it follows [24, 25] by assuming that the metric needed for the BLG action has to be positive definite, a condition that may be relaxed [23] (see also [26] and references therein). We shall not discuss the BLG and related models here; we shall just mention that the original BGL action was subsequently reformulated [20] without using a three-Lie algebra, and that other models for low energy multiple M2 brane dynamics have appeared (albeit with $\mathcal{N}$=6 rather than $\mathcal{N}$=8 supersymmetries) that do not use a FA structure [27] (see footnote 7). This paper will be devoted instead to some purely mathematical aspects of Filippov algebras. Other specific ternary structures, such as Jordan-Okubo triple systems and others will not be discussed here (see [28–31] and references therein).

It is well known that semisimple Lie algebras neither admit non-trivial central extensions (see *e.g.* [35]) nor infinitesimal deformations [32, 33] so that they are rigid or stable. This is so because their cohomology groups $H_0^2(\mathfrak{g})$ and $H_\rho^2(\mathfrak{g})$ are trivial as a result of Whitehead's Lemma (see [34]; in fact, the p-th cohomology groups $H_\rho^p(\mathfrak{g}, V)$, where V is a $\rho(\mathfrak{g})$-module, are zero for ρ non-trivial[2] and $p \geq 0$). The standard proof uses that the Cartan-Killing metric for semisimple algebras can be inverted and then the inverse allows one to show that all p-cocycles are p-coboundaries. We prove in this paper that both the triviality of central extensions and the stability of semisimple Lie algebras under deformations also hold for semisimple Filippov algebras, so that these properties hold true for all $n \geq 2$. We shall show this by using a route that does not require introducing a generalized Cartan-Killing form for the FA. The reason is twofold. First, the Cartan-Killing bilinear form of a Lie algebra $\mathfrak{g}$ is defined on its vector space. In contrast, its n-Lie algebra analogue [15] (see eq. (2.6)), which one might think of using as the Lie algebra Cartan metric to mimic the proof there, is not a bilinear form on $\mathfrak{G}$ but on their fundamental objects[3] $\mathscr{X} \in \wedge^{n-1}\mathfrak{G}$ (Sec. 2 below). However, the semisimplicity criterion for a n-Lie algebra, which states that the $2(n-1)$-linear Killing form generalization k [15] in (2.6) is not degenerate (eq. (2.7) below) does not guarantee that k is non-degenerate as a bilinear form on $\wedge^{n-1}\mathfrak{G}$. Secondly, all simple $n > 2$ Filippov algebras are known [16, 13] and they are few when compared with the plethora of the $n = 2$ Cartan cassification of simple Lie algebras and, further, all of them have the same general structure. Specifically, the only simple real Filippov algebras are the $(n+1)$-dimensional n-Lie algebras of type A_{n+1} (eq. (3.16)) [16, 13], which may be thought of as $n > 2$ generalizations of the $n = 2$ $so(3)$ and $so(1,2)$ ordinary Lie algebras. We shall take advantage of this fact to show first the triviality of the central extensions and deformations of simple n-Lie algebras; then, using that any semisimple Filippov algebra is the direct sum of its simple ideals [16], we shall extend the result to semisimple n-algebras as well.

As already mentioned, the central extension and deformation problems for Lie algebras are formulated in terms of the second Lie algebra cohomology groups (see *e.g.* [35]) for the trivial

than betting the distinction between n-Lie algebras ($\equiv$ FA) and and GLAs on the precise location of a single letter (n-Lie alg. *vs.* Lie n-alg.), we prefer our *higher order Lie algebras* or GLA terminology.

[2] This is no longer true for $\rho = 0$; for a simple compact $\mathfrak{g}$, for instance, the fully antisymmetric structure constants of the Lie algebra always determine a non-trivial three-cocyle.

[3] There is still a parallel, however, if one realizes that in both the $n = 2$ and the arbitrary n cases the Killing metric is a trace form, namely $\mathrm{Tr}(ad_{\mathscr{X}} ad_{\mathscr{Y}})$. This relates it to the Lie algebra of inner derivations of n-Lie algebras ($n \geq 2$) in general.

and the adjoint action respectively. Non-trivial central extensions are characterized by non-trivial two-cocycles in the Lie algebra cohomology group $H_0^2(\mathfrak{g}) = Z_0^2(\mathfrak{g})/B_0^2(\mathfrak{g})$ for the trivial action, whereas non-trivial infinitesimal deformations require $H_\rho^2 \neq 0$ for $\rho = ad$. In the Filippov algebras case, the generalization is not immediate, and in fact we will show that the cocycles responsible for both extensions and deformations may be considered as one-cocycles rather than two-cocycles. The characterization of the cochains and the corresponding cohomology complexes will be given in Secs. 4.1 and 5.1. It will turn out that the cohomology complex adapted to the deformation problem obtained in Sec. 4.1 is essentially equivalent to the one introduced by Gautheron [36] (see also [37–39]), who was the first to consider the full deformation cohomology complex for Nambu algebras.

Nambu algebras are, in fact, a particular case of n-Lie algebras. Their n-bracket is provided by the Jacobian determinant of n functions or Nambu bracket [40], although Nambu did not write the characteristic identity satisfied by his ($n = 3$) bracket, which is none other than the FI. This was done in [13,41–44], and *Nambu-Poisson structures* (N-P) have been much studied since Nambu's original paper [40] and Takhtajan general study [43], see [45,42,46–49,37,50–53] (ref. [37] also considers Nambu superalgebras). In fact, since the earlier considerations of p-branes as gauge theories of volume preserving diffeomorphisms [54], the infinite dimensional FAs given by Nambu brackets have reappeared in applications to brane theory [55] and, in particular, in the Nambu three-bracket realization of the mentioned BLG model as a gauge theory associated with volume preserving diffeormorphisms in a three-dimensional space; see, in particular, [56–59].

Much in the same way the Nambu-Poisson structures follow the pattern of FAs, it is also possible to introduce *generalized Poisson structures* (GPS) [60,1,61] (see further [52,62,63]) whose n-even *generalized Poisson brackets* (GPB) satisfy the GJI and correspond to the GLAs earlier mentioned. This can also be achieved in the graded case, which corresponds to graded GLAs [64]. Note, however, that besides the two properties that each Poisson generalization share respectively with the GLAs and FAs (skewsymmetry of both n-ary Poisson brackets plus the GJI (FI) for the GP (N-P) structures, respectively), the n-ary brackets of both GPS and of N-PS satisfy an additional condition, Leibniz's rule. There has been an extensive discussion since the papers by Nambu [40] and Takhtajan [43] about the difficulties of quantizing the N-P strucures. We shall not touch the point of quantizing n-ary Poisson structures here and will just refer instead to the papers above and *e.g.*, to [55,61,65–67] and references therein.

The plan of this paper is as follows. Sec. 2 reviews some facts on FAs in a (we hope) transparent notation. In Sec. 3 we show that Kasymov's analogue [15] of the Cartan-Killing form for a FA $\mathfrak{G}$ (eq. (2.6) below), when viewed as a *bilinear* form on $\wedge^{n-1}\mathfrak{G}$, is degenerate when $\mathfrak{G}$ is semisimple. This is because the fundamental objects (Sec. 2) $\mathscr{X} \in \wedge^{n-1}\mathfrak{G}$ may involve elements of $\mathfrak{G}$ in different simple ideals (something that cannot happen in the Lie algebra case, where the fundamental objects reduce to single elements $X \in \mathfrak{g}$), although it is non-degenerate when $\mathfrak{G}$ is simple. In Sec 4.1 we derive the conditions for the existence and triviality of a central extension of a Filippov algebra, which allows us to define one-cocycle and one-coboundary conditions respectively. By extending them to higher order cochains, this leads us naturally to the expression of the corresponding cohomology complex, which is given explicitly in that section. Sec. 4.2 contains the proof of the triviality of all central extensions of semisimple Filippov algebras. In Sec. 5.1 we derive the cohomological conditions that govern the infinitesimal deformations of FAs, and subsequently we obtain from them the action of the coboundary operator and the cohomology complex for the non-trivial action, which is the relevant one for deformations of FAs; both the left and the right actions appear naturally in its definition. Sec. 5.2 contains the proof of the rigidity of the semisimple Filippov algebras. Sec. 6 is makes some observations relating the FA and n-Leibniz algebra cohomologies (n-Leibniz algebras share the derivation property of the FI with the FAs but not the total antisymmetry of their n-brackets). Finally, Sec. 7 presents some remarks concerning the extension of the above results to n-Leibniz algebras.

2. Filippov algebras: some basic definitions and properties

We present in this section some salient features of FA in a form that will be convenient for applications later.

A FA algebra $\mathfrak{G}$ is a vector space endowed with a skew-symmetric, n-linear bracket,

$$(X_1, \ldots, X_n) \in \mathfrak{G} \times \cdots \times \mathfrak{G} \mapsto [X_1, \ldots, X_n] \in \mathfrak{G} \tag{2.1}$$

that satisfies the FI,

$$[X_1, \ldots, X_{n-1}, [Y_1, \ldots Y_n]] = \sum_{a=1}^{n} [Y_1, \ldots Y_{a-1}, [X_1, \ldots, X_{n-1}, Y_a], Y_{a+1}, \ldots Y_n] \ , \tag{2.2}$$

which states that the bracket $[X_1, \ldots, X_{n-1}, \ \ ]$, where the last entry is empty, is a derivation of the FA.

Given a basis $\{X_{a_i}\}$ of the $\mathfrak{G}$ vector space, the FA is characterized by its structure constants,

$$[X_{a_1}, \ldots, X_{a_n}] = f_{a_1 \ldots a_n}^{b} X_b \ , \quad a_i = 1, \ldots, n \quad .$$

The properties of n-Lie algebras have been studied, along the lines of Lie algebra theory, by Filippov [13, 44], Kasymov [14, 15], Ling [16] and others. For instance, a subspace $I \subset \mathfrak{G}$ is an *ideal* of $\mathfrak{G}$ if

$$[X_1, \ldots, X_{n-1}, Z] \subset I \quad \forall X \in \mathfrak{G} \, , \forall Z \in I \quad .$$

The above bracket may be rewritten in the form

$$[X_1, \ldots, X_{n-1}, Z] := \mathscr{X} \cdot Z \equiv [\mathscr{X}, Z] \equiv ad_{\mathscr{X}} Z \ , \tag{2.3}$$

where[4] $\mathscr{X} \in \wedge^{n-1}\mathfrak{G}$. The objects $\mathscr{X} \in \wedge^{n-1}\mathfrak{G}$ play an important rôle in the theory of FA, and accordingly we shall call them *fundamental objects.* In fact, the properties of FA are largely determined by them; they also determine derivations that generate an associated Lie algebra. The fundamental objects are characterized by $(n-1)$ elements $(X_1, \ldots, X_{n-1})$ of $\mathfrak{G}$ and are skewsymmetric in them. In terms of these fundamental objects $\mathscr{X}$, the FI may be rewritten as:

$$\mathscr{X} \cdot [Y_1, \ldots, Y_n] = \sum_{a=1}^{n} [Y_1, \ldots, \mathscr{X} \cdot Y_a, \ldots, Y_n] \quad \text{or} \quad ad_{\mathscr{X}} [Y_1, \ldots, Y_n] = \sum_{a=1}^{n} [Y_1, \ldots, ad_{\mathscr{X}} Y_a, \ldots, Y_n] \, . \tag{2.4}$$

Thus, the FI just reflects that $ad_{\mathscr{X}} \equiv \mathscr{X} \cdot \equiv [X_1, \ldots, X_{n-1}, \ \ ]$ is a derivation of the FA (which may be called inner since it is determined by elements of $\mathfrak{G}$). For the particular case of an ordinary Lie algebra $\mathfrak{g}$, $n=2$, $\mathscr{X} = X$ and thus $ad_{\mathscr{X}} \in \mathrm{End}\, \mathfrak{G}$ reduces to the standard adjoint derivative $ad_X \in \mathrm{End}\, \mathfrak{g}$.

[4] The notation $\mathscr{X} \in \wedge^{n-1}\mathfrak{G}$ reflects that the fundamental object $\mathscr{X} = (X_1, \ldots, X_{n-1}) \in \mathfrak{G} \times \overset{n-1}{\cdots} \times \mathfrak{G}$ is antisymmetric in its arguments and does not imply that $\mathscr{X}$ is a $(n-1)$-multivector obtained by the associative wedge product of vector fields.

A FA is *simple* if $[\mathfrak{G},\dots,\mathfrak{G}] \neq \{0\}$ and has no ideals different from the trivial ones, $\{0\}$ and $\mathfrak{G}$. A n-Lie algebra is *semisimple* if it has no solvable ideals, an ideal $I \subset \mathfrak{G}$ being *solvable* [13] if the following sequence of ideals,

$$I^{(0)} := I \;, \quad I^{(1)} := [I^{(0)},\dots,I^{(0)}]\;,\dots,\; I^{(s)} := [I^{(s-1)},\dots,I^{(s-1)}]\;,\dots \tag{2.5}$$

ends *i.e.*, there exists an s for which $I^{(s)} = 0$[5]. Kasymov's generalization [15] of the Cartan criterion then states that a n-Lie algebra is semisimple iff the following $(2n-2)$-linear generalization of the Cartan-Killing form

$$k(\mathscr{X},\mathscr{Y}) = k(X_1,\dots,X_{n-1},Y_1,\dots,Y_{n-1}) := Tr(ad_{\mathscr{X}}ad_{\mathscr{Y}}) \tag{2.6}$$

is non-degenerate, *i.e.* iff

$$k(Z,\mathfrak{G},\overset{n-2}{\dots},\mathfrak{G},\mathfrak{G},\overset{n-1}{\dots},\mathfrak{G}) = 0 \;\Rightarrow\; Z = 0 \tag{2.7}$$

where the $2n-3$ arguments besides Z are arbitrary elements of $\mathfrak{G}$.

It is convenient to introduce a composition law for fundamental objects $\mathscr{X} = (X_1,\dots,X_{n-1})$, $\mathscr{Y} = (Y_1,\dots,Y_{n-1})$, $X_i, Y_i \in \mathfrak{G}$, $i = 1,\dots,(n-1)$. The *composition* $\mathscr{X}\cdot\mathscr{Y}$ is given by the sum of fundamental objects

$$\mathscr{X}\cdot\mathscr{Y} := \sum_{a=1}^{n-1}(Y_1,\dots,Y_{a-1},[X_1,\dots,X_{n-1},Y_a],Y_{a+1},\dots,Y_{n-1})\;. \tag{2.8}$$

For a $n=2$ FA or ordinary Lie algebra, $\mathscr{X}\cdot\mathscr{Y}$ reduces to $X\cdot Y = [X,Y]$. With the above notation, the following Lemma follows from the FI:

Lemma (*Properties of the composition of fundamental objects*)
The dot product of fundamental objects $\mathscr{X}$ of a n-Lie algebra $\mathfrak{G}$ satisfies the relation

$$\mathscr{X}\cdot(\mathscr{Y}\cdot\mathscr{Z}) - \mathscr{Y}\cdot(\mathscr{X}\cdot\mathscr{Z}) = (\mathscr{X}\cdot\mathscr{Y})\cdot\mathscr{Z} \qquad \forall \mathscr{X},\mathscr{Y},\mathscr{Z} \in \wedge^{n-1}\mathfrak{G}\;. \tag{2.9}$$

As a result, the images $ad_{\mathscr{X}}$ of the fundamental objects by the the adjoint map $ad : \wedge^{n-1}\mathfrak{G} \to \mathrm{InDer}\,\mathfrak{G}$ determine (inner) derivations of the FA that satisfy[6]

$$\begin{aligned} &\mathscr{X}\cdot(\mathscr{Y}\cdot Z) - \mathscr{Y}\cdot(\mathscr{X}\cdot Z) = (\mathscr{X}\cdot\mathscr{Y})\cdot Z \quad \text{or} \\ &ad_{\mathscr{X}}ad_{\mathscr{Y}}Z - ad_{\mathscr{Y}}ad_{\mathscr{X}}Z = ad_{\mathscr{X}\cdot\mathscr{Y}}Z \qquad \forall \mathscr{X},\mathscr{Y} \in \wedge^{n-1}\mathfrak{G}, \forall\, Z \in \mathfrak{G} \;. \end{aligned} \tag{2.10}$$

Proof: To prove the assertion, it is sufficient to check eq. (2.9). Let us compute first $\mathscr{X}\cdot(\mathscr{Y}\cdot\mathscr{Z})$. This is given by

$$\begin{aligned} \mathscr{X}\cdot(\mathscr{Y}\cdot\mathscr{Z}) &= \sum_{i=1}^{n-1}\mathscr{X}\cdot(Z_1,\dots,Z_{i-1},[Y_1,\dots,Y_{n-1},Z_i],Z_{i+1},\dots,Z_{i-1}) \\ &= \sum_{i=1}^{n-1}\sum_{j\neq i,j=1}^{n-1}(Z_1,\dots,Z_{j-1},[X_1,\dots,X_{n-1},Z_j],\;Z_{j+1},\dots,Z_{i-1},\;[Y_1,\dots,Y_{n-1},Z_i],Z_{i+1},\dots,Z_{n-1}) \\ &+\sum_{i=1}^{n-1}(Z_1,\dots,Z_{i-1},[X_1,\dots,X_{n-1},[Y_1,\dots,Y_{n-1},Z_i]],Z_{i+1},\dots,Z_{n-1})\;. \end{aligned}$$

[5] The solvability notion for Lie algebras allows for various generalizations when moving to FAs, $n>2$, because the n-bracket has more than two entries. For a n-Lie algebra the notion of k-solvability was introduced by Kasymov [14] (see also [16]) by taking $\mathfrak{G}^{(0,k)} = \mathfrak{G}$, $\mathfrak{G}^{(m,k)} = [\mathfrak{G}^{(m-1,k)},\dots,\mathfrak{G}^{(m-1,k)},\mathfrak{G},\dots,\mathfrak{G}]$, where there are k entries $\mathfrak{G}^{(m-1,k)}$ at the beginning of the n-bracket. Filippov's solvability [13], used above, corresponds to k-solvability for $k=n$; k-solvability is stronger and implies n-solvability for all k [16].

[6] Notice that Z in eq. (2.10) may be replaced in general by $v \in V$, $\mathscr{X}\cdot v := \rho(\mathscr{X})\cdot v$, where ρ is the action that makes the vector space V a $\rho(\mathfrak{G})$-module.

The first term in the $r.h.s$ is symmetric in $\mathscr{X}, \mathscr{Y}$; hence,

$$\begin{aligned}\mathscr{X}\cdot(\mathscr{Y}\cdot\mathscr{Z})-\mathscr{Y}\cdot(\mathscr{X}\cdot\mathscr{Z})=\sum_{j=1}^{n-1} & \Big(Z_1,\ldots,Z_{j-1},\{[X_1,\ldots,X_{n-1},[Y_1,\ldots,Y_{n-1},Z_j]] \\ & -[Y_1,\ldots,Y_{n-1},[X_1,\ldots,X_{n-1},Z_j]]\},Z_{j+1},\ldots,Z_{n-1}\Big).\end{aligned} \tag{2.11}$$

On the other hand, using definition (2.8), we find

$$(\mathscr{X}\cdot\mathscr{Y})\cdot\mathscr{Z}=\sum_{j=1}^{n-1}\sum_{i=1}^{n-1}(Z_1,\ldots,Z_{j-1},[Y_1,\ldots,[X_1,\ldots,X_{n-1},Y_i],\ldots,Y_{n-1},Z_j],Z_{j+1},\ldots,Z_{n-1}). \tag{2.12}$$

Now, using the FI for $[X_1,\ldots,X_{n-1},[Y_1,\ldots,Y_{n-1},Z_j]]$, we see that the above expression reproduces (2.11).

Obviously, the above proof carries forward for the simplest case where the fundamental object $\mathscr{Z}$ in (2.9) is replaced by a FA element Z as in (2.10). It is sufficient to note that, by the FI,

$$\begin{aligned} ad_{\mathscr{X}\cdot\mathscr{Y}}Z &= \sum_{i=1}^{n-1}[Y_1,\ldots,Y_{i-1},[X_1,\ldots,X_{n-1},Y_i],Y_{i+1},\ldots,Y_{n-1},Z] \\ &= [X_1,\ldots,X_{n-1},[Y_1,\ldots,Y_{n-1},Z]]-[Y_1,\ldots,Y_{n-1},[X_1,\ldots,X_{n-1},Z]] \\ &= ad_{\mathscr{X}}ad_{\mathscr{Y}}Z - ad_{\mathscr{Y}}ad_{\mathscr{X}}Z \quad , \end{aligned} \tag{2.13}$$

which completes the proof □

Note that eq. (2.10) shows, by exchanging $\mathscr{X}$ and $\mathscr{Y}$, that $ad_{\mathscr{X}\cdot\mathscr{Y}}Z = -ad_{\mathscr{Y}\cdot\mathscr{X}}Z$ on any $Z\in\mathfrak{G}$, and hence that

$$ad_{\mathscr{X}\cdot\mathscr{Y}} = -ad_{\mathscr{Y}\cdot\mathscr{X}} \qquad \text{or, equivalently,} \qquad (\mathscr{X}\cdot\mathscr{Y})\cdot = -(\mathscr{Y}\cdot\mathscr{X})\cdot \quad , \tag{2.14}$$

where the dots in the last expression should be noted, since the composition of fundamental objects in eq. (2.8) is *not* commutative, $\mathscr{X}\cdot\mathscr{Y}\neq-\mathscr{Y}\cdot\mathscr{X}$. It follows from eqs. (2.14), (2.10) that the inner derivations $ad_{\mathscr{X}}\in \mathrm{End}\,\mathfrak{G}$ of a FA constitute an ordinary Lie algebra; therefore, we have the following

Proposition Let $\mathfrak{G}$ a n-Lie algebra. The inner drivations $ad_{\mathscr{X}}$ associated to the fundamental objects $\mathscr{X}\in\wedge^{n-1}\mathfrak{G}$ determine an ordinary Lie algebra, the *Lie algebra associated with the FA*.

For instance, the Lie algebra associated to the A_4 euclidean algebra (eq. (3.16) below for n=3 and no signs) is $so(4)=so(3)\oplus so(3)$[7]; for the Lorentz case (one $\varepsilon=-1$), an equally simple calculation leads to $so(1,3)$.

[7] Thus, the simple euclidean A_4 determines $SO(4)$ as the gauge group of the A_4-based BLG model. The fact that $SO(4)$ is not semisimple was used [20] to reformulate the BLG action with no reference to A_4, with matter fields taking values in the 'bi-fundamental' representation of the gauge group $SU(2)\otimes SU(2)$, and with the original gauge field replaced by two $SU(2)$-gauge ones. The $\mathcal{N}$=6 model in [27], which describes the low energy limit of the dynamics of N M2 branes, is constructed with scalar and fermion fields taking values in the algebra of $U(N)\otimes U(N)$, with a double set of gauge fields taking values in the adjoint, and does not use a Filippov algebra structure. Its connection with a (non fully skewsymmetric, see Sec. 6) three-bracket structure was elucidated in [21].

3. On Kasymov's analogue of the Cartan-Killing form

To prove that the cohomology groups that govern central extensions and the infinitesimal deformations of semisimple n-Lie algebras are trivial in analogy to the Lie algebra case, it would be convinient to have that the form k defined in (2.6), viewed as a bilinear form on $\wedge^{n-1}\mathfrak{G}$

$$k : \wedge^{n-1}\mathfrak{G} \times \wedge^{n-1}\mathfrak{G} \longrightarrow \mathbb{K} ,$$

be nondegenerate *i.e.*, that $k(\mathscr{X},\mathscr{Y}) = 0 \quad \forall \mathscr{Y} \in \wedge^{n-1}\mathfrak{G} \Rightarrow \mathscr{X} = 0$. In this way one could try repeating for FAs the proof for Lie algebras. But we see immediately that for $n \geq 3$ this is not so. Any semisimple n-Lie algebra is the direct sum of its simple ideals [16],

$$\mathfrak{G} = \bigoplus_{\mathfrak{s}=1}^{k} \mathfrak{G}_{(\mathfrak{s})} = \mathfrak{G}_{(1)} \oplus \cdots \oplus \mathfrak{G}_{(k)} \ . \tag{3.15}$$

As a consequence, the n-bracket $[\ldots, X, \ldots, Y, \ldots] = 0$ whenever X and Y belong to different simple ideals. Then, if one considers, for instance, $\mathscr{X} = X_1 \wedge \cdots \wedge X_{n-2} \wedge Y$, with $X_1, \ldots, X_{n-2}$ and Y in different ideals, it follows that $ad_{\mathscr{X}} \in \mathrm{End}\,\mathfrak{G}$ is identically zero, and so is $k(\mathscr{X},\mathscr{Y})$ for any $\mathscr{Y}$ without $\mathscr{X}$ itself being zero. In contrast, Kasymov criterion for the semisimplicity of n-Lie algebras [15] establishes that a FA is semisimple iff the $2(n-1)$-linear generalization k is nondegenerate in the sense of eq. (2.7).

However, for *simple n-Lie algebras* the form k is nondegenerate on $\wedge^{n-1}\mathfrak{G}$. To show this, we make use of the fact that a *real* simple n-Lie algebra is one of the FA algebras given by [16, 13]

$$[\mathbf{e}_1 \ldots \hat{\mathbf{e}}_i \ldots \mathbf{e}_{n+1}] = (-1)^{i+1}\varepsilon_i \mathbf{e}_i \quad \text{or} \quad [\mathbf{e}_{i_1} \ldots \mathbf{e}_{i_n}] = (-1)^n \sum_{i=1}^{n+1} \varepsilon_i \epsilon_{i_1 \ldots i_n}{}^{i} \mathbf{e}_i \ , \tag{3.16}$$

where $\varepsilon_i = \pm 1$ (no sum over the i of the ε_i factors) just introduce signs[8] that affect the different terms of the sum in i and we have used Filippov's notation to denote the basis $\{\mathbf{e}_i\}$ of $\mathfrak{G}$. Now, the bilinear form k on $\wedge^{n-1}\mathfrak{G}$ is determined by its values on a basis, so taking $\mathscr{X} = (\mathbf{e}_{i_1}, \ldots, \mathbf{e}_{i_{n-1}})$ and $\mathscr{Y} = (\mathbf{e}_{j_1}, \ldots, \mathbf{e}_{j_{n-1}})$, and using eq. (2.3), the action of $ad_{\mathscr{X}} ad_{\mathscr{Y}}$ on a basis vector $\mathbf{e}_j$ is found to be

$$ad_{\mathscr{X}} ad_{\mathscr{Y}}\, \mathbf{e}_j = \sum_{l,s=1}^{n+1} \varepsilon_l \varepsilon_j \epsilon_{j_1 \ldots j_{n-1} j}{}^{l} \epsilon_{i_1 \ldots i_{n-1} l}{}^{s}\, \mathbf{e}_s \ , \tag{3.17}$$

from which we deduce that the trace of $ad_{\mathscr{X}} ad_{\mathscr{Y}}$ is given by

$$k(\mathscr{X},\mathscr{Y}) = \sum_{l,s=1}^{n+1} \varepsilon_l \varepsilon_s \epsilon_{j_1 \ldots j_{n-1} s}{}^{l} \epsilon_{i_1 \ldots i_{n-1} l}{}^{s} \ . \tag{3.18}$$

The matrix appearing on the *r.h.s.* of (3.18), seen as a matrix $k_{(i_1 \cdots {}_{n-1})(j_1 \cdots j_{n-1})}$ with indices $(i_1 \ldots i_{n-1})$ and $(j_1 \ldots j_{n-1})$ determined by the fundamental objects above, is clearly diagonal with non-zero elements on the diagonal, for given $(i_1 \ldots i_{n-1})$, the factor $\epsilon_{i_1 \ldots i_{n-1} l}{}^{s}$ fixes the remaining indices l and s (and $\varepsilon_l \varepsilon_s$) so that $(j_1, \ldots, j_{n-1})$ has to be a reordering of the $(i_1 \ldots i_{n-1})$ indices. For this reason the form k is diagonal with non-zero elements in it and hence non-degenerate.

[8] Note that we might equally well have used the $\epsilon_{i_1 \ldots i_n}{}^{i}$ without signs ε_i in the *r.h.s.* by taking $\epsilon_{i_1 \ldots i_n}{}^{i} = \eta^{ij} \epsilon_{i_1 \ldots i_n j}$ where $\epsilon_{1 \ldots n(n+1)} = +1$ and η is a $(n+1) \times (n+1)$ diagonal metric with $+1$ and -1 in the places indicated by the ε_i's. We shall keep nevertheless the customary ε_i factors above as in *e.g.* [13].

Journal of Physics: Conference Series **175** (2009) 012001 doi:10.1088/1742-6596/175/1/012001

4. Central extensions of n-Lie algebras

4.1. Cohomology and central extensions of Filippov algebras

Given a Filippov algebra $\mathfrak{G}$ with a n-bracket $[\ldots]$, we define a central extension $\widetilde{\mathfrak{G}}$ of $\mathfrak{G}$ by adding a new, central, generator Ξ and modifying the bracket as follows:

$$\begin{aligned} &[\tilde{X}_{a_1}, \ldots, \tilde{X}_{a_n}] := f^{b}_{a_1 \ldots a_n} \tilde{X}_b + \alpha^1(X_1, \ldots, X_n)\Xi \ , \\ &[\tilde{X}_1, \ldots, \tilde{X}_{n-1}, \Xi] = 0 \ , \end{aligned} \tag{4.19}$$

where the $f^{b}_{a_1 \ldots a_n}$ are the structure constants of the unextended algebra $\mathfrak{G}$. One may think of adding more than one central generator, but this will not be needed here for the discussion. Clearly, α^1 has to be a n-linear and fully skewsymmetric map, $\alpha^1 \in \wedge^{n-1}\mathfrak{G}^* \wedge \mathfrak{G}^*$, where $\mathfrak{G}^*$ is the dual of $\mathfrak{G}$; it will be identified with a *one*-cochain. Since the new bracket for the $\tilde{X}_i$ has to satisfy the FI, this gives a condition on α^1. When one of the vectors involved is Ξ, the FI is trivially satisfied; when Ξ is absent it follows that

$$\begin{aligned} &[\tilde{X}_1, \ldots, \tilde{X}_{n-1}, [\tilde{Y}_1, \ldots, \tilde{Y}_n]] = \\ &\qquad \sum_{a=1}^{n} [\tilde{Y}_1, \ldots, \tilde{Y}_{a-1}, [\tilde{X}_1, \ldots, \tilde{X}_{n-1}, \tilde{Y}_a], \tilde{Y}_{a+1}, \ldots, \tilde{Y}_n] \ . \end{aligned} \tag{4.20}$$

Using (4.19) and the FI for the original Filippov algebra, this implies

$$\begin{aligned} &\alpha^1(X_1, \ldots, X_{n-1}, [Y_1, \ldots, Y_n]) - \\ &\qquad \sum_{a=1}^{n} \alpha^1(Y_1, \ldots, Y_{a-1}, [X_1, \ldots, X_{n-1}, Y_a], Y_{a+1}, \ldots, Y_n) = 0 \ . \end{aligned} \tag{4.21}$$

This equation (with $Y_n = Z$), written as $(\delta\alpha^1)(\mathscr{X}, \mathscr{Y}, Z) = 0$, will provide below the condition that characterizes $\alpha^1 \in \wedge^{n-1}\mathfrak{G}^* \wedge \mathfrak{G}^*$, $\alpha^1 : \mathscr{X} \wedge Z \mapsto \alpha^1(\mathscr{X}, Z)$ as a one-cocycle (for the trivial action of $\mathfrak{G}$ on α). It is seen now why becomes natural to call α^1 an one-cochain (rather than a two-cochain, as it would be in the Lie algebra cohomology case) and why we made the split $\alpha^1 \in \wedge^{n-1}\mathfrak{G} \wedge \mathfrak{G}$ explicit rather than simply writing $\alpha^1 \in \wedge^n \mathfrak{G}$: the number of the fundamental objects in the arguments of a cochain determines its order. As we shall see shortly, an arbitrary p-cochain takes $p(n-1)+1$ arguments in $\mathfrak{G}$; a zero-cochain is an element of $\mathfrak{G}^*$.

Let us now construct the cohomology complex relevant for central extensions of FA. Since $\mathfrak{G}$ does not act on $\alpha^1(\mathscr{X}, Z)$, it will be the FA cohomology complex for the trivial action. We define arbitrary p-cochains as elements of $\wedge^{n-1}\mathfrak{G}^* \otimes \cdots \otimes \wedge^{n-1}\mathfrak{G}^* \wedge \mathfrak{G}^*$,

$$\alpha^p : (\mathscr{X}_1, \ldots, \mathscr{X}_p, Z) \mapsto \alpha^p(\mathscr{X}_1, \ldots, \mathscr{X}_p, Z) \ , \tag{4.22}$$

where $\mathscr{X}_1, \ldots, \mathscr{X}_p$ are p fundamental objects. Condition (4.21), which guarantees the consistency of α^1 in eq. (4.19) with the FI (4.20), reads then

$$(\delta\alpha^1)(\mathscr{X}, \mathscr{Y}, Z) = \alpha^1(\mathscr{X}, \mathscr{Y} \cdot Z) - \alpha^1(\mathscr{X} \cdot \mathscr{Y}, Z) - \alpha^1(\mathscr{Y}, \mathscr{X} \cdot Z) = 0 \ , \tag{4.23}$$

where $\mathscr{X} \cdot Z$ and $\mathscr{X} \cdot \mathscr{Y}$ were defined in eqs. (2.3) and (2.8). It is now straightforward to extend (4.23) to a whole cohomology complex; $\delta\alpha^p$ will be a $p+1$ cochain taking arguments on one more fundamental object than α^p. This is done by means of the following

Definition (*FA cohomology complex $(C^\bullet(\mathfrak{G}), \delta)$ adapted to central extensions*)

Let $\alpha^p \in \wedge^{n-1}\mathfrak{G}^* \otimes \cdots \otimes \wedge^{n-1}\mathfrak{G}^* \wedge \mathfrak{G}^*$ be a p-cochain on a FA. The action of the coboundary operator δ on arbitrary p-cochains ($\alpha^p \in C^p(\mathfrak{G})$) is given by (see [61])

$$\begin{aligned}(\delta\alpha)(\mathscr{X}_1, \ldots, \mathscr{X}_{p+1}, Z) = \\ \sum_{1\leq i<j}^{p+1} (-1)^i \alpha(\mathscr{X}_1, \ldots, \hat{\mathscr{X}_i}, \ldots, \mathscr{X}_i \cdot \mathscr{X}_j, \ldots, \mathscr{X}_{p+1}, Z) \\ + \sum_{i=1}^{p+1} (-1)^i \alpha(\mathscr{X}_1, \ldots, \hat{\mathscr{X}_i}, \ldots, \mathscr{X}_{p+1}, \mathscr{X}_i \cdot Z) \,.\end{aligned} \tag{4.24}$$

The proof that $\delta^2 = 0$ is analogous to that for the Lie algebra coboundary operator if we think of $\mathscr{X} \cdot \mathscr{Y}$ as a commutator, in which case eq. (2.9) plays the rôle of a Jacobi identity (we shall come back to this point in Sec. 6).

The p-th cohomology groups are given by $H_0^p(\mathfrak{G}) = Z_0^p(\mathfrak{G})/B_0^p(\mathfrak{G})$, where $Z_0^p(\mathfrak{G})$ is the group (for the natural addition of cochains) of the p-cocycles, $Z_0^p(\mathfrak{G}) = \{\alpha^p \in C^p(\mathfrak{G}) | \delta\alpha = 0\}$, and $B_0^p(\mathfrak{G})$ is the subgroup of the p-coboundaries, $B_0^p(\mathfrak{G}) = \{\alpha^p \in Z_0^p(\mathfrak{G}) | \alpha^p = \delta\alpha^{p-1},\ \alpha^{p-1} \in C^{p-1}(\mathfrak{G})\}$.

A central extension is actually trivial if it is possible to find new generators $\tilde{X}' \in \tilde{\mathfrak{G}}$ from the old ones,

$$\tilde{X}' = \tilde{X} - \beta(X)\Xi \,, \tag{4.25}$$

where $\beta \in \mathfrak{G}^*$ is a zero-cochain, such that they remove Ξ from the *r.h.s.* of eq. (4.19),

$$\begin{aligned}[\tilde{X}'_{a_1}, \ldots, \tilde{X}'_{a_n}] &= f_{a_1\ldots a_n}^b \tilde{X}'_b \\ &= f_{a_1\ldots a_n}^b \tilde{X}_b - \beta([X_{a_1}, \ldots, X_{a_n}])\Xi \,.\end{aligned} \tag{4.26}$$

Comparing the last term above with the original expression in eq. (4.19), we conclude that a central trivial extension is defined by a one-cochain of the form α^1 such that

$$\alpha^1(X_1, \ldots, X_n) = -\beta([X_1, \ldots, X_n]) \,. \tag{4.27}$$

This is tantamount to saying that the one-cocycle α^1 is actually the one-coboundary generated by the zero-cochain β, $\alpha^1(\mathscr{X}, Z) = (\delta\beta)(\mathscr{X}, Z)$, $(\delta\beta)(X_1, \ldots, X_{n-1}, Z) = -\beta([X_1, \ldots, X_{n-1}, Z])$, as it is read from (4.24). Clearly, equivalent extensions correspond to one-cocycles that differ in a coboundary. The different central extensions are thus characterized by the elements of $H_0^1(\mathfrak{G})$, and the trivial extension corresponds to the zero element of $H_0^1(\mathfrak{G})$. This of course recovers the well known Lie algebra cohomology result for $n = 2$: the second cohomology group $H_0^2(\mathfrak{g})$ for a Lie algebra $\mathfrak{g}$ becomes the first one H_0^1 when $\mathfrak{g}$ is viewed as a FA $\mathfrak{G}$, since for $n = 2$ the fundamental objects are single elements $\mathscr{X} = X$ of $\mathfrak{G} = \mathfrak{g}$.

4.2. Triviality of the central extensions of semisimple Filippov algebras

We now show that all the central extensions of a semisimple n-Lie algebra are trivial. To do so, we shall use the explicit form of the simple n-Lie algebras in eq. (3.16) [16, 13] to prove the statement in this case first. Then, the decomposition (3.15) will allow us to extend the result to all semisimple n-Lie algebras.

As a previous example, and since the reasonings below may be considered as a $n > 2$ generalization of the $so(3)$ and $so(1,2)$ Lie algebras, let us consider these $n = 2$ cases first. Their Lie algebra commutators may be jointly expressed as $[X_i, X_j] = \varepsilon_k \epsilon_{ijk} X_k,\ i,j,k = 1,2,3$. The values $\varepsilon_1 = \varepsilon_2 = \varepsilon_3 = 1$ determine $so(3)$,

$$[X_1, X_2] = X_3 \,,\ [X_2, X_3] = X_1 \,, \quad [X_3, X_1] = X_2 \,, \tag{4.28}$$

whereas $so(1,2)$ corresponds to, say, $\varepsilon_1 = \varepsilon_2 = 1$, $\varepsilon_3 = -1$,

$$[X_1, X_2] = -X_3 \ , \ [X_2, X_3] = X_1 \ , \ [X_3, X_1] = X_2 \ . \tag{4.29}$$

Since both $so(3)$ and $so(2,1)$ are simple, they do not have non-trivial central extensions by Whitehead's Lemma. This is easy to check directly. First, any two-cochain α^2 is given by its (skewsymmetric) coordinates $\alpha^2_{i_1 i_2} = \alpha^2(X_{i_1}, X_{i_2})$. But this is also a two-cocycle since it satisfies the two-cocycle condition $\alpha^2([X_{i_1}, X_{i_2}], X_{i_3}) - \alpha^2([X_{i_1}, X_{i_3}], X_{i_2}) + \alpha^2([X_{i_2}, X_{i_3}], X_{i_1}) = 0$ (to check this, it suffices to note that, since the antisymmetrization over four indices is zero, $\epsilon_{[i_1 i_2 l}\alpha^2_{i_3]l} \equiv 0$, which gives $\epsilon_{i_3 i_2 l}\alpha^2_{i_3 l} - \epsilon_{i_1 i_3 l}\alpha^2_{i_2 l} + \epsilon_{i_2 i_3 l}\alpha^2_{i_1 l} = 0$). But then α^2 is a two-coboundary, in fact the two-coboundary generated by the one-cochain β, $\alpha^2 = \delta\beta$, $(\delta\beta)(X_{i_1}, X_{i_2}) = -\beta([X_{i_1}, X_{i_2}]) = \varepsilon_l \epsilon_{i_1 i_2 l}\beta_l$ with $\beta_l = \beta(X_l)$ so that $\alpha^2_{ij} = -\varepsilon_l \epsilon_{i_1 i_2 l}\beta_l$. This may always be satisfied with $\beta_l = \varepsilon_l \frac{1}{2}\epsilon_{ij_l}\alpha^2_{ij}$. Note, however, that this type of argument cannot be extended to other simple algebras, since for all others $\dim \mathfrak{g}$ will be larger than three, and the structure constants will not be given in terms of the three-dimensional skewsymmetric tensor.

With this preliminary remark, ket us now move to the simple n-Lie algebra case.

Lemma: Any one-cochain of a simple n-Lie algebra is a one-coboundary (and thus a trivial one-cocycle).

Proof: Let $\alpha^1 \in \wedge^{n-1}\mathfrak{G}^* \wedge \mathfrak{G}^*$ be a one-cochain and $\mathfrak{G}$ simple. Given a basis $\{\mathbf{e}_i\}_{i=1}^{n+1}$ of $\mathfrak{G}$, α^1 is determined by its coordinates, $\alpha^1_{i_1 \ldots i_n} = \alpha^1(\mathbf{e}_{i_1}, \ldots, \mathbf{e}_{i_n})$. We now show that, in fact, a one-cochain on a simple $\mathfrak{G}$ is a one-coboundary *i.e.*, that there exists a $\beta \in \mathfrak{G}^*$ such that

$$\alpha^1_{i_1 \ldots i_n} = -\beta([\mathbf{e}_{i_1} \ldots \mathbf{e}_{i_n}]) = -\sum_{k=1}^{n+1} \varepsilon_k \epsilon_{i_1 \ldots i_n}{}^k \beta_k \ , \tag{4.30}$$

where $\beta_k = \beta(\mathbf{e}_k)$. Indeed, given α^1, the zero-cochain β given by

$$\beta_k = -\frac{\varepsilon_k}{n!} \sum_{i_1 \ldots i_n = 1}^{n+1} \epsilon^{i_1 \ldots i_n}{}_k \alpha^1_{i_1 \ldots i_n} \tag{4.31}$$

has the desired property (4.30):

$$\begin{aligned} -\beta([\mathbf{e}_{i_1} \ldots \mathbf{e}_{i_n}]) &= -\sum_{k=1}^{n+1} \epsilon_{i_1 \ldots i_n}{}^k \varepsilon_k \beta_k \\ &= \sum_{k=1}^{n+1} \epsilon_{i_1 \ldots i_n}{}^k \frac{\varepsilon_k^2}{n!} \sum_{j_1 \ldots j_n = 1}^{n+1} \epsilon^{j_1 \ldots j_n}{}_k \alpha^1_{j_1 \ldots j_n} \\ &= \frac{1}{n!} \sum_{j_1 \ldots j_n = 1}^{n+1} \epsilon^{j_1 \ldots j_n}_{i_1 \ldots i_n} \alpha^1_{j_1 \ldots j_n} = \alpha^1_{i_1 \ldots i_n} \ , \end{aligned} \tag{4.32}$$

which proves the lemma □

Let now $\mathfrak{G}$ be a semisimple FA, and (3.15) the splitting in its simple components. First, we establish the following simple

Lemma: Let $\alpha^1 \in \wedge^{n-1}\mathfrak{G}^* \wedge \mathfrak{G}^*$ be a one-cocycle on a semisimple n-Lie algebra for the n-Lie algebra trivial action cohomology defined by eq. (4.24). Then, $\alpha(X_1, \ldots, X_{n-2}, Y, Z) = 0$ if Y and Z belong to different ideals.

Proof: Let the different simple ideals be labelled by small gothic letters $\mathfrak{s}$, $\mathfrak{t}$, etc. Let $Z \in \mathfrak{G}_{(\mathfrak{s})}$ and $Y \in \mathfrak{G}_{(\mathfrak{t})}$, $\mathfrak{s} \neq \mathfrak{t}$. Since $\mathfrak{G}_{(\mathfrak{s})}$ is simple, there exist $Z_1, \ldots, Z_n \in \mathfrak{G}_{(\mathfrak{s})}$ such that $[Z_1, \ldots, Z_n] = Z$. We can now use this fact and the cocycle condition in eq. (4.21) to obtain

$$\begin{aligned} \alpha(X_1, \ldots, X_{n-2}, Y, Z) &= \alpha^1(X_1, \ldots, X_{n-2}, Y, [Z_1, \ldots, Z_n]) \\ &= \sum_{k=1}^{n} \alpha^1(Z_1, \ldots [X_1, \ldots, X_{n-2}, Y, Z_k], \ldots, Z_n) = 0 \end{aligned} \tag{4.33}$$

because $[X_1, \ldots, X_{n-2}, Y, Z] = 0$ since $Y \in \mathfrak{G}_{(\mathfrak{t})}$ and $Z \in \mathfrak{G}(\mathfrak{s})$ and $\mathfrak{s} \neq \mathfrak{t}$ □

Using this lemma, we can now prove the main result of this section:

Theorem All central extensions of semisimple n-Lie algebras are trivial.

Proof: Let $\alpha^1 \in \wedge^{n-1}\mathfrak{G}^* \wedge \mathfrak{G}^*$ be a one-cocycle in the FA cohomology for the trivial action, and let $\mathfrak{G}$ be semisimple. Then the theorem follows if α^1 is a one-coboundary.

All $X \in \mathfrak{G}$ can be split in a unique way in the form $X = \sum_{\mathfrak{s}=1}^{k} X_{(\mathfrak{s})}$, where $X_{(\mathfrak{s})}$ is the component of X in the simple ideal labelled by $\mathfrak{s}$. Then we have

$$\alpha^1(X_1, \ldots, X_n) = \sum_{\mathfrak{s}_1 \ldots \mathfrak{s}_n = 1}^{k} \alpha^1(X_{1(\mathfrak{s}_1)}, \ldots, X_{n(\mathfrak{s}_n)}) = \sum_{\mathfrak{s}=1}^{k} \alpha^1(X_{1(\mathfrak{s})}, \ldots, X_{n(\mathfrak{s})}) , \tag{4.34}$$

where the last equality is due to the previous lemma. Every term in the above expression defines a cochain on the simple n-Lie algebra $\mathfrak{G}_{(\mathfrak{s})}$, so they are coboundaries *i.e.* there exist $\beta_{(\mathfrak{s})} \in \mathfrak{G}^*_{(\mathfrak{s})}$ such that $\alpha^1(X_{1(\mathfrak{s})}, \ldots, X_{n(\mathfrak{s})}) = -\beta_{(\mathfrak{s})}([X_{1(\mathfrak{s})}, \ldots, X_{n(\mathfrak{s})}])$. This means that there is a zero-cochain $\beta \in \mathfrak{G}^*$,

$$\beta(X) = \sum_{\mathfrak{s}=1}^{k} \beta_{(\mathfrak{s})}(X_{(\mathfrak{s})}) , \tag{4.35}$$

that generates α^1:

$$\begin{aligned} -\beta([X_1, \ldots, X_n]) &= -\beta\left(\sum_{(\mathfrak{s}_1) \ldots (\mathfrak{s}_n) = 1}^{k} [X_{1(\mathfrak{s}_1)}, \ldots, X_{n(\mathfrak{s}_n)}] \right) \\ &= -\beta\left(\sum_{(\mathfrak{s})=1}^{k} [X_{1(\mathfrak{s})}, \ldots, X_{n(\mathfrak{s})}] \right) = -\sum_{\mathfrak{s}=1}^{k} \beta_{(\mathfrak{s})}([X_{1(\mathfrak{s})}, \ldots, X_{n(\mathfrak{s})}]) \\ &= \sum_{\mathfrak{s}=1}^{k} \alpha^1(X_{1(\mathfrak{s})}, \ldots, X_{n(\mathfrak{s})}) = \alpha^1(X_1, \ldots, X_n) , \end{aligned} \tag{4.36}$$

which concludes the proof □

5. Infinitesimal deformations of n-Lie algebras

5.1. Cohomology complex adapted to deformations of Filippov algebras

In contrast with algebra extensions, which add one or more generators, deformations [32, 33] do not increase the dimension of the algebra. The infinitesimally deformed n-bracket $[\ldots]_t$ may be written in terms of the original one $[\ldots]$ as follows:

$$[X_1, \ldots, X_n]_t = [X_1, \ldots, X_n] + t\alpha^1(X_1, \ldots, X_n) , \tag{5.37}$$

where now α^1 is $\mathfrak{G}$-valued (the added term must belong to $\mathfrak{G}$), and t is the parameter of the infinitesimal deformation [32, 33]. The one-cocycle condition for the deformation problem appears when the deformed n-bracket in (5.37) is made to satisfy the FI so that the deformed bracket does define a FA at order t

$$[X_1, \ldots, X_{n-1}, [Y_1, \ldots, Y_n]_t]_t \\ = \sum_{a=1}^{n} [Y_1, \ldots, Y_{a-1}, [X_1, \ldots, X_{n-1}, Y_a]_t, Y_{a+1}, \ldots, Y_n]_t \,. \tag{5.38}$$

In terms of fundamental objects (and setting $Y_n = Z$) the above condition reads

$$[\mathscr{X}, (\mathscr{Y} \cdot Z)_t]_t = [(\mathscr{X} \cdot \mathscr{Y})_t, Z]_t + [\mathscr{Y}, (\mathscr{X} \cdot Z)_t]_t \tag{5.39}$$

(*cf* eq. (2.10)). Eq. (5.38) implies, keeping only terms linear in t,

$$[X_1, \ldots, X_{n-1}, \alpha^1(Y_1, \ldots, Y_n)] + \alpha^1(X_1, \ldots, X_{n-1}, [Y_1, \ldots, Y_n]) \\ = \sum_{a=1}^{n} [Y_1, \ldots, Y_{a-1}, \alpha^1(X_1, \ldots, X_{n-1}, Y_a), Y_{a+1}, \ldots, Y_n] \\ + \sum_{a=1}^{n} \alpha^1(Y_1, \ldots, Y_{a-1}, [X_1, \ldots, X_{n-1}, Y_a], Y_{a+1}, \ldots, Y_n) \,. \tag{5.40}$$

This expression may be read as the one-cocycle condition $\delta\alpha^1 = 0$ for the $\mathfrak{G}$-valued cochain α^1. In terms of the fundamental objects it may be written, setting again $Y_n = Z$, as

$$(\delta\alpha^1)(\mathscr{X}, \mathscr{Y}, Z) = ad_{\mathscr{X}}\alpha^1(\mathscr{Y}, Z) - ad_{\mathscr{Y}}\alpha^1(\mathscr{X}, Z) - (\alpha^1(\mathscr{X}, \quad) \cdot \mathscr{Y}) \cdot Z \\ -\alpha^1(\mathscr{X} \cdot \mathscr{Y}, Z) - \alpha^1(\mathscr{Y}, \mathscr{X} \cdot Z) + \alpha^1(\mathscr{X}, \mathscr{Y} \cdot Z) = 0 \quad , \tag{5.41}$$

where, for instance for $n = 3$, the term $\alpha^1(\mathscr{X}, \quad) \cdot \mathscr{Y}$ above is the fundamental object defined by

$$\alpha^1(\mathscr{X}, \quad) \cdot \mathscr{Y} := (\alpha^1(\mathscr{X}, \quad) \cdot Y_1, Y_2) + (Y_1, \alpha^1(\mathscr{X}, \quad) \cdot Y_2) \\ = (\alpha^1(\mathscr{X}, Y_1), Y_2) + (Y_1, \alpha^1(\mathscr{X}, Y_2)) \quad , \quad [\alpha^1(\mathscr{X}, \quad) \cdot Y_i := \alpha^1(\mathscr{X}, Y_i)] \quad . \tag{5.42}$$

The general action of the coboundary operator on an arbitrary p-cochain will be given in eq. (5.47) below; we notice at this stage that expression (5.41) involves *both* the left and right actions of $\mathfrak{G}$ on α^1.

An infinitesimal deformation is trivial if there exists a redefinition $X'_i = X_i - t\beta(X_i)$, for some $\mathfrak{G}$-valued zero-cochain β that removes the deformating term in eq. (5.37). If so, the first order deformed bracket in terms of the new, primed generators reads

$$\begin{aligned} [X'_1, \ldots, X'_n]_t &= [X_1, \ldots, X_n]' \\ &\equiv [X_1, \ldots, X_n] - t\beta([X_1, \ldots, X_n]) \,. \end{aligned} \tag{5.43}$$

But, again keeping terms up to order t only, we find that the *l.h.s.* above gives

$$\begin{aligned} [X'_1, \ldots, X'_n]_t &= [X_1, \ldots, X_n]_t \\ &\quad -t\sum_{a=1}^{n} [X_1, \ldots, X_{a-1}, \beta(X_a), X_{a+1}, \ldots, X_n]_t \\ &= [X_1, \ldots, X_n] + t\alpha^1(X_1, \ldots, X_n) \\ &\quad -t\sum_{a=1}^{n} [X_1, \ldots, X_{a-1}, \beta(X_a), X_{a+1}, \ldots, X_n] \,. \end{aligned} \tag{5.44}$$

Therefore, the infinitesimal deformation given by the $\mathfrak{G}$-valued one-cocycle α^1 is trivial if there exists a $\mathfrak{G}$-valued zero-cochain β such that $\alpha^1 = \delta\beta$ where

$$(\delta\beta)(X_1,\ldots,X_n) := -\beta([X_1,\ldots,X_n]) + \sum_{a=1}^{n}[X_1,\ldots,X_{a-1},\beta(X_a),X_{a+1},\ldots,X_n] \ . \tag{5.45}$$

This is the expression that defines the one-coboundary generated by β in the FA cohomology induced by the deformation problem (eq. (5.47) below for $\alpha^0 = \beta$).

The expression of the one-coboundary may again be formulated in terms of the fundamental objects $\mathscr{X}$, as in the central extensions case, which will allow us to generalize the action of the coboundary operator δ on an arbitary cochain $\alpha^p \in C^p(\mathfrak{G},\mathfrak{G})$. Explicitly we find, relabelling the zero-cochain β as α^0,

$$(\delta\alpha^0)(\mathscr{X},Z) = \mathscr{X}\cdot\alpha^0(Z) - \alpha^0(\mathscr{X}\cdot Z) + (\alpha^0(\quad)\cdot\mathscr{X})\cdot Z \ . \tag{5.46}$$

As for Lie algebras, it is the characteristic identity, here the FI, that is responsible for the structure of the whole cohomology complex. This leads to a generalization of the Lie algebra cohomology relative to the adjoint action, and it is given by the following

Definition (*Cohomology complex* $(C^\bullet_{ad}(\mathfrak{G},\mathfrak{G}),\delta)$ *for deformations of FA*)

Let $\alpha^p \in C^p(\mathfrak{G},\mathfrak{G})$ a $\mathfrak{G}$-valued p-cochain, $\alpha^p : \wedge^{(n-1)}\mathfrak{G}\otimes\overset{p}{\cdots}\otimes\wedge^{(n-1)}\mathfrak{G}\wedge\mathfrak{G}\to\mathfrak{G}$. The action of the coboundary operator is given by

$$\begin{aligned}(\delta\alpha^p)&(\mathscr{X}_1,\ldots,\mathscr{X}_p,\mathscr{X}_{p+1},Z) = \\ &\sum_{1\le j<k}^{p+1}(-1)^j\alpha^p(\mathscr{X}_1,\ldots,\widehat{\mathscr{X}_j},\ldots,\mathscr{X}_{k-1},\mathscr{X}_j\cdot\mathscr{X}_k,\mathscr{X}_{k+1},\ldots,\mathscr{X}_{p+1},Z) \\ &+\sum_{j=1}^{p+1}(-1)^j\alpha^p(\mathscr{X}_1,\ldots,\widehat{\mathscr{X}_j},\ldots,\mathscr{X}_{p+1},\mathscr{X}_j\cdot Z) \\ &+\sum_{j=1}^{p+1}(-1)^{j+1}\mathscr{X}_j\cdot\alpha^p(\mathscr{X}_1,\ldots,\widehat{\mathscr{X}_j}\ldots,\mathscr{X}_{p+1},Z) \\ &+(-1)^p(\alpha^p(\mathscr{X}_1,\ldots,\mathscr{X}_p,\quad)\cdot\mathscr{X}_{p+1})\cdot Z\end{aligned} \tag{5.47}$$

where, in the last term,

$$\alpha^p(\mathscr{X}_1,\ldots,\mathscr{X}_p,\quad)\cdot\mathscr{Y} = \sum_{i=1}^{n-1}(Y_1,\ldots,\alpha^p(\mathscr{X}_1,\ldots,\mathscr{X}_p,Y_i),\ldots,Y_{n-1}) \ . \tag{5.48}$$

The above equations define $(C^\bullet_{ad}(\mathfrak{G},\mathfrak{G}),\delta)$ as induced by the deformation problem. It is seen above that both left and right '*ad*' actions enter into the definition of the coboundary operator. The above cohomological complex may be seen essentially equivalent to the one adapted to deformations introduced by Gautheron [36] for the Nambu-Poisson algebras; see also [37, 39, 38]. Since $H^p_{ad}(\mathfrak{G},\mathfrak{G}) = Z^p_{ad}(\mathfrak{G},\mathfrak{G})/B^p_{ad}(\mathfrak{G},\mathfrak{G})$ it follows that the infinitesimal deformations of Filippov algebras are governed by $H^1_{ad}(\mathfrak{G},\mathfrak{G})$.

It is not difficult to check that if one moves to the next order in the deformation of $\mathfrak{G}$ by adding $t^2\alpha^{(2)}(\mathscr{X},X_n)$ to the *r.h.s.* of eq. (5.37), where the superindex two in $\alpha^{(2)}$ refers to the order of the deformation, the FI imposes on the one-cochain $\alpha^{(2)}$ a condition of the form $\gamma(\mathscr{X},\mathscr{Y},Z) = (\delta\alpha^{(2)})(\mathscr{X},\mathscr{Y},Z)$, where $\gamma(\mathscr{X},\mathscr{Y},Z)$ is a two-cocycle which is expressed in terms of α^1. Therefore, if $H^2_{ad}(\mathfrak{G},\mathfrak{G}) \neq 0$, there is an obstruction that prevents extending the infinitesimal deformation to higher orders. As a result, we have the following

Theorem (*Deformations of FA algebras*)

Let $\mathfrak{G}$ be a FA. The first cohomology group $H^1_{ad}(\mathfrak{G},\mathfrak{G})$ for the above complex governs the infinitesimal deformations of $\mathfrak{G}$. The triviality of this cohomology group, $H^1_{ad} = 0$, is a sufficient condition for the rigidity of a FA. The obstruction to expanding an infinitesimal deformation is given by the non-vanishing of the second cohomology group, $H^2_{ad}(\mathfrak{G},\mathfrak{G}) \neq 0$.

Proof: Contained above □

For $n = 2$, and reverting to the notation that characterizes the order of cochains and cocycles by the number of their Lie algebra arguments, the above theorem recovers the standard result [32,33] for Lie algebras: the infinitesimal deformations are characterized by $H^2_{ad}(\mathfrak{g},\mathfrak{g})$ and the obstructions to go to higher orders are in the non-vanishing elements of third cohomology group $H^3_{ad}(\mathfrak{g},\mathfrak{g})$.

5.2. Rigidity of semisimple Filippov algebras

In this section we prove that semisimple n-Lie algebras are also rigid. The pattern of the proof is similar to the case of the central extensions: first we present the proof for simple FAs and then extend it to the semisimple n-Lie algebras. In contrast with the central extensions problem, it is not true now that every one-cochain of a simple n-Lie algebra is a coboundary, so it is necessary to characterize the cocycles for simple n-Lie algebras, eq. (3.16). Taking a basis of $\mathfrak{G}$ $\{\mathbf{e}_i\}_{i=1}^{n+1}$, a one-cochain is defined by

$$\alpha^1(\mathbf{e}_{i_1},\ldots,\mathbf{e}_{i_n}) = \sum_{j=1}^{n+1} (\alpha^1)^j{}_{i_1\ldots i_n}\mathbf{e}_j \ , \qquad (5.49)$$

in terms of its coordinates $(\alpha^1)^j{}_{i_1\ldots i_n}$. It will turn out convenient to define new, dual quantities that are easier to manipulate:

$$(\alpha^1)^{ji} := \frac{1}{n!}\sum_{i_1\ldots i_n=1}^{n+1} \epsilon^{i_1\ldots i_n i}(\alpha^1)^j{}_{i_1\ldots i_n} \ , \quad (\alpha^1)^j{}_{i_1\ldots i_n} = \sum_{i=1}^{n+1}\epsilon_{i_1\ldots i_n i}(\alpha^1)^{ji} \ . \qquad (5.50)$$

Using them we now prove the following

Proposition Let α^1 be a $\mathfrak{G}$-valued one-cochain of a simple n-Lie algebra with coordinates $(\alpha^1)^j{}_{i_1\ldots i_n}$. Then, α^1 is a one-cocycle for the above cohomology complex *iff* $(\alpha^1)^{ji}$ is symmetric, $(\alpha^1)^{ji} = (\alpha^1)^{ij}$.

Proof: First, we write the one-cocycle condition (5.40) in terms of basis elements of $\mathfrak{G}$,

$$\begin{aligned} 0 \ = \ & [\mathbf{e}_{i_1},\ldots,\mathbf{e}_{i_{n-1}},\alpha^1(\mathbf{e}_{j_1},\ldots,\mathbf{e}_{j_{n-1}},\mathbf{e}_k)] \\ & +\alpha^1(\mathbf{e}_{i_1},\ldots,\mathbf{e}_{i_{n-1}},[\mathbf{e}_{j_1},\ldots,\mathbf{e}_{j_{n-1}},\mathbf{e}_k]) \\ & -\sum_{a=1}^{n-1}\alpha^1(\mathbf{e}_{j_1},\ldots,\mathbf{e}_{j_{a-1}},[\mathbf{e}_{i_1},\ldots,\mathbf{e}_{i_{n-1}},\mathbf{e}_{j_a}],\mathbf{e}_{j_{a+1}},\ldots,\mathbf{e}_{j_{n-1}},\mathbf{e}_k) \\ & -\alpha^1(\mathbf{e}_{j_1},\ldots,\mathbf{e}_{j_{n-1}},[\mathbf{e}_{i_1},\ldots,\mathbf{e}_{i_{n-1}},\mathbf{e}_k]) \\ & -\sum_{a=1}^{n-1}[\mathbf{e}_{j_1},\ldots,\mathbf{e}_{j_{a-1}},\alpha^1(\mathbf{e}_{i_1},\ldots,\mathbf{e}_{i_{n-1}},\mathbf{e}_{j_a}),\mathbf{e}_{j_{a+1}},\ldots,\mathbf{e}_{j_{n-1}},\mathbf{e}_k] \\ & -[\mathbf{e}_{j_1},\ldots,\mathbf{e}_{j_{n-1}},\alpha^1(\mathbf{e}_{i_1},\ldots,\mathbf{e}_{i_{n-1}},\mathbf{e}_k)] \ . \end{aligned} \qquad (5.51)$$

Using the commutation relations (3.16) of the simple algebra and that

$$\alpha^1(\mathbf{e}_{i_1},\ldots,\mathbf{e}_{i_{n-1}},\mathbf{e}_k)=\sum_{l=1}^{n+1}\mathbf{e}_l(\alpha^1)^l{}_{i_1\ldots i_{n-1}k}=\sum_{r,s=1}^{n+1}\mathbf{e}_r\,\epsilon_{i_1\ldots i_{n-1}ks}(\alpha^1)^{rs} \tag{5.52}$$

by eq. (5.50), eq. (5.51) gives

$$\begin{aligned}0 \;=\; \sum_{s,l=1}^{n+1}\Bigg(& \varepsilon_r\epsilon_{i_1\ldots i_{n-1}s}{}^r\epsilon_{j_1\ldots j_{n-1}kl}(\alpha^1)^{sl}\\ & +\varepsilon_s\epsilon_{i_1\ldots i_{n-1}sl}\epsilon_{j_1\ldots j_{n-1}k}{}^s(\alpha^1)^{rl}\\ & -\sum_{a=1}^{n-1}\varepsilon_s\epsilon_{i_1\ldots i_{n-1}j_a}{}^s\epsilon_{j_1\ldots j_{a-1}sj_{a+1}\ldots j_{n-1}kl}(\alpha^1)^{rl}\\ & -\varepsilon_s\epsilon_{j_1\ldots j_{n-1}sl}\epsilon_{i_1\ldots i_{n-1}k}{}^s(\alpha^1)^{rl}\\ & -\sum_{a=1}^{n-1}\varepsilon_r\epsilon_{j_1\ldots j_{a-1}sj_{a+1}\ldots j_{n-1}k}{}^r\epsilon_{i_1\ldots i_{n-1}j_al}(\alpha^1)^{sl}\\ & -\varepsilon_r\epsilon_{j_1\ldots j_{n-1}s}{}^r\epsilon_{i_1\ldots i_{n-1}kl}(\alpha^1)^{sl}\Bigg)\,. \end{aligned} \tag{5.53}$$

Since this expression translates the condition that $\delta\alpha^1(\mathscr{X},\mathscr{Y},\mathbf{e}_k)=0$, where $\mathscr{X}\in\wedge^{n-1}\mathfrak{G}$, $\mathscr{Y}\in\wedge^{n-1}\mathfrak{G}$ with $\mathscr{X}=(\mathbf{e}_{i_1},\ldots,\mathbf{e}_{i_{n-1}})$, $\mathscr{Y}=(\mathbf{e}_{j_1},\ldots,\mathbf{e}_{j_{n-1}})$, it follows that it must be antisymmetric in the indices i and also in the indices j and, indeed, it may be checked explicitly in eq. (5.53) that this is so. Using the antisymmetry in $i_1,\ldots,i_{n-1}$ and in $j_1,\ldots,j_{n-1}$, an equivalent expression is obtained by contracting with $\epsilon^{i_1\ldots i_{n+1}}$ and $\epsilon^{j_1\ldots j_{n+1}}$,

$$\sum_{s,l=1}^{n+1}(\epsilon_{sr}^{i_ni_{n+1}}\epsilon_{kl}^{j_nj_{n+1}}-\epsilon_{lr}^{i_ni_{n+1}}\epsilon_{ks}^{j_nj_{n+1}}-\epsilon_{sl}^{i_ni_{n+1}}\epsilon_{kr}^{j_nj_{n+1}})(\alpha^1)^{sl}=0\,. \tag{5.54}$$

Clearly if $(\alpha^1)^{sl}$ is symmetric then (5.54) is satisfied. Conversely, contracting (5.54) with $\delta^k_{j_{n+1}}\delta^r_{i_{n+1}}$ we obtain

$$(n-n^2)\left((\alpha^1)^{i_nj_n}-(\alpha^1)^{j_ni_n}\right)=0\;, \tag{5.55}$$

which means that (5.54) holds if and only if $(\alpha^1)^{ij}$ is symmetric □

Using the previous Proposition it now follows that the simple Filippov n-Lie algebras are stable in the sense of Gerstenhaber [32, 33]

Theorem All infinitesimal deformations of a simple n-Lie algebra are trivial and therefore simple FAs are rigid.

Proof: We have to show that if α^1 is a one-cocycle it is a trivial one. Let the one-cochain be characterized by $(\alpha^1)^{ij}$ as in eq. (5.50). In order to express the one-coboundary condition in terms of $(\alpha^1)^{ij}$, we rewrite the coboundary condition (5.45) in the basis $\{\mathbf{e}_i\}$,

$$\begin{aligned}\alpha^1(\mathbf{e}_{i_1},\ldots,\mathbf{e}_{i_n})=\delta\beta(\mathbf{e}_{i_1},\ldots,\mathbf{e}_{i_n}) &= -\beta([\mathbf{e}_{i_1},\ldots,\mathbf{e}_{i_n}])\\ &+\sum_{a=1}^{n}[\mathbf{e}_{i_1},\ldots,\mathbf{e}_{i_{a-1}},\beta(\mathbf{e}_{i_a}),\mathbf{e}_{i_{a+1}},\ldots,\mathbf{e}_{i_n}]\;,\end{aligned} \tag{5.56}$$

which implies, after using $\beta(\mathbf{e}_j) := \mathbf{e}_i \beta^i{}_j$, $i = 1, \ldots, n+1$, that

$$
\begin{aligned}
(\alpha^1)^r{}_{i_1 \ldots i_n} &= -(-1)^n \sum_{s=1}^{n+1} \varepsilon_s \epsilon_{i_1 \ldots i_n}{}^s \beta^r{}_s \\
&+(-1)^n \sum_{a=1}^{n} \sum_{s=1}^{n+1} \varepsilon_r \epsilon_{i_1 \ldots i_{a-1} s i_{a+1} \ldots i_n}{}^r \beta^s{}_{i_a} \; . \qquad (5.57)
\end{aligned}
$$

If we now contract this equation with $\epsilon^{i_1 \ldots i_n k}$ we find that the coboundary condition may be rewritten as $(\alpha^1)^{rk} = (\delta\beta)^{rk}$ with

$$
(\delta\beta)^{rk} = -(-1)^n (\varepsilon_k \beta^{rk} + \varepsilon_r \beta^{kr}) + (-1)^n \sum_{s=1}^{n+1} \beta^s{}_s \varepsilon_r \delta^{rk} \; . \qquad (5.58)
$$

Let α^1 now be a cocycle. Then, it is generated by the zero-cochain β given by

$$
(\beta)^{jk} = -\frac{(-1)^n}{2} \left[\varepsilon_k (\alpha^1)^{jk} - \frac{1}{n-1} \sum_{s=1}^{n+1} \varepsilon_s (\alpha^1)^s{}_s \delta^{jk} \right] \quad . \qquad (5.59)
$$

Indeed, inserting (5.59) into the $r.h.s$ of (5.58) we get

$$
\begin{aligned}
(\delta\beta)^{jk} &= \frac{1}{2} \Bigg((\alpha^1)^{jk} - \frac{1}{n-1} \varepsilon_k \delta^{jk} \sum_{s=1}^{n+1} \varepsilon_s (\alpha^1)^s{}_s \\
&\quad + (\alpha^1)^{kj} - \frac{1}{n-1} \varepsilon_j \delta^{jk} \sum_{s=1}^{n+1} \varepsilon_s (\alpha^1)^s{}_s \Bigg) \\
&\quad - \frac{1}{2} \varepsilon_j \delta^{jk} \left(\sum_{s=1}^{n+1} \varepsilon_s (\alpha^1)^s{}_s - \frac{n+1}{n-1} \sum_{s=1}^{n+1} \varepsilon_s (\alpha^1)^s{}_s \right) \\
&= (\alpha^1)^{jk} \; , \qquad (5.60)
\end{aligned}
$$

since all sums in the expression cancel among themselves and, because α^1 is a one-cocycle, α^{jk} is symmetric by the previous Proposition. Therefore every one-cocycle of a simple n-Lie algebra is trivial and simple FAs are rigid □

Again, this result can be extended to semisimple n-Lie algebras. By eq. (3.15) every $X \in \mathfrak{G}$ can be written as a sum $X = \sum_{\mathfrak{s}=1}^{l} X_{(\mathfrak{s})}$ of components in the different simple ideals, $X_{(\mathfrak{s})} \in \mathfrak{G}_{(\mathfrak{s})}$. Then, the result of the action of α^1 on n vectors $X_1, \ldots, X_n \in \mathfrak{G}$ may be written as follows:

$$
\begin{aligned}
\alpha^1(X_1, \ldots, X_n) &= \alpha^1 \left(\sum_{\mathfrak{s}_1=1}^{k} X_{1(\mathfrak{s}_1)}, \ldots, \sum_{\mathfrak{s}_n=1}^{k} X_{n(\mathfrak{s}_n)} \right) \\
&= \sum_{\mathfrak{s}_1 \ldots \mathfrak{s}_n=1}^{k} \alpha^1(X_{1(\mathfrak{s}_1)}, \ldots, X_{n(\mathfrak{s}_n)}) \\
&= \sum_{\mathfrak{s}_1 \ldots \mathfrak{s}_n=1}^{k} \sum_{\mathfrak{t}=1}^{k} (\alpha^1)^{(\mathfrak{t})}(X_{1(\mathfrak{s}_1)}, \ldots, X_{n(\mathfrak{s}_n)}) \; . \qquad (5.61)
\end{aligned}
$$

As a result, a one-cochain for a semisimple FA $\mathfrak{G}$ is determined once the components $(\alpha^1)^{(\mathfrak{t})}(X_{1(\mathfrak{s}_1)},\ldots,X_{n(\mathfrak{s}_n)})$, with $(\alpha^1)^{(\mathfrak{t})}(X_{1(\mathfrak{s}_1)},\ldots,X_{n(\mathfrak{s}_n)}) \in \mathfrak{G}_\mathfrak{t}$ and $X_{i(\mathfrak{s}_a)} \in \mathfrak{G}_{\mathfrak{s}_a}$, $a=1,\ldots,n$ are known.

Proposition. If α^1 is a one-cocycle of a semisimple n-Lie algebra, then

$$(\alpha^1)^{(\mathfrak{t})}(X_{1(\mathfrak{s}_1)},\ldots,X_{n(\mathfrak{s}_n)}) = 0$$

when there are at least three indices among the simple ideals $\mathfrak{t},\mathfrak{s}_1,\ldots,\mathfrak{s}_n$ that are different *i.e.*, when the above expression involves components in at least three different ideals.

Proof. Without loss of generality we can choose $\mathfrak{s}_n \neq \mathfrak{t}$. Since the ideal $\mathfrak{G}_{\mathfrak{s}_n}$ is simple, there exist $Y_{1(\mathfrak{s}_n)},\ldots,Y_{n(\mathfrak{s}_n)} \in \mathfrak{G}_{\mathfrak{s}_n}$ such that

$$X_{n(\mathfrak{s}_n)} = [Y_{1(\mathfrak{s}_n)},\ldots,Y_{n(\mathfrak{s}_n)}] \; . \tag{5.62}$$

Then, the projection on the simple ideal $\mathfrak{G}_\mathfrak{t}$ of the cocycle condition (5.40) tells us that

$$\begin{aligned}
&(\alpha^1)^{(\mathfrak{t})}(X_{1(\mathfrak{s}_1)},\ldots,X_{n(\mathfrak{s}_n)}) = \\
&\quad = (\alpha^1)^{(\mathfrak{t})}(X_{1(\mathfrak{s}_1)},\ldots,X_{n-1(\mathfrak{s}_{n-1})},[Y_{1(\mathfrak{s}_n)},\ldots,Y_{n(\mathfrak{s}_n)}]) \\
&= -[X_{1(\mathfrak{s}_1)},\ldots,X_{n-1(\mathfrak{s}_{n-1})},(\alpha^1)^{(\mathfrak{t})}(Y_{1(\mathfrak{s}_n)},\ldots,Y_{n(\mathfrak{s}_n)})] \\
&+\sum_{a=1}^{n}[Y_{1(\mathfrak{s}_n)},\ldots,Y_{a-1(\mathfrak{s}_n)},(\alpha^1)^{(\mathfrak{t})}(X_{1(\mathfrak{s}_1)},\ldots,X_{n-1(\mathfrak{s}_{n-1})},Y_{a(\mathfrak{s}_n)}),Y_{a+1(\mathfrak{s}_n)},\ldots,Y_{n(\mathfrak{s}_n)}] \\
&+\sum_{a=1}^{n}(\alpha^1)^{(\mathfrak{t})}(Y_{1(\mathfrak{s}_n)},\ldots,Y_{a-1(\mathfrak{s}_n)},[X_{1(\mathfrak{s}_1)},\ldots,X_{n-1(\mathfrak{s}_{n-1})},Y_{a(\mathfrak{s}_n)}],Y_{a+1(\mathfrak{s}_n)},\ldots,Y_{n(\mathfrak{s}_n)}) = 0 \, .
\end{aligned} \tag{5.63}$$

The first and third term in the last equality in (5.63) vanish because we are assuming that at least three different ideals are involved and the only possibility for a non-vanishing result is that $\mathfrak{t}=\mathfrak{s}_1=\cdots=\mathfrak{s}_{n-1}$ and $\mathfrak{s}_1=\cdots=\mathfrak{s}_{n-1}=\mathfrak{s}_n$ respectively. The second term vanishes because $\mathfrak{s}_n \neq \mathfrak{t}$ $\square$

The previous Proposition allows us to simplify (5.61) for the one-cocycle α^1 to the case where one or two different ideals are involved. This means that $\alpha^1(X_1,\ldots X_n)$ gives rise to the following terms:

$$\begin{aligned}
\alpha^1(X_1,\ldots,X_n) &= \sum_{\mathfrak{s}=1}^{k}(\alpha^1)^{(\mathfrak{s})}(X_{1(\mathfrak{s})},\ldots,X_{n(\mathfrak{s})}) \\
&+\sum_{\mathfrak{s}\neq\mathfrak{t}}^{k}(\alpha^1)^{(\mathfrak{t})}(X_{1(\mathfrak{s})},\ldots,X_{n(\mathfrak{s})}) \\
&+\sum_{\mathfrak{s}\neq\mathfrak{t}}^{k}\sum_{a=1}^{n}(\alpha^1)^{(\mathfrak{s})}(X_{1(\mathfrak{s})},\ldots,X_{a-1(\mathfrak{s})},X_{a(\mathfrak{t})},X_{a+1(\mathfrak{s})},\ldots,X_{n(\mathfrak{s})}) \; .
\end{aligned} \tag{5.64}$$

Proposition. Let $X_{n(\mathfrak{t})} = [Y_{1(\mathfrak{t})},\ldots,Y_{n(\mathfrak{t})}]$. If $\mathfrak{t}\neq\mathfrak{s}$, then

$$(\alpha^1)^{(\mathfrak{s})}(X_{1(\mathfrak{s})},\ldots,X_{n-1(\mathfrak{s})},X_{n(\mathfrak{t})}) = -[X_{1(\mathfrak{s})},\ldots,X_{n-1(\mathfrak{s})},(\alpha^1)^{(\mathfrak{s})}(Y_{1(\mathfrak{t})},\ldots,Y_{n(\mathfrak{t})})] \; . \tag{5.65}$$

Proof. First, we notice that all the terms in the last line of (5.64) have the structure of the *l.h.s.* of (5.65), and therefore eq. (5.65) will apply to all of them. To prove now this relation, we again make use of eq. (5.63), particularized to the case where the a priori different ideals labelled $\mathfrak{t}, \mathfrak{s}_1, \ldots, \mathfrak{s}_n$ are such that the first n are equal to $\mathfrak{s}$, say, and the $(n+1)$-th one $\mathfrak{s}_n$ is different, say $\mathfrak{t}$. Then,

$$\begin{aligned}
&(\alpha^1)^{(\mathfrak{s})}(X_{1(\mathfrak{s})}, \ldots, X_{n-1(\mathfrak{s})}, X_{n(\mathfrak{t})}) = \\
&= (\alpha^1)^{(\mathfrak{s})}(X_{1(\mathfrak{s})}, \ldots, X_{n-1(\mathfrak{s})}, [Y_{1(\mathfrak{t})}, \ldots, Y_{n(\mathfrak{t})}]) \\
= \; - \; & [X_{1(\mathfrak{s})}, \ldots, X_{n-1(\mathfrak{s})}, (\alpha^1)^{(\mathfrak{s})}(Y_{1(\mathfrak{t})}, \ldots, Y_{n(\mathfrak{t})})] \\
+ \; & \sum_{a=1}^{n} [Y_{1(\mathfrak{t})}, \ldots, Y_{a-1(\mathfrak{t})}, (\alpha^1)^{(\mathfrak{s})}(X_{1(\mathfrak{s})}, \ldots, X_{n-1(\mathfrak{s})}, Y_{a(\mathfrak{t})}), Y_{a+1(\mathfrak{t})}, \ldots, Y_{n(\mathfrak{t})}] \\
+ \; & \sum_{a=1}^{n} (\alpha^1)^{(\mathfrak{s})}(Y_{1(\mathfrak{t})}, \ldots, Y_{a-1(\mathfrak{t})}, [X_{1(\mathfrak{s})}, \ldots, X_{n-1(\mathfrak{s})}, Y_{a(\mathfrak{t})}], Y_{a+1(\mathfrak{t})}, \ldots, Y_{n(\mathfrak{t})}) \; , \qquad (5.66)
\end{aligned}$$

and we see that the terms in the last two lines vanish, which proves eq. (5.65) □

Using the two previous propositions, we now prove the following

Theorem. Let (α^1) be a one-cocycle for the deformation cohomology of a semisimple n-Lie algebra $\mathfrak{G}$, eq. (5.47). Then there exists a zero-cochain β such that

$$\begin{aligned}
\alpha^1(X_1, \ldots, X_n) &= (\delta\beta)(X_1, \ldots, X_n) \\
&= -\beta([X_1, \ldots, X_n]) + \sum_{a=1}^{n} [X_1, \ldots, X_{a-1}, \beta(X_a), X_{a+1}, \ldots X_n] \; . \qquad (5.67)
\end{aligned}$$

Therefore, any semisimple FA is rigid.

Proof. Let us first consider the $\mathfrak{G}_\mathfrak{s}$-valued one-cochain with arguments in $\mathfrak{G}_\mathfrak{s}$ given by $(\alpha^1)^{(\mathfrak{s})}(X_{1(\mathfrak{s})}, \ldots, X_{n(\mathfrak{s})})$ which corresponds to the first line in (5.64). If α^1 is a cocycle, then so is $(\alpha^1)^{(\mathfrak{s})}$. But since $\mathfrak{G}_\mathfrak{s}$ is simple there exists a $\mathfrak{G}_\mathfrak{s}$-valued zero-cochain that takes arguments in $\mathfrak{G}_\mathfrak{s}$, ${\beta^{(\mathfrak{s})}}_{(\mathfrak{s})}$, such that eq. (5.45) reads

$$\begin{aligned}
(\alpha^1)^{(\mathfrak{s})}(X_{1(\mathfrak{s})}, \ldots, X_{n(\mathfrak{s})}) &= -{\beta^{(\mathfrak{s})}}_{(\mathfrak{s})}([X_{1(\mathfrak{s})}, \ldots, X_{n(\mathfrak{s})}]) \\
&+ \sum_{a=1}^{n} [X_{1(\mathfrak{s})}, \ldots, X_{a-1(\mathfrak{s})}, {\beta^{(\mathfrak{s})}}_{(\mathfrak{s})}(X_{a(\mathfrak{s})}), X_{a+1(\mathfrak{s})}, \ldots X_{n(\mathfrak{s})}] \; . \qquad (5.68)
\end{aligned}$$

Consider now the $\mathfrak{G}_\mathfrak{t}$-valued one-cochain with arguments in $\mathfrak{G}_\mathfrak{s}$ ($\mathfrak{t} \neq \mathfrak{s}$) given by $(\alpha^1)^{(\mathfrak{t})}(X_{1(\mathfrak{s})}, \ldots, X_{n(\mathfrak{s})})$ (eq. (5.64), second line). This part also satisfies the one-cocycle condition since α^1 is a cocycle, but the one cocycle condition (5.40) reduces to the terms

$$\begin{aligned}
&(\alpha^1)^{(\mathfrak{t})}(X_{1(\mathfrak{s})}, \ldots, X_{n-1(\mathfrak{s})}, [Y_{1(\mathfrak{s})}, \ldots, Y_{n(\mathfrak{s})}]) \\
&= \sum_{a=1}^{n} (\alpha^1)^{(\mathfrak{t})}(Y_{1(\mathfrak{s})}, \ldots, Y_{a-1(\mathfrak{s})}, [X_{1(\mathfrak{s})}, \ldots, X_{n-1(\mathfrak{s})}, Y_{a(\mathfrak{s})}], Y_{a+1(\mathfrak{s})}, \ldots, Y_{n(\mathfrak{s})}) \; . \qquad (5.69)
\end{aligned}$$

This is the one-cocycle condition that we considered already for the trivial action (eq. (4.21)), and we know that then α^1 may be generated by a zero-cochain β, namely

$$(\alpha^1)^{(\mathfrak{t})}(X_{1(\mathfrak{s})}, \ldots, X_{n(\mathfrak{s})}) = (\delta {\beta^{(\mathfrak{t})}}_{(\mathfrak{s})})(X_{1(\mathfrak{s})}, \ldots, X_{n(\mathfrak{s})}) = -{\beta^{(\mathfrak{t})}}_{(\mathfrak{s})}([X_{1(\mathfrak{s})}, \ldots, X_{n(\mathfrak{s})}]) \; , \qquad (5.70)$$

where the $\mathfrak{G}_{\mathfrak{t}}$-valued zero-cochain takes arguments in $\mathfrak{G}_{\mathfrak{s}}$, hence the notation $\beta^{(\mathfrak{t})}{}_{(\mathfrak{s})}$.

Using these zero-cochains, we check that $\alpha^1 = \delta\beta$ with

$$\beta(X) := \sum_{\mathfrak{s},\mathfrak{t}=1}^{k} \beta^{(\mathfrak{t})}{}_{(\mathfrak{s})}(X_{(\mathfrak{s})}) \ . \tag{5.71}$$

Indeed, we have, from the definition of $\delta\beta$ (eq. (5.45)),

$$\begin{aligned}
\delta\beta(X_1,\dots,X_n) &= -\beta([X_1,\dots,X_n]) \\
&\quad+\sum_{a=1}^{n}[X_1,\dots,X_{a-1},\beta(X_a),X_{a+1},\dots X_n] \\
&= -\sum_{\mathfrak{s},\mathfrak{t}=1}^{k} \beta^{(\mathfrak{t})}{}_{(\mathfrak{s})}([X_1,\dots,X_n]_{(\mathfrak{s})}) \\
&\quad+\sum_{a=1}^{n}\sum_{\mathfrak{s},\mathfrak{t}=1}^{k}[X_1,\dots,X_{a-1},\beta^{(\mathfrak{t})}{}_{(\mathfrak{s})}(X_{a(\mathfrak{s})}),X_{a+1},\dots X_n] \\
&= -\sum_{\mathfrak{s},\mathfrak{t}=1}^{k} \beta^{(\mathfrak{t})}{}_{(\mathfrak{s})}([X_{1(\mathfrak{s})},\dots,X_{n(\mathfrak{s})}]) \\
&\quad+\sum_{a=1}^{n}\sum_{\mathfrak{s},\mathfrak{t}=1}^{k}[X_{1(\mathfrak{t})},\dots,X_{a-1(\mathfrak{t})},\beta^{(\mathfrak{t})}{}_{(\mathfrak{s})}(X_{a(\mathfrak{s})}),X_{a+1(\mathfrak{t})},\dots X_{n(\mathfrak{t})}] \ .
\end{aligned} \tag{5.72}$$

The proof is now completed by checking that the coboundary (5.71) generates the one-cocycle α^1 in (5.64) since, as argued there, all other cases give zero. This is achieved by means of the Proposition expressed by eq. (5.65). Using eq. (5.65) in the last terms of (5.64), eq. (5.68) in the first term of (5.64) and eq. (5.70) both in the second term plus in the terms which have now appeared from the *r.h.s* of (5.65) we obtain that the one-cocycle α^1 gives rise to

$$\begin{aligned}
\alpha^1(X_1,\dots,X_n) &= -\sum_{\mathfrak{s}=1}^{k} \beta^{(\mathfrak{s})}{}_{(\mathfrak{s})}([X_{1(\mathfrak{s})},\dots,X_{n(\mathfrak{s})}]) \\
&\quad+\sum_{\mathfrak{s}=1}^{k}\sum_{a=1}^{n}[X_{1(\mathfrak{s})},\dots,X_{a-1(\mathfrak{s})},\beta^{(\mathfrak{s})}{}_{(\mathfrak{s})}(X_{a(\mathfrak{s})}),X_{a+1(\mathfrak{s})},\dots X_{n(\mathfrak{s})}] \\
&\quad-\sum_{\mathfrak{s}\neq\mathfrak{t}}^{k} \beta^{(\mathfrak{t})}{}_{(\mathfrak{s})}([X_{1(\mathfrak{s})},\dots,X_{n(\mathfrak{s})}]) \\
&\quad+\sum_{\mathfrak{s}\neq\mathfrak{t}}^{k}\sum_{a=1}^{n}[X_{1(\mathfrak{s})},\dots,X_{a-1(\mathfrak{s})},\beta^{(\mathfrak{s})}{}_{(\mathfrak{t})}([Y_{a1(\mathfrak{t})},\dots,Y_{an(\mathfrak{t})}]),X_{a+1(\mathfrak{s})},\dots X_{n(\mathfrak{s})}] \\
&= -\sum_{\mathfrak{s},\mathfrak{t}=1}^{k} \beta^{(\mathfrak{t})}{}_{(\mathfrak{s})}([X_{1(\mathfrak{s})},\dots,X_{n(\mathfrak{s})}]) \\
&\quad+\sum_{\mathfrak{s},\mathfrak{t}=1}^{k}\sum_{a=1}^{n}[X_{1(\mathfrak{s})},\dots,X_{a-1(\mathfrak{s})},\beta^{(\mathfrak{s})}{}_{(\mathfrak{t})}(X_{a(\mathfrak{t})}),X_{a+1(\mathfrak{s})},\dots X_{n(\mathfrak{s})}] \ ,
\end{aligned} \tag{5.73}$$

since, again, $[Y_{a1(\mathfrak{t})},\dots,Y_{an(\mathfrak{t})}] = X_{a(\mathfrak{t})}$, which allows us to join the two double sums in the second and fourth lines above into a single one. This reproduces eq. (5.72) so that $\alpha^1 = \delta\beta$, which completes the proof □

6. An observation on n-Leibniz algebras, FAs, and cohomology

Consider the case of ordinary or $n = 2$ *Leibniz algebras* $\mathscr{L}$ [68–71]. These algebras share with the Lie algebras a form of the JI identity, but not the anticommutativity of the two-bracket. As a result, their characteristic identity is the *Leibniz identity*, which retains the aspect (b) of the JI for Lie algebras mentioned in the Introduction. Specifically, a *Leibniz algebra* is a vector space $\mathscr{L}$ endowed with a bilinear operation $\mathscr{L} \times \mathscr{L} \to \mathscr{L}$ that satisfies the *Leibniz identity*

$$[X,[Y,Z]] = [[X,Y],Z] + [Y,[[X,Z]] \qquad \forall X,Y,Z \in \mathscr{L} \quad , \tag{6.74}$$

Actually, this defines a *left* Leibniz algebra. There is a right counterpart when the equation above is modified to correspond to a right derivation: the right Leibniz identity reads $[[X,Y],Z] = [[X,Z],Y] + [X,[Y,Z]]$ and accordingly defines a right Leibniz algebra. Obviously, when the bracket is anticommutative, the Leibniz algebra $\mathscr{L}$ becomes a Lie algebra $\mathfrak{g}$ and both the left and right Leibniz identities become one and the same Lie algebra JI.

Similarly, n-Leibniz algebras $\mathfrak{L}$ [37, 72] (see also [38]) share with the FAs the derivation property expressed by the n-Leibniz identity, which follows the pattern of the FI of eq. (2.2), where now the brackets are n-Leibniz brackets and thus not fully antisymmetric[9]. Thus, their characteristic identity has still the structure of eq. (2.9) which, to stress the similarity with the left character of the Leibniz identity above we shall write in the form

$$\mathscr{X} \cdot (\mathscr{Y} \cdot \mathscr{Z}) = (\mathscr{X} \cdot \mathscr{Y}) \cdot \mathscr{Z} + \mathscr{Y} \cdot (\mathscr{X} \cdot \mathscr{Z}) \qquad \forall \mathscr{X}, \mathscr{Y}, \mathscr{Z} \in \otimes^{n-1}\mathfrak{L} \, , \tag{6.75}$$

where the n-bracket involved in the definition of the composition of fundamental objects (eq. (2.8)) is now the n-Leibniz bracket in $\mathfrak{L}$. It is clear that the above defines a (left) n-Leibniz algebra. FAs $\mathfrak{G}$ may be viewed as as a particular case of n-Leibniz algebras $\mathfrak{L}$, namely those with a fully antisymmetric n-bracket and fundamental objects in $\wedge^{n-1}\mathfrak{G}$ rather than in $\otimes^{n-1}\mathfrak{L}$.

When we were discussing the various FA cohomology complexes, the essential ingredients in their definition were the FI and the specific action of $\mathfrak{G}$ on the cochains; in particular, the skewsymmetry of the FA n-bracket was not needed for the nilpotency of δ in, say, eq. (4.24). As a result, this δ also defines a coboundary operator for the cohomology of a n-Leibniz algebra $\mathfrak{L}$, in which the p-cochains are now elements $\alpha \in (\otimes^{(n-1)}\mathfrak{L}^*) \otimes \overset{p}{\cdots} \otimes (\otimes^{(n-1)}\mathfrak{L}^*) \otimes \mathfrak{L}^* = \otimes^{p(n-1)+1}\mathfrak{L}^*$. The key ingredient that guarantees the nilpotency of the coboundary operator is still the (left) identity that follows from eq. (2.2), which implies eqs. (2.9), (2.10), etc. The essential difference between the n-Lie algebra and n-Leibniz algebra cohomology complexes for the trivial representation, unimportant for the nilpotency of δ, is that $\mathscr{X} \in \wedge^{n-1}\mathfrak{G}$ for a n-Lie algebra and $\mathscr{X} \in \otimes^{n-1}\mathfrak{L}$ for a Leibniz algebra, and the fact the brackets that appear in the composition of fundamental objects in eq. (2.8) are, respectively, n-Lie and n-Leibniz brackets.

Thus, with the appropriate changes in the definition of the p-cochain spaces C^p, the coboundary operator in eq. (4.24) defines the corresponding cohomologies for n-Lie $\mathfrak{G}$ and n-Leibniz $\mathfrak{L}$ algebras adapted to the central extension problem, which corresponds to the trivial action. Similar considerations apply to the n-Leibniz algebra cohomology adapted to the deformation problem already discussed for the FAs (eqs. (5.47), (5.48)), but we shall not consider this further here. For an ordinary Leibniz algebra $\mathcal{L}$, for instance, generalizing to the case where the action is given through a representation [68, 71] ρ on $\mathscr{A}$ and reverting (since $n = 2$) to the notation where p indicates the number of algebra elements on which α^p takes

[9] This is the case of the three algebras used in [22] in the context of the BLG model.

Journal of Physics: Conference Series **175** (2009) 012001 doi:10.1088/1742-6596/175/1/012001

arguments, $\alpha^p \in C^p(\mathscr{L},\mathscr{A}) = \mathrm{Hom}(\otimes^p\mathscr{L},\mathscr{A})$, eq. (5.47) leads to

$$\begin{aligned}(\delta\alpha^p)(X_1,\ldots,X_p,X_{p+1}) = \\ \sum_{1\leq j<k}^{p+1} (-1)^j\alpha^p(X_1,\ldots,\widehat{X_j},\ldots,X_{k-1},[X_j,X_k],X_{k+1},\ldots,X_{p+1}) \\ +\sum_{j=1}^{p}(-1)^{j+1}\rho(X_j)\cdot\alpha^p(X_1,\ldots,\widehat{X_j}\ldots,X_{p+1}) \\ +(-1)^{p+1}\alpha^p(X_1,\ldots,X_p)\cdot\rho(X_{p+1}) \quad ,\end{aligned} \tag{6.76}$$

which coincides with the coboundary operator for the Leibniz algebra cohomology complex $(C^\bullet(\mathscr{L},\mathscr{A}),\delta)$ [69,68,72] (there given for right Lebiniz algebras). We note that if ρ is a symmetric representation [68, 71], the $\rho(X_{p+1})$ in the last term above may be moved to the left and the resulting contribution may then be added as one more term to the third line by enlarging the range of the sum. The resulting expression has then the same form of the expression that gives the action of the Lie algebra cohomology coboundary operator on p-cochains valued on a $\rho(\mathfrak{g})$-module (see, *e.g.* [35]).

Thus, the FA cohomologies defined by the coboundary operators (4.24) that define the FA cohomology complexes (and the corresponding homology) constitute simply the translation of the n-Leibniz algebra $\mathfrak{L}$ cohomology complexes to the n-Lie algebra $\mathfrak{G}$ case. In fact, the previous discussion shows that since Leibniz algebras largely underly the structural cohomological properties of the FAs, the FA cohomology complexes could also have been found from those for the n-Leibniz algebras by demanding full skewsymmetry for the n-Leibniz bracket to become the n-bracket of a FA, and by modifying accordingly the definition of the cochains to account for the skewsymmetry of the fundamental objects of a FA.

We conclude this section with a remark. Let $\mathfrak{L}$ be a n-Leibniz algebra. The composition $\mathscr{X}\cdot\mathscr{Y}$ of fundamental objects, rewritten as $[\mathscr{X},\mathscr{Y}]$, has the properties of a (non-antisymmetric) Leibniz algebra commutator. Indeed, eq. (2.9) also holds for $\mathfrak{L}$, and with the notation $\mathscr{X}\cdot\mathscr{Y}\equiv[\mathscr{X}\,,\,\mathscr{Y}]$ it takes the form

$$[\mathscr{X}\,,[\mathscr{Y}\,,\,\mathscr{Z}]] = [[\mathscr{X}\,,\,\mathscr{Y}],\,\mathscr{Z}] + [\mathscr{Y}\,,[\mathscr{X}\,,\,\mathscr{Z}]] \quad , \quad \mathscr{X}\,,\mathscr{Y}\,,\mathscr{Z}\in\otimes^{n-1}\mathfrak{L} \quad , \tag{6.77}$$

where $[\mathscr{X}\,,\,\mathscr{Y}] \neq -[\mathscr{Y}\,,\,\mathscr{X}]$ is a non-antisymmetric two-bracket. Comparing with eq. (6.74), we see that (6.77) defines an ordinary (left) Leibniz algebra where the two entries in $[\ ,\ ]$ are fundamental objects $\mathscr{X}\in\otimes^{n-1}\mathfrak{L}$. Hence, given a n-Leibniz algebra $\mathfrak{L}$, the linear space of the fundamental objects endowed with the dot operation (2.8) which defines a non-antisymmetric two-bracket $[\mathscr{X},\mathscr{Y}]\equiv\mathscr{X}\cdot\mathscr{Y}$, becomes a (here left) Leibniz algebra [37], the *ordinary Leibniz algebra* $\mathscr{L}$ *associated with a n-Leibniz algebra* $\mathfrak{L}$.

When the fundamental objects are those of a FA, $\mathscr{X},\mathscr{Y}\in\wedge^{n-1}\mathfrak{G}$, and the n-bracket involved in the definition of $\mathscr{X}\cdot\mathscr{Y}$ is therefore the fully antisymmetric bracket of a n-Lie algebra, the resulting $\mathscr{L}$ is the *Leibniz algebra associated with the Filippov algebra* $\mathfrak{G}$. For n=2, the Leibniz algebra $\mathscr{L}$ associated with the $n=2$ FA $\mathfrak{G}$ is in fact an ordinary Lie algebra $\mathfrak{g}$ since the FA bracket is skewsymmetric and the FA itself is an ordinary Lie algebra.

7. Final comments

We have proved the analogue of Whitehead's lemma for n-Lie algebras: semisimple Filippov algebras cannot be centrally extended in a non-trivial way and furthermore they are rigid because $H^1(\mathfrak{G},\mathfrak{G})=0$. Actually, Witehead's lemma for ordinary Lie algebras is more general since it states that the Lie algebra cohomology groups for a non-trivial action ($\rho\neq 0$) are trivial,

$H^p_\rho(\mathfrak{g}, V) = 0 \ \forall p \geq 0$, when $\mathfrak{g}$ is semisimple. The analogous proof for Filippov algebras $\mathfrak{G}$, using our procedure, would require proving first the triviality of the corresponding higher order cohomology groups for simple n-Lie algebras. This is much more involved, but the fact that the analogue [15] of the Cartan-Killing form for a simple $\mathfrak{G}$ is non-degenerate when considered as a bilinear form on $\wedge^{n-1}\mathfrak{G}$, could help in the analysis of all these higher order cohomology groups, which we conjecture to be also trivial.

Another extension of our results would be to consider the case of Leibniz's n-algebras $\mathfrak{L}$ mentioned in Sec. 6, since their cohomological structure, being based on the n-Leibniz identity, is similar. However, since the proof of the triviality of the relevant FA cohomology groups relies heavily on the skewsymmetry of the structure constants of the simple FAs, one would not expect *e.g.* their rigidity property to extend automatically to the corresponding Leibniz algebra case. In fact, it has been shown [73] that the simple euclidean three-Lie algebra A_4 may be infinitesimally deformed as a three-Leibniz algebra of a specific type. This is not surprising in the light of the above discussion, and indeed this increase of possibilities for deformations has been observed already for ordinary Lie algebras when they are treated as Leibniz algebras [74]; in other words, the Leibniz algebra deformations of a Lie algebra are richer. However, the mentioned Leibniz deformability of the simple A_4 euclidean algebra may not extend to the other simple FAs (the case n=3 is special), which may be rigid under this type of n-Leibniz algebra deformations [75].

Acknowledgments
This work has been partially supported by research grants from the Spanish Ministry of Science and Innovation (FIS2008-01980, FIS2005-03989), the Junta de Castilla y León (VA013C05) and EU FEDER funds.

References

[1] J. A. de Azcárraga, A. M. Perelomov and J. C. Pérez Bueno, *The Schouten-Nijenhuis bracket, cohomology and generalized Poisson structures*, J. Phys. **A29**, 7993-8010 (1996) [arXiv:hep-th/9605067].
[2] J. A. de Azcárraga and J. C. Pérez-Bueno, *Higher-order simple Lie algebras*, Commun. Math. Phys. **184**, 669-681 (1997) [arXiv:hep-th/9605213].
[3] J. A. de Azcárraga, J. M. Izquierdo and J. C. Pérez Bueno, *An introduction to some novel applications of Lie algebra cohomology in mathematics and physics*, Rev. Real Acad. Cien. Exactas Fís. Nat., Ser. A Mat., **95**, 225-248 (2001) [arXiv:physics/9803046].
[4] P. Hanlon and M. Wachs, *On Lie k-algebras*, Adv. in Math. **113**, 206-236 (1995)
[5] J.-L. Loday, *La renaissance of opérades*, Sem. Bourbaki **792**, 47-74 (1994-95).
[6] V. Gnedbaye, *Les algèbres k-aires el leurs opérads*, C. R. Acad. Sci. Paris, Série I **321**, 147-152 (1995)
[7] P.W. Michor and A. M. Vinogradov, *n-ary Lie and associative algebras*, Rend. Sem. Mat. Univ. Pol. Torino **53**, 373-392 (1996) [arXiv:math.QA/9801087].
[8] V. Gnedbaye, *Operads of k-ary algebras*, Contemp. Math. **202**,83-114 (1997)
[9] T. Lada and J. Stasheff, *Introduction to SH Lie algebras for physicists*, Int. J. Theor. Phys. **32**, 1087-1104 (1993) [arXiv:hep-th/9209099].
[10] T. Lada and M. Markl, *Strongly homotopy Lie algebras*, Commun. in Alg. **23**, 2147-2161 (1995).
[11] J. Stasheff, *Deformation theory and the Batalin-Vilkovisky master equation*, in *Deformation theory and symplectic geometry*, D. Sternheimer *et al.* eds., Kluwer Acad. Publ., p. 271-284, 1997 [arXiv:q-alg/9702012].
[12] K. Bering and T. Lada, *Examples of homotopy Lie algebras*, arXiv:0903.5433v1 [math.QA]
[13] V. Filippov, *n-Lie algebras*, Sibirsk. Mat. Zh. **26**, 126-140 (1985) (English transl.: Siberian Math. J. **26**, 879-891 (1985)).
[14] S. M. Kasymov, *Theory of n-lie algebras*, Algebra i Logika **26**, 277-297 (1987) (English transl.: Algebra and Logic **26**, 155-166 (1988)).
[15] S. M. Kasymov, *Analog of the Cartan criteria for n-lie algebras*, Algebra i Logika **34**, 274-287 (1995) (English transl.: Algebra and Logic **34**, 147-154 (1995)).
[16] W. X. Ling, *On the structure of n-Lie algebras*, PhD thesis, Siegen, 1993.
[17] J. Bagger and N. Lambert, *Modelling multiple M2's*, *Phys. Rev.* **D75**, 045020 (2007) arXiv:hep-th/0611108.

[18] J. Bagger and N. Lambert, *Gauge Symmetry and Supersymmetry of Multiple M2-Branes*, Phys. Rev. **D77**, 065008 (2008) [arXiv:0711.0955 [hep-th]].
[19] A. Gustavsson, *One-loop corrections to Bagger-Lambert theory*, Nucl. Phys. **B807**, 315-333 (2009) arXiv:0805.4443 [hep-th].
[20] M. Van Raamsdonk, *Comments on the Bagger-Lambert theory and multiple M2-branes*, JHEP **05**, 105 (2008) arXiv:0803.3803 [hep-th].
[21] J. Bagger and N. Lambert, *Three-Algebras and N=6 Chern-Simons Gauge Theories*, Phys. Rev. **D79**, 025002 (2009) [arXiv:0807.0163 [hep-th]].
[22] S. A. Cherkis and C. Sämann, *Multiple M2-branes and Generalized 3-Lie algebras*, Phys. Rev. **D78**, 066019 (2008) [arXiv:0807.0808 [hep-th]].
[23] J. Gomis, G. Milanesi and J. G. Russo, *Bagger-Lambert Theory for General Lie Algebras*, JHEP **0806**, 075 (2008) [arXiv:0805.1012 [hep-th]].
[24] J. P. Gauntlett and J. B. Gutowski, *Constraining maximally supersymmetric membrane actions*, arXiv:0804.3078 [hep-th].
[25] G. Papadopoulos, *M2-branes, 3-Lie algebras and Plücker relations*, JHEP **0805**, 054 (2008) [arXiv:0804.2662 [hep-th]].
[26] P. de Medeiros, J. Figueroa-O'Farrill, E. Méndez-Escobar and P. Ritter, *Metric 3-Lie algebras for unitary Bagger-Lambert theories*, JHEP **0904**, 037 (2009) [arXiv:0902.4674 [hep-th]].
[27] O. Aharony, O. Bergman, D. L. Jafferis and J. Maldacena, *N=6 superconformal Chern-Simons-matter theories, M2-branes and their gravity duals*, JHEP **0810**, 091 (2008) [arXiv:0806.1218 [hep-th]].
[28] N. Jacobson, *Lie and Jordan triple systems*, Amer. J. Math., **71**, 149-170, (1949)
[29] S. Okubo and N. Kamiya, *Jordan-Lie superalgebra and Jordan-Lie triple system*, Univ. Rocherster preprint UR-1467 (1996)
[30] S. Okubo, *Construction of Lie superalgebras from triple product systems*, AIP Conf. Proc. **687**,33 (2003).
[31] R. Kerner, *Ternary and non-associative structures*, Int. J. of Geom. Meth. in Mod. Phys. **5**, 1265-1294 (2008)
[32] M. Gerstenhaber, *On the deformation of rings and algebras*, Annals Math. **79**, 59-103 (1964).
[33] A. Nijenhuis and R. W. Richardson Jr., *Deformation of Lie algebra structures*, J. Math. Mech. **171**, 89-105 (1967)
[34] N. Jacobson, *Lie algebras*. Dover Pub., N.Y., 1979.
[35] J. A. de Azcárraga and J. M. Izquierdo, *Lie groups, Lie algebras, cohomology and some applications in physics*, Cambridge University Press, Cambridge, UK, 1995.
[36] P. Gautheron, *Some remarks concerning Nambu mehcanics*, Lett. Math. Phys. **37**, 103-116 (1996).
[37] Y. L. Daletskii and L. Takhtajan, *Leibniz and Lie algebra structures for Nambu algebra*, Lett. Math. Phys. **39** 127-141 (1997).
[38] M. Rotkiewicz, *Cohomology ring of n-Lie algebras*, Extracta Math. **20**, 219-232 (2005).
[39] L. Takhtajan, *A higher order analog of the Chevalley-Eilenberg complex and the deformation theory of n-algebras*, St. Petersburg Math. J. **6**, (1994) (English transl. **6**, 429-437 (1995)
[40] Y. Nambu, *Generalized Hamiltonian dynamics*, Phys. Rev. **D7**, 2405-2414 (1973).
[41] D. Sahoo and M. C. Valsakumar, *Nambu mechanics and its quantization*, Phys. Rev. **A46**, 4410-4412 (1992)
[42] D. Sahoo and M. C. Valsakumar, *Algebraic structure of Nambu mechanics*, Pramana **40**, 1-16 (1993).
[43] L. Takhtajan, *On Foundation of the generalized Nambu mechanics*, Commun. Math. Phys. **160**, 295-316 (1994), [arXiv:hep-th/9301111].
[44] V. Filippov, *On n-Lie algebra of jacobians*, Sibirsk. Mat. Zh. **39**, no. 3, 660-669, (1998) (English transl.: Siberian Math. J. **39**, 573-581 (1998)).
[45] N. Mukunda and E. G. C. Sudarshan, *Relation between Nambu and Hamiltonian mechanics*, Phys. Rev. **D13**, 2846-2850 (1976).
[46] R. Chatterjee, *Dynamical symmetries and Nambu mechanics*", Lett. Math. Phys. **36**, 117-126 (1996) [arXiv:hep-th/9501141].
[47] R. Chatterjee and L. Takhtajan, *Aspects of classical and quantum Nambu mechanics*, Lett. Math. Phys.**37**, 475-482 (1996) [arXiv:hep-th/9507125].
[48] D. Alekseevsky and P. Guha, *On decomposability of Nambu-Poisson tensor*, Acta Math. Univ. Comenianae **LXV**, 1-9 (1996)
[49] J. Hietarinta, *Nambu tensors and commuting vector fields*, J. Phys.**A30**, L27-L23 (1997)
[50] G. Marmo, G. Vilasi and A. M. Vinogradov, *The local structure of n-Poisson and n-Jacobi manifolds*, J. Geom. Phys. **25**, 141 (1998) [arXiv:physics/9709046].
[51] I. Vaisman, *A survey on Nambu-Poisson brackets*, Acta Math. Univ. Comenianae **LXVIII**, 213-241 (1999) [arXiv:math/9901047].
[52] P.W. Michor and A. M. Vinogradov, *A note on n-ary Poisson brackets*, Rend. Circ. Mat. di Palermo Ser. II, suppl. **63**, 165-172 (2000) [arXiv:math.SG/9901117].

[53] T. Curtright, C. K. Zachos, *Classical and quantum Nambu mechanics*, Phys. Rev. **D68**, 085001-1-29 (2003) [arXiv:hep-th/0212267].

[54] E. Bergshoeff, E. Sezgin, Y. Tanii and P. K. Townsend, *Super p-Branes as gauge theories of volume preserving diffeomorphisms*, Annals Phys. **199**, 340-365 (1990).

[55] J. Hoppe, *On M-Algebras, the quantisation of Nambu-mechanics, and volume preserving diffeomorphisms*, Helv. Phys. Acta **70**, 302-317 (1997) [arXiv:hep-th/9602020].

[56] P. M. Ho, R. C. Hou and Y. Matsuo, *Lie 3-algebra and multiple M2-branes*, JHEP **0806**, 020 (2008) [arXiv:0804.2110 [hep-th]].

[57] C. Sochichiu, *On Nambu-Lie 3-algebra representations*, arXiv:0806.3520 [hep-th].

[58] P. M. Ho and Y. Matsuo, *M5 from M2* JHEP **0806**, 105 (2008) [arXiv:0804.3629 [hep-th]].

[59] I. A. Bandos and P. K. Townsend, *SDiff gauge theory and the M2 condensate*, JHEP **0902**, 013 (2009) [arXiv:0808.1583 [hep-th]].

[60] J. A. de Azcárraga, A. M. Perelomov, and J. C. Pérez Bueno, *New generalized Poisson structures*, J. Phys. **A29**, L151-L157 (1996) [arXiv:q-alg/9601007].

[61] J. A. de Azcárraga, J. M Izquierdo and J. C. Pérez Bueno, *On the generalizations of Poisson structures*, J. Phys. **A30**, L607-L616 (1997) [arXiv:hep-th/9703019].

[62] R. Ibáñez M. de León, J. C. Marrero and and D. Martín de Diego, *Dynamics of generalized Poisson and Nambu-Poisson brackets*, J. Math. Phys. **38**, 2332-2344 (1997)

[63] R. Ibáñez M. de León, J. C. Marrero, *Homology and cohomology on generalized Poisson manifolds*, J. Phys. **A31**, 1253-1266 (1998)

[64] J. A. de Azcárraga, J. M. Izquierdo, A. M. Perelomov and J. C. Pérez Bueno, *The Z_2-graded Schouten-Nijenhuis bracket and generalized super-Poisson structures*, J. Math. Phys. **38**, 3735 (1997) [arXiv:hep-th/9612186].

[65] D. Sternheimer, *Deformation quantization: twenty years after*, AIP Conf. Proc., **453**, 107-145 (1998) [arXiv:math/9809056].

[66] T. Curtright and C. K. Zachos, *Quantizing Dirac and Nambu brackets*, AIP Conf. Proc. **672**, 165-182 (2003) [arXiv:hep-th/0303088].

[67] T. Curtright, D. Fairlie, X. Jin, L. Mezincescu and C. Zachos, *Classical and Quantal Ternary Algebras*, arXiv:0903.4889 [hep-th].

[68] J.-L. Loday, *Une version non-commutative des algèbres de Lie*, L'Ens. Math. **39**, 269-293 (1993).

[69] J. L. Loday and T. Pirashvili, *Universal enveloping algebras of Leibniz algebras and (co)homology*, Mat. Annalen **296**, 139-158 (1993).

[70] C. Cuvier, *Algèbres de Leibnitz: définitions, propriétés*, Ann. Scient. Éc. Norm. Sup. **27**, 1-45 (1994)

[71] J.-L. Loday and T. Pirashvili, *Leibniz representations of Lie algebras*, J. Alg. **181**, 414-425 (1996)

[72] J. M. Casas, J.-L. Loday and T. Pirashvili, *Leibniz n-algebras*, Forum Math. **14**,189-207 (2002).

[73] J. M. Figueroa-O'Farrill, *Three lectures on 3-algebras*, arXiv:0812.2865 [hep-th].

[74] A. Fialowski and A. Mandal, *Leibniz algebra deformations of a Lie algebra*, J. Math. Phys. **49**, 093511-1-11 (2008) [arXiv:0802.1263 [math.KT]]

[75] J. A. de Azcárraga and J. M. Izquierdo, *On Leibniz deformations and rigidity of simple n-Lie algebras*, in preparation.

The homogeneity conjecture for supergravity backgrounds

José Miguel Figueroa-O'Farrill
Maxwell Institute for Mathematical Sciences & School of Mathematics, The University of Edinburgh, King's Buildings, Edinburgh EH9 3JZ, United Kingdom
E-mail: j.m.figueroa@ed.ac.uk

Abstract. These notes record three lectures given at the workshop "Higher symmetries in Physics", held at the Universidad Complutense de Madrid in November 2008. In them we explain how to construct a Lie (super)algebra associated to a spin manifold, perhaps with extra geometric data, and a notion of privileged spinors. The typical examples are supersymmetric supergravity backgrounds; although there are more classical instances of this construction. We focus on two results: the geometric constructions of compact real forms of the simple Lie algebras of type B_4, F_4 and E_8 from S^7, S^8 and S^{15}, respectively; and the construction of the Killing superalgebra of eleven-dimensional supergravity backgrounds. As an application of this latter construction we show that supersymmetric supergravity backgrounds with enough supersymmetry are necessarily locally homogeneous.

1. Geometric construction of exceptional Lie algebras

The Killing–Cartan classification of complex simple Lie algebras consists of four infinite series of *classical* Lie algebras: $A_{n\geq 1}$, $B_{n\geq 2}$, $C_{n\geq 3}$ and $D_{n\geq 4}$, and a small number of *exceptional* Lie algebras: E_6, E_7, E_8, F_4 and G_2, where the subscripts denote the ranks and in the classical case they have been chosen so as to avoid low rank isomorphisms. The classical Lie algebras are well-understood: they are matrix algebras and their compact real forms are the Lie algebras of (special) unitary matrices over $\mathbb{R}$ (B and D), $\mathbb{C}$ (A) and $\mathbb{H}$ (C). In contrast, the exceptional Lie algebras result from "baroque" constructions involving octonions or else from constructions involving spinors, as explained by Adams in his posthumous lecture notes [1] and, for the case of E_8, also in [2]. It is this latter construction which we will geometrise in today's lecture, using a device well-known in supergravity and which will be subject of the next two lectures: the so-called *Killing superalgebra*.

The basic idea of these lectures is to assign to a spin manifold a 2-graded algebra. In this first lecture we will consider the particular example of the exceptional Hopf fibration

$$S^7 \longrightarrow S^{15} \longrightarrow S^8 , \tag{1}$$

where S^n stands for the unit n-sphere in $\mathbb{R}^{n+1}$. If we think of $S^{15} \subset \mathbb{O} \oplus \mathbb{O}$ and $S^7 \subset \mathbb{O}$, then $S^8 \cong \mathbb{OP}^1$ is the octonionic projective plane. Applying the Killing superalgebra construction to the spaces in the above fibration we will obtain compact (or split) real forms of the simple Lie algebras of type B_4, E_8 and F_4, respectively. We have been unable thus far to pinpoint the relation between these Lie algebras which is suggested by the Hopf fibration relating the corresponding spaces. This first lecture is based on [3].

1.1. Clifford algebras, spin group and spinor representations

We start with a flash review of Clifford algebras, the spin group and the spinor representations. For more details, see the books [4] or [5].

Let $E, \langle -, - \rangle$ be a euclidean vector space. For instance, we could take $E = \mathbb{R}^n$ with the standard "dot" product. We define the **Clifford algebra** $C\ell(E)$ to be the associative algebra obtained by quotienting the tensor algebra $T(E)$ by the two-sided ideal generated by elements of the form $x \otimes x + \langle x, x \rangle 1$; in symbols,

$$C\ell(E) = T(E)/\left\{x \otimes x + \langle x, x \rangle 1\right\} . \tag{2}$$

Since the ideal is not homogeneous, the Clifford algebra is not graded, but only filtered. The associated graded algebra is the exterior algebra ΛE, whence we may think of the Clifford algebra as a quantisation of the exterior algebra. Since the ideal has even parity, the Clifford algebra inherits a 2-grading which agrees under the "classical limit" with the 2-grading of ΛE into odd and even forms. When $E = \mathbb{R}^n$ relative to the usual dot product, we call the corresponding Clifford algebra $C\ell(n)$. Up to isomorphism we have the following table of euclidean Clifford algebras:

n	0	1	2	3	4	5	6	7	$k+8$
$C\ell(n)$	$\mathbb{R}$	$\mathbb{C}$	$\mathbb{H}$	$\mathbb{H}\oplus\mathbb{H}$	$\mathbb{H}(2)$	$\mathbb{C}(4)$	$\mathbb{R}(8)$	$\mathbb{R}(8)\oplus\mathbb{R}(8)$	$C\ell(k)\otimes_{\mathbb{R}}\mathbb{R}(16)$

(3)

where we use the notation $\mathbb{K}(m)$ for the algebra of $m \times m$ matrices with entries in $\mathbb{K}$, and where the last column goes by the name of **Bott periodicity**.

The subspace $\Lambda^2 E \subset C\ell(E)$ is a Lie subalgebra under the Clifford commutator isomorphic to $\mathfrak{so}(E)$. Exponentiating inside the (associative) Clifford algebra, gives (for $n > 2$) a simply-connected Lie group $\mathrm{Spin}(E) \subset C\ell(E)^{\text{even}}$, called the **spin group** of E. Conjugating with $\mathrm{Spin}(E)$ preserves $E \subset C\ell(E)$ and defines a two-to-one group homomorphism

$$\mathrm{Spin}(E) \to \mathrm{SO}(E) . \tag{4}$$

When $E = \mathbb{R}^n$ with the standard dot product, we denote the spin group by $\mathrm{Spin}(n)$.

The Clifford algebra $C\ell(n)$ is isomorphic either to a matrix algebra or to two copies of a matrix algebra, and as such has either one or two inequivalent irreducible representations. This follows from the fact that $\mathbb{R}(n)$ and $\mathbb{H}(n)$ have up to isomorphism a unique irreducible representation isomorphic to $\mathbb{R}^n$ and $\mathbb{H}^n$, respectively, whereas $\mathbb{C}(n)$ has two non-isomorphic irreducible representations: $\mathbb{C}^n$ and its complex conjugate representation. Similarly, $\mathbb{R}(n)\oplus\mathbb{R}(n)$ and $\mathbb{H}(n) \oplus \mathbb{H}(n)$ have two inequivalent irreducible representations, isomorphic to $\mathbb{R}^n$ or $\mathbb{H}^n$, respectively. These are called the **pinor** representations of $C\ell(E)$. Restricting them to $\mathrm{Spin}(E)$ one obtains (perhaps reducible) representations called **spinor** representations and denoted $S(E)$. The type ($\mathbb{R}$, $\mathbb{C}$ or $\mathbb{H}$) of $S(E)$ follows from the fact that $\mathrm{Spin}(E) \subset C\ell(E)^{\text{even}}$ and that $C\ell(n)^{\text{even}} \cong C\ell(n-1)$. For example, the spinor representations of $\mathrm{Spin}(n)$ for small n, have types $\mathbb{H}$ for $n = 3,4,5$, $\mathbb{C}$ for $n = 2,6$ and $\mathbb{R}$ for $n = 7,8,9$. This is consistent with the low-dimensional isomorphisms $\mathrm{Spin}(2) \cong \mathrm{U}(1)$, $\mathrm{Spin}(3) = \mathrm{Sp}(1)$, $\mathrm{Spin}(4) = \mathrm{Sp}(1) \times \mathrm{Sp}(1)$, $\mathrm{Spin}(5) = \mathrm{Sp}(2)$, $\mathrm{Spin}(6) = \mathrm{SU}(4)$.

We will be using the fact that $S(E)$ admits a $C\ell(E)$-invariant inner product $(-,-)$ obeying

$$(x \cdot \psi_1, \psi_2) = -(\psi_1, x \cdot \psi_2) , \tag{5}$$

for all $\psi_i \in S(E)$ and $x \in E \subset C\ell(E)$. This means that $(-,-)$ is also $\mathrm{Spin}(E)$-invariant. For E euclidean, as we have been assuming, the spinor inner product is also positive-definite. However if we allow for E to have arbitrary signature, then all seven types of elementary inner products appear among the spinor inner products.

The transpose of the Clifford action $E \otimes S(E) \to S(E)$ defines a real bilinear map $S(E) \otimes S(E) \to E$ which in the cases of interest in this lecture will be skewsymmetric, whence defines a map $\Lambda^2 S(E) \to E$. This will form part of a Lie bracket on a 2-graded algebra whose odd subspace will be isomorphic to (a subspace of) $S(E)$.

1.2. Globalisation: spin geometry

Let (M^n, g) be a riemannian manifold. At every point $x \in M$ we can consider the orthonormal frames for the tangent space $T_x M$. This is the fibre at x of a principal fibre bundle $\mathrm{O}(M)$ called, unsurprisingly, the bundle of orthonormal frames. If M is oriented, and restricting to *oriented* frames, we obtain a subbundle $\mathrm{SO}(M)$. The obstruction to the existence of $\mathrm{SO}(M)$ is the triviality of $\det(TM)$ which is captured by the first Stiefel–Whitney class $w_1(TM) \in H^1(M;\mathbb{Z}_2)$. Hence roughly speaking half the manifolds are orientable. Assuming that M is oriented, a spin structure on M is a lift $\mathrm{Spin}(M) \to \mathrm{SO}(M)$ restricting fibrewise to the two-to-one homomorphism $\mathrm{Spin}(n) \to \mathrm{SO}(n)$. The obstruction is now captured by the second Stiefel–Whitney class $w_2(TM) \in H^2(M;\mathbb{Z}_2)$, whence roughly speaking one quarter of all manifolds are spin. If $w_2(TM) = 0$, the set of inequivalent spin structures are in one-to-one correspondence with $H^1(M;\mathbb{Z}_2) \cong \mathrm{Hom}(\pi_1(M), \mathbb{Z}_2)$, so roughly speaking to the assignment of a sign to every noncontractible loop.

For example, if $S^n \subset \mathbb{R}^{n+1}$ is the unit sphere, then $T_x S^n$ is the perpendicular complement of the line in $\mathbb{R}^{n+1}$ through the origin and x. An oriented orthonormal basis for $T_x S^n$ is then an oriented orthonormal frame for $x^\perp \cong \mathbb{R}^n$ and hence in one-to-one correspondence with the points in $\mathrm{SO}(n)$. Adding x itself we obtain an oriented orthonormal frame for $\mathbb{R}^{n+1}$ and hence an element of $\mathrm{SO}(n+1)$: the element which takes the standard orthonormal basis to that one. Conversely, we have a map $\mathrm{SO}(n+1) \to S^n$ sending the matrix $g \in \mathrm{SO}(n+1)$ to its first column, say x, which is a unit vector in $\mathbb{R}^{n+1}$. The fibre of this map consists of the remaining n columns, which form an oriented frame in the n-dimensional subspace perpendicular to x. In other words, $\mathrm{SO}(S^n) = \mathrm{SO}(n+1)$. The spin cover $\mathrm{Spin}(S^n)$ is precisely the spin group $\mathrm{Spin}(n+1)$ and since $\pi_1(S^n) = \{1\}$ (for $n > 1$), there is a unique such spin structure. For $n = 1$ there are two spin structures, which physicists like to call Neveu–Schwarz and Ramond [2].

Let $\rho : \mathrm{Spin}(n) \to \mathrm{GL}(S)$ be a spinor representation of $\mathrm{Spin}(n)$ and define the associated vector bundle

$$\$ = \mathrm{Spin}(M) \times_\rho S \,, \tag{6}$$

called a bundle of spinors. Its sections are called spinor fields. (Had the lectures been given in Spanish, the spinor bundle would have been called €!)

The tangent space $(T_x M, g_x)$ to M at x is a euclidean vector space and gives rise to a Clifford algebra $C\ell(T_x M)$. As x varies, this globalises to a bundle $C\ell(TM)$ of Clifford algebras. As a vector bundle, we have a natural isomorphism $C\ell(TM) \cong \Lambda T^*M$. We will always think of spinor representations as the restriction to the spin group of a pinor representation of the Clifford algebra. Globalising, this means that the spinor bundle \$ will always admit an action of the Clifford bundle $C\ell(TM)$, making it into a bundle of Clifford modules. In this way, differential forms, which are sections of ΛT^*M, will be able to act on spinor fields via the natural isomorphism $\Lambda T^*M \cong C\ell(TM)$ and the action of $C\ell(TM)$ on \$.

The Levi-Cività connection on $\mathrm{SO}(M)$ lifts to a connection on $\mathrm{Spin}(M)$ and hence defines a connection on any associated vector bundle. In particular we have a covariant derivative ∇ on sections of \$: for all vector fields $X \in \mathfrak{X}(M)$ and spinor fields $\psi \in \Gamma(\$)$, $\nabla_X \psi \in \Gamma(\$)$. The covariant derivative ∇_X along X is linear and obeys the Leibniz rule

$$\nabla_X (f\psi) = (Xf)\psi + f\nabla_X \psi \,, \tag{7}$$

for all functions $f \in C^\infty(M)$.

1.3. Killing spinors and the cone construction

The Levi-Cività connection may be used to write down natural equations on spinors, whose solutions define privileged notions of spinor fields:

- **parallel spinors**: $\nabla\psi = 0$. By the holonomy principle, the holonomy group of ∇ must be included in the stabilizer of a spinor. The determination of which manifolds admit parallel spinors was thus solved by Wang [6] using Berger's holonomy classification. The irreducible holonomy groups of manifolds admitting parallel spinors are $\mathrm{SU}(n)$, $\mathrm{Sp}(n)$, G_2 and $\mathrm{Spin}(7)$.
- **Killing spinors**: $\nabla_X\psi = \lambda X \cdot \psi$ for all $X \in \mathfrak{X}(M)$ for some nonzero constant $\lambda \in \mathbb{C}$, called the **Killing constant**. Iterating the definition of a Killing spinor, we find that M is Einstein with scalar curvature proportional to λ^2, which means that $\lambda^2 \in \mathbb{R}$ and hence $\lambda \in \mathbb{R}^\times \cup i\mathbb{R}^\times$. This gives rise to two separate notions of **real** or **imaginary** Killing spinors, according to whether λ is real or imaginary, respectively.

In this lecture we will concentrate on the case of real Killing spinors. Moreover by rescaling the metric, if necessary, we may always take $\lambda = \pm\frac{1}{2}$. Therefore such a manifold M is Einstein with positive scalar curvature and, if complete, is compact by the Bonnet–Myers theorem. The question of which complete spin manifolds admit real Killing spinors was solved by Bär [7] via his celebrated cone construction by mapping the problem to the problem of determining which manifolds admit parallel spinors.

Indeed, given a spin manifold (M, g) we define its **metric cone** $C(M) = \mathbb{R}^+ \times M$, with metric

$$g_C = dr^2 + r^2 g \ , \tag{8}$$

where $r > 0$ is the parameter of the $\mathbb{R}^+$. For example, if $M = S^n$, then $C(M) = \mathbb{R}^{n+1} \setminus \{0\}$. In this case, and in this case alone, the metric extends smoothly to the origin and it is the flat metric on $\mathbb{R}^{n+1}$ written in spherical polar coordinates. In all other cases, the metric has a conical singularity at $r = 0$. Bär's penetrating observation was that $\nabla_X\psi = \pm\frac{1}{2}X \cdot \psi$ on M becomes the condition $\widetilde{\nabla}\widetilde{\psi} = 0$ on the cone, where the tilded objects live on the cone. The sign in the Killing spinor equation is reflected either in the chirality of the parallel spinor in the case of n odd, or in the embedding $C\ell(n) \subset C\ell(n+1)$ if n is even. This says that the existence of real Killing spinors is again a holonomy problem, albeit in an auxiliary manifold one dimension higher. A theorem of Gallot [8] says that if M is complete, then its metric cone is either flat — so that M is the round sphere — or irreducible. In this latter case, we may use Wang's classification of holonomy groups leaving a spinor invariant. In this way one arrives at Table 1 of (types of) complete manifolds admitting real Killing spinors.

Table 1. Manifolds admitting real Killing spinors

n	hol $\widetilde{\nabla}$	$C(M)$	M
n	$\{1\}$	$\mathbb{R}^{n+1}$	S^n
$2m-1$	$\mathrm{SU}(m)$	Calabi–Yau	Sasaki–Einstein
$4m-1$	$\mathrm{Sp}(m)$	hyperkähler	3-Sasaki
6	G_2		nearly Kähler (non-Kähler)
7	$\mathrm{Spin}(7)$		weak G_2 holonomy

Such manifolds play an important rôle in the AdS/CFT correspondence, as pointed out originally in [9].

1.4. The Killing superalgebra

Let (M, g) be a spin manifold and $\$ \to M$ a bundle of $C\ell(TM)$-modules. We will assume that M admits real Killing spinors and, without loss of generality (i.e., rescaling the metric and reversing orientation, if necessary), assume that the Killing constant $\lambda = \frac{1}{2}$. We define a 2-graded vector space $\mathfrak{g} = \mathfrak{g}_0 \oplus \mathfrak{g}_1$, where $\mathfrak{g}_0$ is the vector space of Killing vector fields on M and

$$\mathfrak{g}_1 = \left\{\psi \in \Gamma(\$) \middle| \nabla_X \psi = \tfrac{1}{2} X \cdot \psi \ \forall X \in \mathfrak{X}(M)\right\} \tag{9}$$

is the vector space of Killing spinors. Remember we have a real bilinear map $\mathfrak{g}_1 \times \mathfrak{g}_1 \to \mathfrak{X}(M)$ by transposing the Clifford action. Explicitly, given $\psi_1, \psi_2 \in \mathfrak{g}_1$, we let $[\psi_1, \psi_2] \in \mathfrak{X}(M)$ be defined by

$$g([\psi_1, \psi_2], X) = (\psi_1, X \cdot \psi_2) \tag{10}$$

for all $X \in \mathfrak{X}(M)$. The following result explains the terminology.

Lemma 1. *$[\psi_1, \psi_2]$ is a Killing vector field.*

Proof. This is a simple calculation. For all $X, Y \in \mathfrak{X}(M)$, we have

$$\begin{aligned} g(\nabla_X[\psi_1, \psi_2], Y) &= X\, g([\psi_1, \psi_2], Y) - g([\psi_1, \psi_2], \nabla_X Y) \\ &= X\,(\psi_1, Y \cdot \psi_2) - (\psi_1, \nabla_X Y \cdot \psi_2) \\ &= (\nabla_X \psi_1, Y \cdot \psi_2) + (\psi_1, Y \cdot \nabla_X \psi_2) \\ &= \tfrac{1}{2}(X \cdot \psi_1, Y \cdot \psi_2) + \tfrac{1}{2}(\psi_1, Y \cdot X \cdot \psi_2) \\ &= \tfrac{1}{2}(\psi_1, (Y \cdot X - X \cdot Y) \cdot \psi_2) \ , \end{aligned}$$

whence

$$g(\nabla_X[\psi_1, \psi_2], Y) + g(\nabla_Y[\psi_1, \psi_2], X) = 0 \ . \tag{11}$$

□

Let $K \in \mathfrak{g}_0$ be a Killing vector field. Then $A_K : TM \to TM$, defined by $A_K(Y) = -\nabla_Y K$, is a skewsymmetric endomorphism of TM. Define the following **Lie derivative** on spinor fields:

$$\mathcal{L}_K = \nabla_K + \varrho(A_K) \ , \tag{12}$$

where $\varrho : \mathfrak{so}(TM) \to \mathfrak{gl}(\$)$ is the spinor representation at the level of the Lie algebra.

Taking for ϱ any other representation of $\mathfrak{so}(TM)$ defines equally well a Lie derivative on sections of the corresponding associated vector bundle. For example, if we take ϱ to be the defining representation on TM, we have

$$\mathcal{L}_K X = \nabla_K X + A_K X = \nabla_K X - \nabla_X K = [K, X] \ , \tag{13}$$

which is the standard Lie derivative of vector fields.

Proposition 2. *The Lie derivative $\mathcal{L}_K$ on $\Gamma(\$)$ obeys a number of properties:*

(i) $[\mathcal{L}_{K_1}, \mathcal{L}_{K_2}]\psi = \mathcal{L}_{[K_1,K_2]}\psi$,

(ii) $\mathcal{L}_K(f\psi) = (K\,f)\psi + f\mathcal{L}_K\psi$,

(iii) $\mathcal{L}_K(X \cdot \psi) = [K, X] \cdot \psi + X \cdot \mathcal{L}_K\psi$, *and*

(iv) $\mathcal{L}_K \nabla_X \psi = \nabla_X \mathcal{L}_K \psi + \nabla_{[K,X]}\psi$,

for all Killing vector fields K, K_1, K_2 and all $f \in C^\infty(M)$, $X \in \mathfrak{X}(M)$ and $\psi \in \Gamma(\$)$.

We remark that properties (1) and (2) justify calling $\mathcal{L}_K$ a Lie derivative, whereas (3) and (4) say that $\mathcal{L}_X$ leaves invariant the Clifford action and ∇, respectively. The definition of $\mathcal{L}_K$ goes back to Lichnerowicz and appears in the thesis of Kosmann-Schwarzbach [10]. It appeared also more recently, in a supergravity context, in [11].

The last two properties in Proposition 2 imply that if $\psi \in \mathfrak{g}_1$ is a Killing spinor, then so is $\mathcal{L}_K\psi$ for all Killing vector fields K. Indeed,

$$\begin{aligned}
\nabla_X \mathcal{L}_K \psi &= \mathcal{L}_K \nabla_X \psi - \nabla_{[K,X]}\psi && \text{by (4) in Proposition 2}\\
&= \tfrac{1}{2}\mathcal{L}_K(X\cdot\psi) - \tfrac{1}{2}[K,X]\cdot\psi && \text{since } \psi\in\mathfrak{g}_1\\
&= \tfrac{1}{2}X\cdot\mathcal{L}_K && \text{by (3) in Proposition 2}
\end{aligned}$$

whence $\mathcal{L}_K\psi \in \mathfrak{g}_1$. This defines a real bilinear map $\mathfrak{g}_0 \times \mathfrak{g}_1 \to \mathfrak{g}_1$, denoted $(K,\psi) \mapsto [K,\psi] := \mathcal{L}_K\psi$.

We now have defined a 2-graded multiplication on $\mathfrak{g} = \mathfrak{g}_0 \oplus \mathfrak{g}_1$, denoted by a bracket $[-,-]$ anticipating the fact that in some cases it will be a Lie (super)algebra. To wit, we have a map $\Lambda^2\mathfrak{g}_0 \to \mathfrak{g}_0$, given by the Lie bracket of vector fields, the above-defined map $\mathfrak{g}_0\otimes\mathfrak{g}_1 \to \mathfrak{g}_1$ given by the spinorial Lie derivative, and the map $\mathfrak{g}_1\otimes\mathfrak{g}_1 \to \mathfrak{g}_0$ given by transposing the Clifford action. This last map is either symmetric or skewsymmetric depending on dimension. This suggests that it may define a Lie (super)algebra structure on $\mathfrak{g}$. This requires satisfying the relevant Jacobi identity. Being a 2-graded algebra, the *jacobator*, the element in $\mathrm{Hom}(\mathfrak{g}^{\otimes 3},\mathfrak{g})$ whose vanishing implies the Jacobi identity, breaks up into four components depending on whether we have three, two, one or no elements in $\mathfrak{g}_0$:

- all elements in $\mathfrak{g}_0$: this is simply the Jacobi identity of the Lie bracket of vector fields;
- two elements in $\mathfrak{g}_0$ and one in $\mathfrak{g}_1$:

$$\begin{aligned}
[K_1,[K_2,\psi]] - [K_2,[K_1,\psi]] &= \mathcal{L}_{K_1}\mathcal{L}_{K_2}\psi - \mathcal{L}_{K_2}\mathcal{L}_{K_1}\psi\\
&= [\mathcal{L}_{K_1},\mathcal{L}_{K_2}]\psi\\
&= \mathcal{L}_{[K_1,K_2]}\psi && \text{by (1) in Proposition 2}\\
&= [[K_1,K_2],\psi]\ ;
\end{aligned}$$

- one element in $\mathfrak{g}_0$ and two in $\mathfrak{g}_1$: this is property (3) in Proposition 2; and
- all elements in $\mathfrak{g}_1$: this does *not* follow from the formalism and has to be checked case by case. For Lie algebras, it lives in $\left(\Lambda^3\mathfrak{g}_1^*\otimes\mathfrak{g}_1\right)^{\mathfrak{g}_0}$, whereas for Lie superalgebras it lives in $\left(S^3\mathfrak{g}_1^*\otimes\mathfrak{g}_1\right)^{\mathfrak{g}_0}$. In some cases, representation theory shows that such spaces are 0, and hence this last component of the jacobator vanishes. In other cases, such spaces are not 0, but the jacobator vanishes all the same. In most cases, however, this last component of the jacobator will not vanish. Hence the generic situation is a 2-graded **$\frac{3}{4}$-Lie (super)algebra**.

1.5. *Some examples of 2-graded Lie algebras*

Consider now the unit spheres $S^7 \subset \mathbb{R}^8$, $S^8 \subset \mathbb{R}^9$ and $S^{15} \subset \mathbb{R}^{16}$, thought of as riemannian manifolds with the canonical spin structure given by the spin groups. In all these cases, the spinor inner product is real symmetric and positive-definite. Since Clifford action is skewsymmetric, so is its transpose, whence the odd-odd bracket is similarly skewsymmetric, defining a map $\Lambda^2\mathfrak{g}_1 \to \mathfrak{g}_0$. In other words, $\mathfrak{g}$ is a 2-graded (possibly) Lie algebra. Notice that $[\mathfrak{g}_1,\mathfrak{g}_1]$ is an ideal of $\mathfrak{g}_0$: this does not use the vanishing of the last component of the jacobator. Since $\mathfrak{g}_0$, the isometry Lie algebra of the above spheres, is simple, we see that $[\mathfrak{g}_1,\mathfrak{g}_1] = \mathfrak{g}_0$ in this case.

To determine $\mathfrak{g}_1$ as an $\mathfrak{g}_0$-module we use the cone construction and the fact, proved in [11], that this construction is equivariant under the action of $\mathfrak{g}_0$, which is naturally a Lie subalgebra of the isometries of the cone. This means that $[K, \psi] = \mathcal{L}_K \psi$ can be lifted and calculated on the cone:

$$\mathcal{L}_{\widetilde{K}} \widetilde{\psi} = \widetilde{\nabla}_{\widetilde{K}} \widetilde{\psi} + \widetilde{\varrho}(A_{\widetilde{K}}) \widetilde{\psi} \,, \tag{14}$$

but $\widetilde{\psi}$ is parallel and (relative to flat coordinates on the cone) $\widetilde{\psi}$ and $A_{\widetilde{K}}$ are constant because $\widetilde{K}$ is a linear vector field. Therefore this is the standard action of $\mathfrak{g}_0 = \mathfrak{so}(n+1)$ ($n = 7, 8, 15$) on (positive-chirality, when applicable) spinors: $S(8)_+$, $S(9)$ and $S(15)_+$, all of which are irreducible representations.

In all cases, a roots-and-weights calculation (made less painful by using LiE [12]) shows that

$$\left(\Lambda^3 \mathfrak{g}_1^* \otimes \mathfrak{g}_1\right)^{\mathfrak{g}_0} = 0 \,, \tag{15}$$

whence the Jacobi identity is satisfied and $\mathfrak{g}$ becomes a 2-graded Lie algebra. To identify the Lie algebras in question we simply observe that $\mathfrak{g}_0$ being simple and $\mathfrak{g}_1$ being irreducible, implies that $\mathfrak{g}$ is simple. The dimensions are easy to compute and the Lie algebras are thus easy to recognise from the Killing–Cartan classification. The results are summarised in Table 2.

Table 2. Killing Lie algebras of some spheres

M	$\mathfrak{g}_0$	$\dim \mathfrak{g}_0$	$\mathfrak{g}_1$	$\dim \mathfrak{g}_1$	$\dim \mathfrak{g}$	$\mathfrak{g}$
S^7	$\mathfrak{so}(8)$	28	$S(8)_+$	8	36	B_4
S^8	$\mathfrak{so}(9)$	36	$S(9)$	16	52	F_4
S^{15}	$\mathfrak{so}(16)$	120	$S(16)_+$	128	248	E_8

Since the inner products on $\mathfrak{g}_0$ and $\mathfrak{g}_1$ are invariant and positive-definite, $\mathfrak{g}$ is a compact real form of the corresponding complex simple Lie algebra. By the usual device of taking $i\mathfrak{g}_1$ instead of $\mathfrak{g}_1$, we may obtain the maximally split real forms.

2. Supergravity backgrounds

Supergravity is an extension of Einstein (or Einstein–Maxwell) theory. At the level of its solutions, it is given by some geometric data $(g, F, \ldots)$, where g is a local lorentzian metric and $F, \ldots$ stand for extra fields, all subject to partial differential equations of the form

- *Einstein*

$$\mathrm{Ric}(g) - \tfrac{1}{2} R g = T(F, \ldots) \tag{16}$$

- *"Maxwell"*

$$dF = 0 \qquad \text{and} \qquad d \star F = \cdots \tag{17}$$

The details depend on the supergravity theory in question and have hence kept purposefully vague in the above description. We will consider here only so-called *Poincaré* supergravities. There are other supergravities: massive, gauged,... Supergravity theories are dictated by the representation theory of the Poincaré superalgebras. There are (physically interesting) supergravity theories in dimension $d \leq 11$ and lorentzian signature, meaning that the local metric g is lorentzian. Supergravity theories are among the jewels of twentieth century theoretical physics and a good review of the structure of supergravity theories from the representation theory point of view can be found in [13].

2.1. Eleven-dimensional supergravity

My favourite, and to some extent the simplest yet nontrivial, supergravity theory is the unique eleven-dimensional supergravity theory. Its existence was conjectured by Nahm [14], whereas it was constructed by Cremmer, Julia and Scherk [15]. Its field content is a lorentzian eleven-dimensional metric g and a closed 4-form F. We can motivate this as follows.

Supergravity is a theory invariant under local supersymmetry, hence the spectrum should carry a representation of the corresponding supersymmetry algebra. In the case of eleven-dimensional supergravity this is the eleven-dimensional Poincaré superalgebra $(\mathfrak{so}(1,10) \oplus \mathbb{R}^{1,10}) \oplus S(1,10)$, where $S(1,10)$ is the spinor representation of $\mathrm{Spin}(1,10)$. It is not hard to show, using Bott periodicity, that $C\ell(1,10) = \mathbb{R}(32) \oplus \mathbb{R}(32)$ and hence $S(1,10) \cong \mathbb{R}^{32}$. The supertranslation ideal generated by $\mathbb{R}^{1,10} \oplus S(1,10)$ has as nonzero brackets the projection $S^2S(1,10) \to \mathbb{R}^{1,10}$ of the symmetric square of the spinor representation $S^2S(1,10) \cong \mathbb{R}^{1,10} \oplus \Lambda^2\mathbb{R}^{1,10} \oplus \Lambda^5\mathbb{R}^{1,10}$ onto the vector representation of $\mathrm{Spin}(1,10)$.

Irreducible unitary representations of the Poincaré superalgebra are induced by representations of the supertranslation ideal generated by $S(1,10) \oplus \mathbb{R}^{1,10}$. This is done by first fixing a character of the abelian translation ideal $\mathbb{R}^{1,10}$; that is, a momentum $p \in \left(\mathbb{R}^{1,10}\right)^*$. Since we are interested in a theory of gravity, which we expect even in eleven dimensions to be a long range force, we require a massless representations, whence $p^2 = 0$, but $p \neq 0$. The *little group* of p, which is the maximal compact subgroup of the stabiliser of p in $\mathrm{Spin}(1,10)$ is isomorphic to $\mathrm{Spin}(9)$. Once a momentum p has been fixed, the supertranslation ideal takes the form of a Clifford algebra

$$[Q_1, Q_2] = -2\,(Q_1, p \cdot Q_2)\,1\ , \tag{18}$$

where $Q_i \in S(1,10)$ and the spinor inner product is symplectic in this signature, whence the bracket here is symmetric. The bilinear form defining the Clifford algebra, $\langle Q_1, Q_2 \rangle = (Q_1, p \cdot Q_2)$ is degenerate because $p^2 = 0$. In fact, it has rank 16. This is shown by exhibiting a "dual" momentum q such that $q^2 = 0$ and $p \cdot q = 1$. Then $S(1,10) = \ker p \oplus \ker q$, where the kernel refers to the Clifford action. The common stabiliser of p and q in $\mathrm{Spin}(1,10)$ is a $\mathrm{Spin}(9)$-subgroup, which we can identify with the little group of either p or q. The degenerate Clifford algebra (18) becomes an honest Clifford algebra on $\ker q \subset S(1,10)$ isomorphic to $C\ell(16)$, and in fact, as $\mathrm{Spin}(9)$-module, $\ker q$ is the spinor module $S(9)$. There is, up to isomorphism, a unique irreducible representation of $C\ell(16)$ and it is real and of dimension 256. Indeed, by Bott periodicity,

$$C\ell(16) \cong C\ell(8) \otimes_{\mathbb{R}} \mathbb{R}(16) \cong \mathbb{R}(16) \otimes_{\mathbb{R}} \mathbb{R}(16) \cong \mathbb{R}(16^2)\ . \tag{19}$$

As a $\mathrm{Spin}(9)$-module, this is nothing but $S(S(9))$; that is, spinors of spinors! A roots-and-weights calculation shows that as a representation of $\mathrm{Spin}(9)$ we have

$$S(S(9)) \cong S^2_0(\mathbb{R}^9) \oplus \Lambda^3\mathbb{R}^9 \oplus \mathrm{RS}(\mathbb{R}^9)\ , \tag{20}$$

where $\mathbb{R}^9$ stands for the vector representation of $\mathrm{Spin}(9)$, S^2_0 denotes traceless symmetric tensors and RS stands for the Rarita–Schwinger representation, which is the subrepresentation of $\mathbb{R}^9 \otimes S(9)$ consisting of the kernel of the Clifford action $\mathbb{R}^9 \otimes S(9) \to S(9)$. Counting dimensions, we see that for the bosonic part of the representation

$$\dim S^2_0(\mathbb{R}^9) + \dim \Lambda^3\mathbb{R}^9 = 44 + 84 = 128\ , \tag{21}$$

whereas for the fermionic part of the representation

$$\dim \mathrm{RS}(\mathbb{R}^9) = \dim \mathbb{R}^9 \otimes S(9) - \dim S(9) = 16 \times 9 - 16 = 128\ , \tag{22}$$

whence the physical degrees of freedom match, as expected. In terms of fields, $S^2_0(\mathbb{R}^9)$ parametrise the fluctuations of a metric tensor g, whereas $\Lambda^3\mathbb{R}^9$ parametrises the fluctuations of a (locally defined) 3-form potential A, and $\mathrm{RS}(\mathbb{R}^9)$ parametrises the fluctuations of a gravitino.

The supergravity action, ignoring terms involving the gravitino, consists of three terms: an Einstein–Hilbert term, a Maxwell term and a Chern–Simons term. The lagrangian density is given by

$$R\,\mathrm{dvol}_g - \tfrac{1}{4} F \wedge \star F + \tfrac{1}{12} F \wedge F \wedge A\ , \tag{23}$$

where $F = dA$ locally. Although A appears explicitly in the above lagrangian, the Euler–Lagrange equations only involve F. The equations are of Einstein–Maxwell type with a twist provided by the Chern–Simons term; namely, the Maxwell equation is nonlinear:

$$d \star F = -\tfrac{1}{2} F \wedge F\ . \tag{24}$$

2.2. Supersymmetric supergravity backgrounds

We define a (bosonic) **eleven-dimensional supergravity background** to be an eleven-dimensional lorentzian spin manifold $(M, g, \$)$ and a closed 4-form $F \in \Omega^4(M)$ subject to the Einstein–Maxwell equations derived from the lagrangian (23).

The lagrangian (23) admits a supersymmetric completion by adding extra terms involving the gravitino $\Psi \in \Omega^1(M, \$)$. The variation of the gravitino under supersymmetry defines a connection D on the spinor bundle $\$$:

$$D_X \psi := \nabla_X \psi + \tfrac{1}{6} \iota_X F \cdot \psi + \tfrac{1}{12} X^\flat \wedge F \cdot \psi\ , \tag{25}$$

for all $\psi \in \Gamma(\$)$ and $X \in \mathfrak{X}(M)$ and where $X^\flat \in \Omega^1(M)$ is the one-form such that $X^\flat(Y) = g(X, Y)$ for all $Y \in \mathfrak{X}(M)$.

The connection D is the fundamental object in this game, as it encodes virtually all the information of the theory. For example, the Einstein–Maxwell equations can be recovered by demanding the vanishing of the Clifford-trace of its curvature. More explicitly, let e_i be a pseudo-orthonormal frame for M and let e^i denote the dual frame, defined by $g(e^i, e_j) = \delta^i_j$. Then, as shown in [16], the field equations defining the notion of a supergravity background are equivalent to

$$\sum_i e^i \cdot R^D(e_i, X) = 0 \quad \forall\ X \in \mathfrak{X}(M)\ . \tag{26}$$

A nonzero spinor field $\psi \in \Gamma(\$)$ which is D-parallel is called a **(supergravity) Killing spinor**. Although this seems a priori to be a generalisation of the notion of a parallel spinor, it is in fact the original notion of a Killing spinor. The geometrical notion in the first lecture is a special case of the supergravity Killing spinor equation for a particular Ansatz for (M, g, F), known as a *Freund–Rubin background* [17].

Being a linear equation, Killing spinors form a vector space, which anticipating the construction of the Killing superalgebra, will be denoted $\mathfrak{g}_1$. Being defined by a parallel condition, a Killing spinor is determined by its value at a point, whence the dimension of $\mathfrak{g}_1$ is bounded above by the rank of the spinor bundle; that is, $\dim \mathfrak{g}_1 \leq 32$. The ratio

$$\nu = \frac{\dim \mathfrak{g}_1}{32} \tag{27}$$

is called the **supersymmetry fraction** of the background (M, g, F). If $\nu > 0$, (M, g, F) is said to be **supersymmetric**.

2.3. Examples

A large number of supersymmetric backgrounds are known. Maximally supersymmetric backgrounds – those with $\nu = 1$ — have been classified in [18, 19, 20]. For such backgrounds, D

is flat and the equations of motions are automatically satisfied. These backgrounds are related as follows:

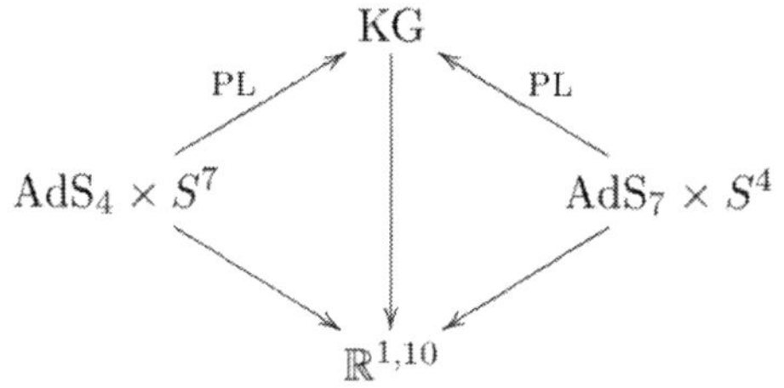

where AdS_n is the n-dimensional anti de Sitter spacetime — i.e., lorentzian hyperbolic space —, KG is a special type of plane wave [21] whose geometry is described by a lorentzian symmetric space of Cahen–Wallach type [22], and $\mathbb{R}^{1,10}$ is Minkowski spacetime with $F = 0$. The arrows labelled "PL" are Penrose limits, described in this context in [23, 24], but tracing their origin to work of Güven [25] and, of course, Penrose [26]. The undecorated arrows are zero-curvature limits.

The $AdS_4 \times S^7$ and $AdS_7 \times S^4$ backgrounds depend on a parameter, interpreted as the scalar curvature of the eleven-dimensional geometry. The radii of curvature of the factors are in a ratio of 2 : 1, whence these backgrounds do not describe realistic compactifications as they once were thought to do. They are known as Freund–Rubin backgrounds. The Killing spinors of the Freund–Rubin backgrounds are $\otimes$ of geometric Killing spinors on the two factors: real on the riemannian factor and imaginary on the lorentzian factor. One can substitute either factor by an Einstein manifold with the same scalar curvature and admitting the relevant kind of Killing spinors. In particular one can consider $AdS_4 \times X^7$, where X is a riemannian manifold admitting real Killing spinors, whence its cone has holonomy contained in Spin(7). Whenever X is not a sphere, the resulting background has a smaller fraction ν of supersymmetry. They can be understood as *near-horizon* geometries of M2-branes, to which we now turn.

The **M2-brane** is a interesting background with $\nu = \frac{1}{2}$, discovered in [27] and interpreted as an *interpolating soliton* in [28]. It is described as follows:

$$\begin{aligned} g &= H^{-2/3} ds^2(\mathbb{R}^{1,2}) + H^{1/3}\left(dr^2 + r^2 ds^2(S^7)\right) \\ F &= \mathrm{dvol}(\mathbb{R}^{1,2}) \wedge dH^{-1} \\ H(r) &= \alpha + \frac{\beta}{r^6}\,, \end{aligned} \tag{28}$$

where $ds^2(\mathbb{R}^{1,2})$ and $\mathrm{dvol}(\mathbb{R}^{1,2})$ are the metric and volume of 3-dimensional Minkowski spacetime, $ds^2(S^7)$ is the metric on the unit sphere in $\mathbb{R}^8$ and H is a two-parameter harmonic function on $\mathbb{R}^8$. If we take $\beta \to 0$ while keeping $\alpha \neq 0$ fixed, we obtain eleven-dimensional Minkowski spacetime with $F = 0$, but taking $\alpha \to 0$ while keeping $\beta \neq 0$ fixed, one obtains $AdS_4 \times S^7$ with scalar curvature depending on β. Therefore the M2-brane interpolates between these two maximally supersymmetric backgrounds. The Killing spinors are given by

$$\psi = H^{1/6}\psi_\infty\,, \tag{29}$$

where ψ_∞ is a parallel spinor in the asymptotic Minkowski spacetime obeying the projection condition

$$\mathrm{dvol}(\mathbb{R}^{1,2}) \cdot \psi_\infty = \psi_\infty\,. \tag{30}$$

Since $\mathrm{dvol}(\mathbb{R}^{1,2})^2 = 1$ and the parallel spinors of $\mathbb{R}^{1,10}$ split into two half-dimensional eigenspaces of $\mathrm{dvol}(\mathbb{R}^{1,2})$. As a result the solution has $\nu = \frac{1}{2}$.

As observed in [9], replacing S^7 by another seven-dimensional manifold admitting real Killing spinors — that is, weak G_2-holonomy, Sasaki-Einstein or 3-Sasaki manifolds — we obtain an

M2-brane at a conical singularity in an 8-dimensional manifold with Spin(7), SU(4) or Sp(2) holonomy, respectively.

To this day a large class class of backgrounds with various values of ν are known to exist. General local metrics with minimal supersymmetry have been written down in [16, 29]. To date, the only fraction which has been ruled out is $\nu = \frac{31}{32}$ [30, 31].

3. The Killing superalgebra of supergravity backgrounds

In the first lecture we saw that from a spin manifold admitting Killing spinors one could define (in the good cases) a 2-graded Lie algebra and in this way we recovered the compact real forms of the simple Lie algebras of types B_4, F_4 and E_8. At its most basic, what we have is a spin manifold with a privileged subspace of spinor fields which then generates a 2-graded algebra with the spinors being the odd-subspace.

In the second lecture we saw how supersymmetric (eleven-dimensional) supergravity backgrounds gave rise to precisely such a situation: an eleven-dimensional *lorentzian* spin manifold with a privileged notion of spinor: the supergravity Killing spinors, which are parallel with respect to a connection D on the spinor bundle. Unlike the spin connection ∇, the connection D is *not* induced from a connection on the tangent bundle: it is genuinely a spinor connection. The spinor bundle is a real rank-32 symplectic vector bundle, but D does *not* preserve the symplectic structure. In fact, as shown by Hull [32], the holonomy algebra of D is generically contained in $\mathfrak{sl}(32,\mathbb{R})$ since only the 'determinant' is preserved.

In this third and last lecture we will see that to every supersymmetric background of eleven-dimensional supergravity one can assign a Lie superalgebra by the techniques in the first lecture and using it we will show that if ν is sufficiently large, the background is forced to be homogeneous. This lecture is based on [33].

3.1. The Killing superalgebra

Let (M, g, F) be a supersymmetric eleven-dimensional supergravity background. Following the method in the first lecture, let us define a 2-graded vector space $\mathfrak{g} = \mathfrak{g}_0 \oplus \mathfrak{g}_1$, where

$$\mathfrak{g}_0 = \{X \in \mathfrak{X}(M) | \mathcal{L}_X g = 0 = \mathcal{L}_X F\} \tag{31}$$

is the Lie algebra of F-preserving isometries of the background, and

$$\mathfrak{g}_1 = \{\psi \in \Gamma(\$) | D\psi = 0\} \tag{32}$$

is the space of Killing spinors. Clearly $\mathfrak{g}$ is finite-dimensional, since as mentioned above $\dim \mathfrak{g}_1 \leq 32$ and $\dim \mathfrak{g}_0 \leq 66$, which is the maximum dimension of the isometry algebra of an eleven-dimensional lorentzian manifold. It is only for $\mathbb{R}^{1,10}$ with $F = 0$ that both of these upper bounds are realised.

Given $\psi \in \Gamma(\$)$, we may define $[\psi, \psi] \in \mathfrak{X}(M)$ by transposing the Clifford action. This is nonzero because this bilinear product is now symmetric, since the inner product on \$ is symplectic. In fact, it is not difficult to show that $[\psi, \psi]$ is always causal; that is, it has non-positive minkowskian norm.

Lemma 3. *If* $\psi \in \mathfrak{g}_1$ *then* $[\psi, \psi] \in \mathfrak{g}_0$.

Proof. We first show that $[\psi, \psi]$ is Killing. We first have that

$$\begin{aligned}
g(\nabla_X[\psi,\psi], Y) &= Xg([\psi,\psi], Y) - g([\psi,\psi], \nabla_X Y) && \text{since } \nabla g = 0\\
&= X(\psi, Y\cdot\psi) - (\psi, \nabla_X Y \cdot \psi) && \text{by definition of } [\psi,\psi]\\
&= (\nabla_X\psi, Y\cdot\psi) + (\psi, Y\cdot\nabla_X\psi)\\
&= 2(\nabla_X\psi, Y\cdot\psi) && \text{since } (\psi_1, Y\cdot\psi_2) = (\psi_2, Y\cdot\psi_1)\\
&= -2(Y\cdot\nabla_X\psi, \psi) \ .
\end{aligned}$$

Now since $D\psi = 0$,

$$\nabla_X \psi = -\tfrac{1}{6}\iota_X F \cdot \psi - \tfrac{1}{12} X^\flat \wedge F \cdot \psi ,$$

whence

$$\begin{aligned} g(\nabla_X[\psi,\psi], Y) &= \tfrac{1}{3}\left(Y \cdot \iota_X F \cdot \psi, \psi\right) + \tfrac{1}{6}\left(Y \cdot (X^\flat \wedge F) \cdot \psi, \psi\right) \\ &= \tfrac{1}{3}\left((Y^\flat \wedge \iota_X F - \iota_Y \iota_X F) \cdot \psi, \psi\right) \\ &\quad + \tfrac{1}{6}\left((Y^\flat \wedge X^\flat \wedge F + g(X,Y)F - X^\flat \wedge \iota_Y F) \cdot \psi, \psi\right) \\ &= -\tfrac{1}{3}\left(\iota_Y \iota_X F \cdot \psi, \psi\right) + \tfrac{1}{6}\left(Y^\flat \wedge X^\flat \wedge F \cdot \psi, \psi\right) , \end{aligned}$$

where we have used that for every 4-form $\Phi \in \Omega^4(M)$,

$$(\Phi \cdot \psi, \psi) = 0 .$$

It follows that

$$g(\nabla_X[\psi,\psi], Y) + g(\nabla_Y[\psi,\psi], X) = 0 ,$$

whence $[\psi,\psi]$ is a Killing vector. One can also prove that $\mathfrak{L}_{[\psi,\psi]}F = 0$. Indeed, since $dF = 0$, $\mathfrak{L}_{[\psi,\psi]}F = d\iota_{[\psi,\psi]}F$ and it is just a calculation to show that

$$\iota_{[\psi,\psi]}F = -dB ,$$

where $B \in \Omega^2(M)$ is the 2-form in the square of ψ:

$$B(X,Y) = \left(\psi, X^\flat \wedge Y^\flat \cdot \psi\right) .$$

□

This result explains why ψ is a called a Killing spinor, since it is the "square root" of a Killing vector. It follows by the usual polarisation trick that if $\psi_1, \psi_2 \in \mathfrak{g}_1$, then $[\psi_1, \psi_2] \in \mathfrak{g}_0$. We therefore have a symmetric bilinear map $\mathfrak{g}_1 \times \mathfrak{g}_1 \to \mathfrak{g}_0$ denoted by $(\psi_1, \psi_2) \mapsto [\psi_1, \psi_2]$.

We now define a bilinear map $\mathfrak{g}_0 \times \mathfrak{g}_1 \to \mathfrak{g}_1$ using the spinorial Lie derivative $\mathfrak{L}_X$ defined in equation (12). Let $X \in \mathfrak{g}_0$. It follows from Proposition 2 and the fact that X preserves F, that $\mathfrak{L}_X$ preserves D; that is,

$$[\mathfrak{L}_X, D_Y]\psi = D_{[X,Y]}\psi \quad \forall\, \psi \in \Gamma(\$) , \tag{33}$$

whence if $D\psi = 0$, also $D\mathfrak{L}_X\psi = 0$. Therefore $[X,\psi] = \mathfrak{L}_X\psi$ defines the desired bilinear map. Together with the Lie bracket of vector fields, under which $\mathfrak{g}_0$ becomes a Lie algebra, we have on $\mathfrak{g} = \mathfrak{g}_0 \oplus \mathfrak{g}_1$ the structure of a superalgebra.

In checking the Jacobi identity, one again sees as in the first lecture that 3/4 of the jacobator is identically zero because of properties of the Lie derivative $\mathfrak{L}_X$. The fourth component of the jacobator vanishes if and only if for all $\psi \in \mathfrak{g}_1$,

$$[[\psi,\psi],\psi] = 0 \qquad \text{or equivalently} \qquad \mathfrak{L}_{[\psi,\psi]}\psi = 0 . \tag{34}$$

Representation theory is not useful here, since we are interested in a general result for unspecified $\mathfrak{g}_0$ and $\mathfrak{g}_1$. An explicit calculation (made less painful with Mathematica or Maple) shows that this is indeed the case. Therefore we have a Lie superalgebra called the **symmetry superalgebra** of the background. The ideal generated by $\mathfrak{g}_1$, $\mathfrak{k} = [\mathfrak{g}_1, \mathfrak{g}_1] \oplus \mathfrak{g}_1$ is called the **Killing superalgebra** of the background. Some (but not all, see [34]) backgrounds are such that their Killing superalgebra admits an extension $\mathfrak{m} = \mathfrak{m}_0 \oplus \mathfrak{g}_1$ which is "maximal" in the sense that

$$\mathfrak{m}_0 = [\mathfrak{g}_1, \mathfrak{g}_1] \cong S^2\mathfrak{g}_1 ,$$

where the isomorphism is one of vector spaces. When it exists, it is called the **maximal superalgebra** of the background.

3.2. Examples

Let us consider some examples. The simplest is of course the Minkowski maximally supersymmetric background $\mathbb{R}^{1,10}$ with $F = 0$. The symmetry superalgebra is the Poincaré superalgebra, whereas the Killing superalgebra is the supertranslation ideal, as explained in Section 2.1. The maximal superalgebra is obtained by taking $\mathfrak{m}_0 = S^2\mathfrak{g}_1$ and declaring $\mathfrak{m}_0$ to be central. The extra elements in $\mathfrak{m}_0$ not in the Killing superalgebra can be understood in terms of *brane charges*, as explained, for example, in [35].

Backgrounds with $F = 0$ are said to be *purely gravitational.* The Killing spinors are parallel with respect to the Levi-Civitá connection ∇. This means that the holonomy of ∇ is contained in the stabiliser of a spinor in $\mathrm{Spin}(1,10)$. There are two types of spinor orbits in $S(1,10)$ and hence two stabilisers, up to isomorphism. As shown by Bryant [36] and the author [37], the orbits are labelled by the value of a quartic polynomial q, whose value $q(\psi)$ at $\psi \in S(1,10)$ is the minkowskian norm of the vector $[\psi,\psi]$, which as mentioned above is always non-negative. If $q(\psi) = 0$ we must distinguish between the $\psi = 0$ and a 25-dimensional orbit with stabiliser isomorphic to $(\mathrm{Spin}(9) \ltimes \mathbb{R}^8) \times \mathbb{R} \subset \mathrm{Spin}(1,10)$, whereas if $q(\psi) < 0$, the stabiliser is isomorphic to $\mathrm{SU}(5)$. This dichotomy gives rise to two types of supersymmetric purely gravitational backgrounds: one generalising the M-wave [38], where $[\psi,\psi]$ is a lightlike parallel vector and thus the geometry is described by a Brinkmann metric, and another generalising the Kaluza–Klein monopole [39, 40, 41]. This latter class gives rise to reducible geometries of the form $\mathbb{R} \times N$, where N is a riemannian ten-dimensional manifold with holonomy contained in $\mathrm{SU}(5)$. The Kaluza–Klein monopole is the case $N = \mathbb{R}^6 \times K$, with K a 4-dimensional hyperkähler manifold. Because ψ is parallel with respect to ∇, so is $[\psi,\psi]$ and the resulting Killing superalgebras are of the supertranslation type, sketchily $[Q,Q] = P_+$, where P_+ is the parallel null vector in the case of the waves, or else $[Q,Q] =$ translation in the flat factor, for the generalised Kaluza–Klein monopoles.

For backgrounds with nonzero F, as in the M2-brane discussed in the second lecture, the Killing superalgebra is still of the supertranslation type, where now $[Q,Q] =$ translations along the brane worldvolume.

For the maximally supersymmetric Freund–Rubin backgrounds the symmetry superalgebra is simple and isomorphic to $\mathfrak{osp}(8|4)$ in the case of $\mathrm{AdS}_4 \times S^7$ and to $\mathfrak{osp}(6,2|2)$ for $\mathrm{AdS}_7 \times S^4$. Simplicity implies that the Killing superalgebra agrees with the symmetry superalgebra. In [34] we showed via an explicit geometric construction that the maximal superalgebra is isomorphic to $\mathfrak{osp}(1|32)$.

It was shown in [24] that the Penrose limit contracts the Killing superalgebra, whence for the maximally supersymmetric KG background, obtained via the Penrose limit from the above Freund–Rubin backgrounds, the Killing superalgebra is a contraction of either of the orthosymplectic superalgebras $\mathfrak{osp}(8|4)$ or $\mathfrak{osp}(6,2|2)$. This contraction was performed explicitly in [42] obtaining the Killing superalgebra previously computed in [18].

3.3. The homogeneity conjecture

Because squaring Killing spinors one obtains Killing vectors, it stands to reason that the more supersymmetric a background is, the more isometries it has. It is therefore tempting to conjecture that there given sufficient supersymmetry — that is, a sufficiently large value of the fraction ν — the background might be homogeneous, where we say that a supergravity background (M,g,F) is **homogeneous** if it admits the transitive action of a Lie group via F-preserving isometries. So the question is whether there is some critical fraction ν_c such that if a background has a fraction $\nu > \nu_c$, then it is forced to be homogeneous. All maximally supersymmetric backgrounds discussed in the second lecture are symmetric spaces, whence in particular homogeneous. On the other hand, the M2-brane, which has $\nu = \frac{1}{2}$, has cohomogeneity one: with orbits labelled by the radial coordinate r in the solution. This suggests that $\nu_c \geq \frac{1}{2}$.

Furthermore, inspecting the catalogue of known solution with $\nu > \frac{1}{2}$, one sees that they are always homogeneous. This prompted Patrick Meessen to state the

Homogeneity Conjecture. *All supergravity backgrounds with* $\nu > \frac{1}{2}$ *are homogeneous.*

In fact, we have to be a little careful because in practice we have that g, F are only locally defined in some open neighbourhood of $\mathbb{R}^{11}$, so that a more relevant notion is that of *local homogeneity*, which is implied by *local transitivity*, by which we mean that around every point there is a local frame consisting of F-preserving Killing vectors.

In [33] we proved something weaker and at the same stronger than the homogeneity conjecture. We proved that if a background has $\nu > \frac{3}{4}$ then it is locally homogeneous, but we proved that already $[\mathfrak{g}_1, \mathfrak{g}_1]$ acts locally transitively. In other words, local homogeneity is a direct consequence of supersymmetry.

3.4. Status of the conjecture

The conjecture does not just make reference to eleven-dimensional supergravity, but in fact to any Poincaré supergravity theory. Concentrating for definiteness on the ten-dimensional supergravity theories, similar results exist for these theories as well. In [43] we showed that any background of either type IIA or IIB supergravity with $\nu > \frac{3}{4}$ is locally homogeneous, whereas any background of type I/heterotic supergravity with $\nu > \frac{1}{2}$ is locally homogeneous. This latter result benefited from the classification of parallelisable backgrounds [44], which in turn was made possible by the fact that the Killing spinors are defined by the lift to the spin bundle of a metric connection with torsion. We believe that the conjecture is true as stated, but proving this for type II and eleven-dimensional supergravities will require a better understanding of the connection D.

Acknowledgments

These notes are a reasonably faithful transcription of three lectures delivered at the Universidad Complutense de Madrid in November 2008 as part of the *Workshop on Higher Symmetries in Physics*. I would like to extend my gratitude to Marco Castrillón López for the invitation. The notes are an expanded version of two lectures given at the Universität Bielefeld in May 2008 as part of the *Meeting on Geometry and Supersymmetry* organised by Andriy Haydys, to whom I am grateful for the invitation.

The results detailed in the first lecture had their origin in lectures delivered in May 2007 at the Dipartamento di Matematica "U. Dini" of the Università degli Studi di Firenze. It is a pleasure to take this opportunity to thank Dmitri Alekseevsky and Andrea Spiro for arranging that visit, Luigi Mangiarotti for support, Cristina Giannotti for the many delicious meals and Suor Tarcisia of *le Suore "Stabilite nella Carità" della Villa Agape* for keeping me off the streets at night.

Most of the work described here is based on collaborations with a number of people, whom it is my pleasure to thank: Bobby Acharya, Matthias Blau, Emily Hackett-Jones, Chris Hull, Patrick Meessen, George Moutsopoulos, George Papadopoulos, Simon Philip, Hannu Rajaniemi, Joan Simón and Bill Spence.

Finally, these notes were prepared while on sabbatical at the Universitat de València, supported under research grant FIS2008-01980. I am grateful to José de Azcárraga for making this visit possible.

References

[1] Adams J F 1996 *Lectures on exceptional Lie groups* (The University of Chicago Press) edited by Zafer Mahmud and Mamoru Mimura

[2] Green M, Schwarz J and Witten E 1987 *Superstring Theory* (Cambridge, UK: Cambridge University Press) 2 vols.

[3] Figueroa-O'Farrill J M 2008 *Comm. Math. Phys.* **283** 663–674 (*Preprint* 0706.2829)
[4] Harvey F R 1990 *Spinors and calibrations* (Academic Press)
[5] Lawson H and Michelsohn M 1989 *Spin geometry* (Princeton University Press)
[6] Wang M 1989 *Ann. Global Anal. Geom.* **7** 59–68
[7] Bär C 1993 *Comm. Math. Phys.* **154** 509–521
[8] Gallot S 1979 *Ann. Sci. École Norm. Sup.* **12** 235–267
[9] Acharya B S, Figueroa-O'Farrill J M, Hull C M and Spence B 1998 *Adv. Theor. Math. Phys.* **2** 1249–1286 (*Preprint* hep-th/9808014)
[10] Kosmann Y 1972 *Annali di Mat. Pura Appl. (IV)* **91** 317–395
[11] Figueroa-O'Farrill J M 1999 *Class. Quant. Grav.* **16** 2043–2055 (*Preprint* hep-th/9902066)
[12] van Leeuwen M A A 1994 *Euromath Bull.* **1** 83–94 ISSN 1359-4346
[13] Van Proeyen A (*Preprint* hep-th/0301005)
[14] Nahm W 1978 *Nucl. Phys.* **B135** 149–166
[15] Cremmer E, Julia B and Scherk J 1978 *Phys. Lett.* **76B** 409–412
[16] Gauntlett J P and Pakis S 2003 *J. High Energy Phys.* **04** 039 (*Preprint* hep-th/0212008)
[17] Freund P and Rubin M 1980 *Phys. Lett.* **B97** 233–235
[18] Figueroa-O'Farrill J M and Papadopoulos G 2001 *J. High Energy Phys.* **06** 036 (*Preprint* hep-th/0105308)
[19] Figueroa-O'Farrill J M 2001 *Special geometric structures in String Theory* ed Alekseevsky D V, Cortés V, Devchand C and Proeyen A V (Bonn: EMIS) (*Preprint* math.DG/0109162)
[20] Figueroa-O'Farrill J M and Papadopoulos G 2003 *J. High Energy Phys.* **03** 048 (*Preprint* hep-th/0211089)
[21] Kowalski-Glikman J 1984 *Phys. Lett.* **134B** 194–196
[22] Cahen M and Wallach N 1970 *Bull. Am. Math. Soc.* **76** 585–591
[23] Blau M, Figueroa-O'Farrill J M, Hull C M and Papadopoulos G 2002 *Class. Quant. Grav.* **19** L87–L95 (*Preprint* hep-th/0201081)
[24] Blau M, Figueroa-O'Farrill J M and Papadopoulos G 2002 *Class. Quant. Grav.* **19** 4753–4805 (*Preprint* hep-th/0202111)
[25] Güven R 2000 *Phys. Lett.* **B482** 255–263 (*Preprint* hep-th/0005061)
[26] Penrose R 1976 *Differential geometry and relativity* (Dordrecht: Reidel) pp 271–275
[27] Duff M and Stelle K 1991 *Phys. Lett.* **253B** 113–118
[28] Duff M, Gibbons G and Townsend P K 1994 *Phys. Lett.* **332B** 321–328 (*Preprint* hep-th/9405124)
[29] Gauntlett J P, Gutowski J and Pakis S 2003 *J. High Energy Phys.* **12** 049 (*Preprint* hep-th/0311112)
[30] Gran U, Gutowski J, Papadopoulos G and Roest D 2007 *J. High Energy Phys.* **02** 043 (*Preprint* hep-th/0610331)
[31] Figueroa-O'Farrill J M and Gadhia S 2007 *J. High Energy Phys.* **06** 043 (*Preprint* hep-th/0702055)
[32] Hull C M (*Preprint* hep-th/0305039)
[33] Figueroa-O'Farrill J M, Meessen P and Philip S 2005 *Class. Quant. Grav.* **22** 207–226 (*Preprint* hep-th/0409170)
[34] Figueroa-O'Farrill J M, Hackett-Jones E, Moutsopoulos G and Simón J 2008 (*Preprint* 0809.5034)
[35] Townsend P K (*Preprint* hep-th/9712004)
[36] Bryant R L 2000 *Global analysis and harmonic analysis (Marseille-Luminy, 1999)* (*Sémin. Congr.* vol 4) ed Bourgignon J P, Branson T and Hijazi O (Paris: Soc. Math. France) pp 53–94 (*Preprint* math.DG/0004073)
[37] Figueroa-O'Farrill J M 2000 *Class. Quant. Grav.* **17** 2925–2947 (*Preprint* hep-th/9904124)
[38] Hull C M 1984 *Phys. Lett.* **139B** 39–41
[39] Sorkin R 1983 *Phys. Rev. Lett.* **51** 87–90
[40] Gross D and Perry M 1983 *Nucl. Phys.* **B226** 29
[41] Han S and Koh I 1985 *Phys. Rev.* **D31** 2503
[42] Hatsuda M, Kamimura K and Sakaguchi M 2002 *Nucl. Phys.* **B637** 168–176 (*Preprint* hep-th/0204002)
[43] Figueroa-O'Farrill J M, Hackett-Jones E and Moutsopoulos G 2007 *Class. Quant. Grav.* **24** 3291–3308 (*Preprint* hep-th/0703192)
[44] Figueroa-O'Farrill J M, Kawano T and Yamaguchi S 2003 *J. High Energy Phys.* **10** 012 (*Preprint* hep-th/0308141)

Some Results on cubic and higher order extensions of the Poincaré algebra

M. Rausch de Traubenberg
IPHC-DRS, UdS, CNRS, IN2P3; 23 rue du Loess, 67037 Strasbourg, France
E-mail: `Michel.Rausch@IReS.in2p3.fr`

Abstract. In these lectures we study some possible higher order (of degree greater than two) extensions of the Poincaré algebra. We first give some general properties of Lie superalgebras with some emphasis on the supersymmetric extension of the Poincaré algebra or Supersymmetry. Some general features on the so-called Wess-Zumino model (the simplest field theory invariant under Supersymmetry) are then given. We further introduce an additional algebraic structure called Lie algebras of order F, which naturally comprise the concepts of ordinary Lie algebras and superalgebras. This structure enables us to define various non-trivial extensions of the Poincaré algebra. These extensions are studied more precisely in two different contexts. The first algebra we are considering is shown to be an (infinite dimensional) higher order extension of the Poincaré algebra in $(1+2)-$dimensions and turns out to induce a symmetry which connects relativistic anyons. The second extension we are studying is related to a specific finite dimensional Lie algebra of order three, which is a cubic extension of the Poincaré algebra in $D-$space-time dimensions. Invariant Lagrangians are constructed.

1. Introduction

Describing the laws of physics in terms of underlying symmetries has always been a powerful tool. For instance the Casimir operators of the Poincaré algebra are related to the mass and the spin of elementary particles. Moreover, it has been understood that all the fundamental interactions (electromagnetic, weak and strong) are related to the Lie algebra $\mathfrak{u}(1)_Y \times \mathfrak{su}(2)_L \times \mathfrak{su}(3)_c$, in the so-called Standard Model. The Standard Model is then described by the Lie algebra $I\mathfrak{so}(1,3) \times \mathfrak{u}(1)_Y \times \mathfrak{su}(2)_L \times \mathfrak{su}(3)_c$, where $I\mathfrak{so}(1,3)$ is related to space-time symmetries and $\mathfrak{u}(1)_Y \times \mathfrak{su}(2)_L \times \mathfrak{su}(3)_c$ to internal symmetries. The elementary particles, namely, the quarks, the leptons and the Higgs boson are in appropriate representations of $\mathfrak{u}(1)_Y \times \mathfrak{su}(2)_L \times \mathfrak{su}(3)_c$. Even if the Standard Model is the physical theory were the confrontation between experimental results and theoretical predictions is in an extremely good agreement, there are strong arguments (which cannot be summarised here) that it is not the final theory.

Thus, to understand the properties of elementary particles, it is then interesting to study the kind of symmetries which are allowed in space-time. Within the framework of Quantum Field Theory (unitarity of the S matrix *etc.*), S. Coleman and J. Mandula have shown [1] that if the symmetries are described in terms of Lie algebras, only trivial extensions of the Poincaré algebra can be obtained. Namely, the fundamental symmetries are based on $I\mathfrak{so}(1,3) \times \mathfrak{g}$ with $\mathfrak{g} \supseteq \mathfrak{u}(1)_Y \times \mathfrak{su}(2)_L \times \mathfrak{su}(3)_c$ a compact Lie algebra describing the fundamental interactions and $[I\mathfrak{so}(1,3), \mathfrak{g}] = 0$. Several algebras, in relation to phenomenology, have been investigated (see *e.g.* [2]) such as $\mathfrak{su}(5), \mathfrak{so}(10), \mathfrak{e}_6$ *etc.* Such theories are usually refer to "Grand-Unified-Theories" or GUT, *i.e.* theories which unify all the fundamental interactions. The fact that elements of $\mathfrak{g}$ and $I\mathfrak{so}(1,3)$ commute means that we have a trivial extension of the Poincaré algebra.

Then, R. Haag, J. T. Lopuszanski and M. F. Sohnius [3] understood that is was possible to extend in a non-trivial way, *i.e.* to introduce new generators which are not Lorentz scalars but spinors, the symmetries of space-time within the framework of Lie superalgebras (see Definition 2.1). This extension is unique (up to the number of supercharges) and called supersymmetry. For a discussion of the Coleman & Mandula theorem see also the Appendix B of [4].

In the context of the two above mentioned theorems, two mathematical structures emerge naturally in the description of the symmetries in physics: the Lie algebras and the Lie superalgebras. Since, the properties of elementary particles are dictated by its symmetries, it should be interesting to know, whether or not other types of symmetries might be possible. The main feature of the above descriptions of physical symmetries are based upon algebras, *i.e.* vector spaces equipped with a binary multiplication, the commutator or the anticommutator. This observation leads immediately to the simple question: would it be possible to obtain higher order algebras in Quantum Field Theory? Of course these new types of symmetries should not be in conflict with the basic principle of Quantum Field Theory. Furthermore, since we do not want to leave aside the basic principles of Relativity this new structure, if acceptable, should in some sense reproduce the Poincaré symmetries. Ternary algebras and higher order algebras have been considered in physics only occasionally (see for instance [5, 6, 7, 8, 9, 10, 11, 12] and references therein). For some mathematical references one can see [13, 14, 15, 16]. Recently they was some revival of interest in ternary algebras when it has been realised that a ternary algebra defined by a fully antisymmetric product appears in the description of multiple M2-branes [17] (see also [18]).

In a series of papers [19, 20, 21] a specific $F-$ary algebra, called Lie algebra of order F was introduced. This algebra presents the interesting feature to have a Lie algebra as a subalgebra. It is thus a kind of hybrid mathematical structure with two products: one binary and the second of order F. Furthermore, Lie algebras of order F can be seen as a possible generalisation of Lie (super)algebras. A Lie algebra of order F admits a $\mathbb{Z}_F-$grading, the zero-graded part being a Lie algebra. An $F-$fold symmetric product (playing the role of the anticommutator in the case $F = 2$) expresses the zero graded part in terms of the non-zero graded part. In the same manner than the Lie (super)algebras lead to description of (extended) space-time symmetries, Lie algebras of order F allow to construct higher order extensions of the Poincaré algebras. See also [22] for different extensions of the Poincaré algebra.

These lectures are devoted to give some examples of higher order extensions of the Poincaré algebra. These extensions will be studied with many details, and a collection of useful technical points will be given in some appendices. We also give some emphasise on the way we by-pass the no-go theorems. Since these types of structures can be seen as a possible "generalisations" of supersymmetry, and in order to stress on the analogy on both structures, we give in Section 2 of this lecture some well-known results on supersymmetry. Some results upon the Wess-Zumino model are given. Section three is the main subject of these lectures. We firstly give the precise definition of Lie algebra of order F that we apply in two different contexts to extend the Poincaré symmetries. It is known that small dimensional space-times present exceptional properties. In particular, in three space-time dimensions there exists states which are neither bosons nor fermions but anyons. We construct a higher order extension of the Poincaré algebra in three space-time dimensions [23]. Studying its representations explicitly shows that this extension induces a symmetry between relativistic anyons in a straight analogy with supersymmetry, which is a Fermi-Bose symmetry. In the next subsection, we construct a cubic extension of the Poincaré algebra in any space-time dimensions. In this case, the cubic extension considered turns out to induce a symmetry on generalised gauge fields or $p-$forms. The transformation laws have a geometrical interpretation in terms of the natural operations on $p-$forms. Then invariant Lagrangians are constructed [24]. In a series of appendices useful technical details are given, in order to illustrate some points, or to give all the needed definitions and identities to prove the results. In Appendix A we give our conventions for spinors in four space-times dimensions. In Appendix B, some emphasis on relativistic wave equations for anyons in three space-times dimensions is given. The main point of our algebras is that they are partially $F-$ary. This means that studying their representations goes along the same lines of studying the representations of supersymmetric algebras, but in the context of Clifford algebras of polynomial's instead of Clifford algebras. In Appendix C, we give some results on Clifford algebras of polynomial's. In Appendix D, infinite dimensional representations of Lie algebras are studied with some emphasis on Verma modules or indecomposable representations. Appendix E is a collection of useful identities on spinors and $p-$forms in $4n-$dimensional space-time.

2. Lie superalgebras and supersymmetry

The purpose of this section is to give some general features of Lie superalgebras and to show its implementation in Quantum Field Theory. The basic idea is very simple. It is a consequence of the Noether and spin-statistics theorems. The Noether theorem establishes a deep relation between symmetries and conservation laws. The spin-statistics theorem is related to the quantisation of fields. It is known that, when the space-time dimension is higher than three, there are two types of fields, the bosons of integer spin and the fermions of half-integer spin. The former are quantised using commutation relations whilst the latter are quantised with anticommutation relations. Thus, if we assume that we have some conserved charges of integer spin $X_1, \cdots, X_m$ and of half-integer spin $Y_1, \cdots, Y_n$ that form a *closed algebra*, the only allowed possibility, is given by an algebraic structure involving commutation relations and anticommutation relations. Furthermore, since both sides of the equality have to behave in the same manner with respect to the Poincaré algebra, the only possibility is given by

$$[X_i, X_j] = f_{ij}{}^k X_k, \quad [X_i, Y_a] = R_{ia}{}^b Y_b, \quad \{Y_a, Y_b\} = C_{ab}{}^i X_i. \tag{1}$$

The mathematical structure hidden behind relations (1) is called a Lie superalgebra. In fact it is the discovery of supersymmetry in Quantum Field Theory [25, 26] which gave rise to the concept of Lie superalgebra and its subsequent classification [27].

Definition 2.1 *A Lie (complex or real) superalgebra is a $\mathbb{Z}_2-$graded vector space $\mathfrak{g} = \mathfrak{g}_0 \oplus \mathfrak{g}_1$ endowed with the following structure*

(i) $\mathfrak{g}_0$ *is a Lie algebra, we denote by* $[\ ,\ ]$ *the bracket on* $\mathfrak{g}_0$ ($[\mathfrak{g}_0, \mathfrak{g}_0] \subseteq \mathfrak{g}_0$)*;*

(ii) $\mathfrak{g}_1$ *is a representation of* $\mathfrak{g}_0$ ($[\mathfrak{g}_0, \mathfrak{g}_1] \subseteq \mathfrak{g}_1$)*;*

(iii) there exits a $\mathfrak{g}_0-$*equivariant mapping* $\{\ ,\ \} : S^2(\mathfrak{g}_1) \to \mathfrak{g}_0$ *where* $S^2(\mathfrak{g}_1)$ *denotes the two-fold symmetric product of* $\mathfrak{g}_1$ ($\{\mathfrak{g}_1, \mathfrak{g}_1\} \subseteq \mathfrak{g}_0$)*;*

(iv) The following Jacobi identities hold ($\forall\ b_1, b_2, b_3 \in \mathfrak{g}_0, \forall\ f_1, f_2, f_3 \in \mathfrak{g}_1$)

$$\begin{aligned}
&[[b_1, b_2], b_3] + [[b_2, b_3], b_1] + [[b_3, b_1], b_2] = 0 \\
&[[b_1, b_2], f_3] + [[b_2, f_3], b_1] + [[f_3, b_1], b_2] = 0 \\
&[b_1, \{f_2, f_3\}] - \{[b_1, f_2], f_3\} - \{f_2, [b_1, f_3]\} = 0 \\
&[f_1, \{f_2, f_3\}] + [f_2, \{f_3, f_1\}] + [f_3, \{f_1, f_2\}] = 0.
\end{aligned} \tag{2}$$

In the definition above, the generators of zero (resp. one) gradation are called the bosonic (resp. fermionic) generators or $\mathfrak{g}_0$ (resp. $\mathfrak{g}_1$), is called the bosonic (resp. fermionic) part of the Lie superalgebra. The first Jacobi identity is the usual Jacobi identity for Lie algebras, the second says that $\mathfrak{g}_1$ is a representation of $\mathfrak{g}_0$, the third identity is the equivariance of $\{\ ,\ \}$. These identities are just consequences of 1., 2. and 3., respectively. However, the fourth Jacobi identity, which is an extra constraint, is just the $\mathbb{Z}_2-$graded Leibniz rule. This means that in Definition 2.1 we could have avoided the three first Jacobi identities. Let us mention however that, for some authors the Jacobi identities are given in the definition of Lie superalgebras, and as a consequence they deduce that $\mathfrak{g}_0$ is a Lie algebra and $\mathfrak{g}_1$ is a representation of $\mathfrak{g}_0$. Finally all the Jacobi identities unify. Indeed, if for homogeneous elements X, Y, Z of $\mathfrak{g}$ we denote by $|X|$ *etc* their $\mathbb{Z}_2-$grade and define the graded-commutator by $[X,Y]_\pm = -(-1)^{|X||Y|}[Y,X]_\pm$, we have

$$(-1)^{|Z||X|}[X,[Y,Z]_\pm]_\pm + (-1)^{|X||Y|}[Y,[Z,X]_\pm]_\pm + (-1)^{|Y||Z|}[Z,[X,Y]_\pm]_\pm = 0.$$

2.1. The super-Poincaré algebra

The supersymmetric extension of the Poincaré algebra is constructed, in the framework of Lie superalgebras, by adjoining to the Poincaré generators anticommuting elements, called supercharges (we denote Q), which belong to the spinor representation of the Poincaré algebra. Thus the supersymmetric algebra is a Lie superalgebra $\mathfrak{g} = \mathfrak{Iso}(1, d-1) \oplus \mathcal{S}$ with brackets

$$[L,L] = L,\ \ [L,P] = P,\ \ [L,Q] = Q,\ \ [P,Q] = 0,\ \ \{Q,Q\} = P, \tag{3}$$

with (L,P) the generators of the Poincaré algebra that belong to the bosonic part of the algebra and Q the fermionic part of the algebra which is called the superchages. This extension is non-trivial, because the supercharges Q are spinors, and thus do not commute with the generators of the Lorentz algebra. In these lectures we are considering the simplest supersymmetric extension of the Poincaré algebra, *i.e.* where there is only one spinor supercharge. For $N-$extended supersymmetry ($N \le 8$), *i.e.* where N spinor-supercharges are introduced, the reader is referred to the literature. There are many good text books on the subject. For historical references one can see [28, 29, 30, 31]. For a more modern presentation see *e.g.* [32], and for the construction of supersymmetric models in particle physics see for instance [33, 34, 35].

We now set up our conventions. We recall that the left- (right-)handed spinors are respectively in the $\mathbf{2}$ (resp. $\overline{\mathbf{2}}$) representation of $SL(2,\mathbb{C})$. We denote $\mathbf{2} = \langle \psi_\alpha, \alpha = 1,2\rangle$ and $\overline{\mathbf{2}} = \langle \bar\chi^{\dot\alpha}, \dot\alpha = 1,2\rangle$ the two-dimensional representation of $SL(2,\mathbb{C})$ and its complex conjugate representation. A Majorana spinor is given by

$$\Psi_M = \begin{pmatrix} \psi_\alpha \\ \bar\psi^{\dot\alpha} \end{pmatrix},$$

with $(\psi_\alpha)^* = \bar\psi_{\dot\alpha}$ (where * denotes the complex conjugation). See Appendix Appendix A for our conventions and useful identities. With the notations above the supersymmetric algebra is given by $\mathfrak{g} = \mathfrak{Iso}(1,3) \oplus \mathbf{2} \oplus \overline{\mathbf{2}}$, with $\mathfrak{Iso}(1,3) = \langle L_{\mu\nu} = -L_{\nu\mu}, P_\mu,\ \mu,\nu = 0,\cdots,3\rangle$ the Poincaré algebra and $\mathbf{2} = \langle Q_\alpha, \alpha = 1,2\rangle$, $\overline{\mathbf{2}} = \langle \bar Q^{\dot\alpha}, \dot\alpha = 1,2\rangle$ the supercharges that we take Majorana $(Q_\alpha)^\dagger = \bar Q_{\dot\alpha}$. We now determine the various brackets of the supersymmetric algebra:

(i) The even-even part of the algebra is just the Lie algebra structure of the Poincaré algebra

$$\begin{aligned} [L_{\alpha\beta}, L_{\mu\nu}] &= \eta_{\beta\nu}L_{\mu\alpha} - \eta_{\alpha\nu}L_{\mu\beta} + \eta_{\beta\mu}L_{\alpha\nu} - \eta_{\alpha\mu}L_{\beta\nu}, \\ [L_{\alpha\beta}, P_\mu] &= \eta_{\beta\mu}P_\alpha - \eta_{\alpha\mu}P_\beta, \\ [P_\alpha, P_\beta] &= 0. \end{aligned} \tag{4}$$

with $\eta = \text{diag}(1,-1,-1,-1)$ the Minkowski metric. We have to emphasise that with our conventions, since our structure constants are real, for a unitary representation, the operators $L_{\mu\nu}$ and P_μ are antihermitian. The usual quadri-momentum and angular momentum of physical applications are given by $-iP_\mu, -iL_{\mu\nu}$, and are thus hermitian for unitary representations.

(ii) The odd-even part of the algebra is given by the action of the Poincaré algebra on the spinors Q_α and $\bar{Q}^{\dot\alpha}$. The action of the Lorentz generator is known. For the action of P on Q we have *a priori*

$$[P_\mu, Q_\alpha] = p\sigma_{\mu\alpha\dot\alpha}\bar{Q}^{\dot\alpha}, \quad [P_\mu, \bar{Q}^{\dot\alpha}] = p'\bar{\sigma}_\mu{}^{\dot\alpha\alpha}Q_\alpha.$$

To write the R.H.S. of the equation above we have identified the only tensors with the appropriate index structure. But the Jacobi identity with (P_μ, P_ν, Q_α) gives $pp' = 0$, and since Q is the complex conjugate of $\bar{Q}$, $p = 0$ automatically implies that $p' = 0$ and conversely. Thus we have the non-vanishing brackets

$$\begin{aligned} [L_{\mu\nu}, Q_\alpha] &= \sigma_{\mu\nu\alpha}{}^\beta Q_\beta, \\ [L_{\mu\nu}, \bar{Q}^{\dot\alpha}] &= \bar{\sigma}_{\mu\nu}{}^{\dot\alpha}{}_{\dot\beta}\bar{Q}^{\dot\beta}. \end{aligned} \tag{5}$$

We have to pose for a while with this part of the algebra. Indeed, the relations (5) seem to be incompatible with the fact that we are dealing with a real Lie superalgebra since the structure constants are complex. However, it is known that when the space-time dimension is four one can find a representation where the Dirac $\Gamma-$matrices are purely imaginary ($\Gamma_0^M = \sigma_2 \otimes \sigma_1, \Gamma_1^M = i\sigma_3 \otimes \sigma_0, \Gamma_2^M = -i\sigma_2 \otimes \sigma_2, \Gamma_3^M = -i\sigma_1 \otimes \sigma_0$), the Majorana representation. This means that in this representation the matrices $\Gamma_{\mu\nu}^M$ are real. Thus we clearly see, in this representation, that the structure constants become real.

(iii) The odd-odd part of the algebra has to close on the even part of the algebra. Since $\mathbf{2} \otimes \mathbf{2} = \mathbf{1} \oplus \mathbf{3}_+$, $\bar{\mathbf{2}} \otimes \bar{\mathbf{2}} = \mathbf{1} \oplus \mathbf{3}_-$ and $\mathbf{2} \otimes \bar{\mathbf{2}} = \mathbf{4}$ with $\mathbf{1}$ the scalar representation, $\mathbf{3}_\pm$ the (anti-)self-dual two-forms and $\mathbf{4}$ the vector representation of $SL(2,\mathbb{C})$, we have *a priori*

$$\begin{aligned} \{Q_\alpha, Q_\beta\} &= a\sigma^{\mu\nu}{}_{\alpha\beta}L_{\mu\nu}, \\ \left\{\bar{Q}_{\dot\alpha}, \bar{Q}_{\dot\beta}\right\} &= b\bar{\sigma}^{\mu\nu}{}_{\dot\alpha\dot\beta}L_{\mu\nu}, \\ \{Q_\alpha, Q_\beta\} &= -ic\sigma^\mu{}_{\alpha\dot\alpha}P_\mu. \end{aligned} \tag{6}$$

The right handed part of (6) is obtained by means of the natural tensors acting on the spinor space. Now using the Jacobi identity with (P, Q, Q) gives $a = 0$, since P commute with Q and do not commute with L. Similarly we have $b = 0$. We identify now the constant c. The relation above gives $\left\{Q_1, \bar{Q}_1\right\} + \left\{Q_2, \bar{Q}_2\right\} = -2icP_0$. Now if we assume that the supercharges act on a Hilbert space, since $\bar{Q}_{\dot\alpha} = Q^\dagger_\alpha$, for any element $|\psi>$ we have $\sum_\alpha <\psi|\left\{Q^\dagger_\alpha, Q_\alpha\right\}|\psi> = -2ic <\psi|P_0|\psi>$. Thus $-2ic <\psi|P_0|\psi>$ is positive. Assuming $-iP_0$ is a positive operator (the energy with our convention) gives $c > 0$. The conventional normalisation is to take $c = 2$. The odd-odd part of the algebra is then

$$\left\{Q_\alpha, \bar{Q}_{\dot\alpha}\right\} = -2i\sigma^\mu_{\alpha\dot\alpha}P_\mu. \tag{7}$$

The relation above seems to be in conflict with the reality of the Lie superalgebra. To solve this discrepancy, write the relations (7) in the four dimensional representation

$$\left\{ \begin{pmatrix} Q_\alpha \\ \bar{Q}^{\dot{\alpha}} \end{pmatrix}, \begin{pmatrix} Q_\beta & \bar{Q}^{\dot{\beta}} \end{pmatrix} \right\} = -2i \begin{pmatrix} 0 & \sigma_{\mu\alpha\dot{\gamma}} \\ \bar{\sigma}_\mu{}^{\dot{\alpha}\gamma} & 0 \end{pmatrix} \begin{pmatrix} -\epsilon_{\gamma\beta} & 0 \\ 0 & -\bar{\epsilon}^{\dot{\gamma}\dot{\beta}} \end{pmatrix} P^\mu, \tag{8}$$

introduce

$$C = -i\Gamma_0\Gamma_2 = \begin{pmatrix} -\varepsilon & 0 \\ 0 & -\bar{\varepsilon} \end{pmatrix}$$

the charge conjugation matrix. Denoting Q_a the components of the four dimensional Majorana spinor, the relations (8) become

$$\{Q_a, Q_b\} = -2i(\Gamma^\mu C)_{ab} P_\mu.$$

Now, we write the equation above in the Majorana representation for which the Dirac $\Gamma-$matrices and C are purely imaginary. If we denote Q_a^M the supercharge in the Majorana representation and substitute $Q_a^M \longrightarrow S_a = e^{i\frac{\pi}{4}} Q_a^M$, then we obtain

$$\{S_a, S_b\} = -2(\Gamma_\mu^M C^M)_{ab} P^\mu,$$

and the structure constant becomes real. We however prefer to write the odd-odd part of the algebra in the form (7).

The algebra given by equation (4), (5) and (7) defines the simplest supersymmetric extension of the Poincaré algebra. Extended supersymmetric algebras with a number of spinor charges $N \leq 8$ [3, 26, 36, 37] or supersymmetric algebras in any space-times dimensions $D \leq 11$ [38] can however be defined. As we are concerned, we just focus on the simplest extension of the Poincaré symmetry (4), (5) and (7). We will however not follow the standard approach where irreducible representations of the Poincaré superalgebra were systematically investigated [37] leading to the implementation of supersymmetry in Quantum Field Theory and then to construction of supersymmetric extensions of the Standard Model of Particle Physics [33, 34, 35]. We simply quote some results that can be found in the text book given in references: (i) we clearly see that $P_\mu P^\mu$ is a Casimir operator, this means that in any representation all states have the same mass; (ii) for any representation of the supersymmetric algebra there is an equal number of bosonic and fermionic degrees of freedom. Following the method of Wigner induced representations we identify to types of relevant representations the massive representations $P^\mu P_\mu < 0$ (recall that with our conventions the mass operator is given by $M^2 = -P_\mu P^\mu$) and the massless representations $P^\mu P_\mu = 0$. When supersymmetry is implemented in Quantum Field Theory, it turns out that it is a new symmetry that mixes bosons and fermions. Beyond its purely mathematical interest this has also some phenomenological interesting consequences. This was probably why supersymmetry becomes so popular in the description of physical laws in particle physics (see for instance [33, 34]). Finally, observing that the symmetric product of two supersymmetric transformations gives rise to a space-time translation, the construction of local supersymmetric theory is necessarily a theory of gravity called supergravity. Several authors introduced independently supergravity [39].

2.2. Finite dimensional matrix representation of the supersymmetric algebra

We now give a finite dimensional matrix representation of the algebra above. At first, we introduce a five dimensional representation of the Poincaré algebra

$$L_{\mu\nu} \longrightarrow L_{5\mu\nu} = \left(\begin{array}{c|c} J_{\mu\nu} & 0 \\ \hline 0 & 0 \end{array} \right) \quad P_\mu \longrightarrow P_{5\mu} = \left(\begin{array}{c|c} 0 & \delta_\mu \\ \hline 0 & 0 \end{array} \right), \quad \begin{pmatrix} 4\times 4 & 4\times 1 \\ 1\times 4 & 1\times 1 \end{pmatrix} \tag{9}$$

where δ_μ is a column vector with components $\delta_\mu{}^\nu$. In this representation, the group element's read

$$(\Lambda, T) = e^{\frac{1}{2}\omega^{\alpha\beta}J_{5\alpha\beta}}e^{t^\alpha P_{5\alpha}} = \left(\begin{array}{c|c}\Lambda & t\\ \hline 0 & 1\end{array}\right),$$

and the quadrivector x is embedded in a five dimensional rep. $X_5 = \begin{pmatrix}x\\1\end{pmatrix}$ such that in the Poincaré transformation above we have

$$X' = \begin{pmatrix}x'\\1\end{pmatrix} = \left(\begin{array}{c|c}\Lambda & t\\ \hline 0 & 1\end{array}\right)\begin{pmatrix}x\\1\end{pmatrix} = \begin{pmatrix}\Lambda x + t\\1\end{pmatrix}.$$

In a similar manner, we can define a nine-dimensional matrix representation of the supersymmetric algebra. We define a nine-dimensional vector with indices $I = (\mu, 4, \alpha, \dot\alpha)$ given by

$$X_9{}^I = \begin{pmatrix}x^\mu\\1\\ \theta^\alpha\\ \bar\theta^{\dot\alpha}\end{pmatrix}.$$

Since we consider θ^α instead of θ_α we have for its transformation law under a Lorentz transformation

$$\delta_\omega\psi_\alpha = \frac{1}{2}\omega^{\mu\nu}\sigma_{\mu\nu\alpha}{}^\beta\psi_\beta \quad\Rightarrow\quad \delta_\omega\psi^\alpha = -\frac{1}{2}\omega^{\mu\nu}\sigma_{\mu\nu}{}^\alpha{}_\beta\psi^\beta$$

but we have $\sigma_{\mu\nu}{}^\alpha{}_\beta = \sigma_{\mu\nu\beta}{}^\alpha$.

We define now the 9×9 matrices (to simplify notations we keep the same symbols)

$$L_{\mu\nu} = \left(\begin{array}{c|c|c|c} J_{\mu\nu} & 0 & 0 & 0\\ \hline 0 & 0 & 0 & 0\\ \hline 0 & 0 & -\sigma^t_{\mu\nu} & 0\\ \hline 0 & 0 & 0 & \bar\sigma_{\mu\nu}\end{array}\right), \quad P_\mu = \left(\begin{array}{c|c|c|c} 0 & \delta_\mu & 0 & 0\\ \hline 0 & 0 & 0 & 0\\ \hline 0 & 0 & 0 & 0\\ \hline 0 & 0 & 0 & 0\end{array}\right), \quad Q = \begin{pmatrix}0 & Q_{01}\\ Q_{10} & 0\end{pmatrix} = \begin{pmatrix}5\times 5 & 5\times 4\\ 4\times 5 & 4\times 4\end{pmatrix}$$

with $J_{\mu\nu}$ the Lorentz generators in the defining vector representation with matrix elements $(J_{\mu\nu})^\alpha{}_\beta = \delta_\mu{}^\alpha\eta_{\nu\beta} - \delta_\nu{}^\alpha\eta_{\mu\alpha}$. We introduce $Q_{01}{}^M{}_A, Q_{10}{}^A{}_M$ with $M = 0, 1, 2, 3, 4$ and $A = 1, 2, \dot 1, \dot 2$ the components of the supercharges. If we note $I = (\mu\beta\dot\beta)$ and $J = (\nu\gamma\dot\gamma)$ the indices of lines and columns of the following matrices, we get

$$\begin{aligned}
(Q_{01\alpha})^M{}_A &= \begin{pmatrix}0 & i\sigma^\mu{}_{\alpha\dot\gamma}\\ 0 & 0\end{pmatrix}, & (\bar Q_{01\dot\alpha})^M{}_A &= \begin{pmatrix}-i\sigma^\mu{}_{\gamma\dot\alpha} & 0\\ 0 & 0\end{pmatrix}, & &\begin{pmatrix}4\times 2 & 4\times 2\\ 1\times 2 & 1\times 2\end{pmatrix},\\
(Q_{10\alpha})^A{}_M &= \begin{pmatrix}0 & \delta^\beta{}_\alpha\\ 0 & 0\end{pmatrix}, & (\bar Q_{10\dot\alpha})^A{}_M &= \begin{pmatrix}0 & 0\\ 0 & -\delta^{\dot\beta}{}_{\dot\alpha}\end{pmatrix}, & &\begin{pmatrix}2\times 4 & 2\times 1\\ 2\times 4 & 2\times 1\end{pmatrix}.
\end{aligned} \tag{10}$$

A direct calculation shows that the matrices (10) satisfy the algebra (4), (5) and (7). For the even-odd part we use the relation

$$(J_{\alpha\beta})^\mu{}_\nu\Gamma^\nu = -\begin{pmatrix}0 & \sigma_{\alpha\beta}\sigma^\mu - \sigma^\mu\bar\sigma_{\alpha\beta}\\ \bar\sigma_{\alpha\beta}\bar\sigma^\mu - \bar\sigma^\mu\sigma_{\alpha\beta} & 0\end{pmatrix}.$$

A matrix representation of the supersymmetric algebra was also given in [40], but with different notations. The transformations are now parameterised by

$$(\Lambda, t, e) = e^{\frac{1}{2}\omega^{\alpha\beta}L_{\alpha\beta}}e^{t^\alpha P_\alpha}e^{\varepsilon^\alpha Q_\alpha + \bar{Q}_{\dot\alpha}\varepsilon^{\dot\alpha}},$$

with $\varepsilon^\alpha, \bar{\varepsilon}^{\dot\alpha}$ Majorana spinors. And since $(\varepsilon^\alpha Q_\alpha + \bar{Q}_{\dot\alpha}\varepsilon^{\dot\alpha})^2 = 0$, we have for a supersymmetric transformation

$$X' = \begin{pmatrix} x'^\mu \\ 1 \\ \theta'^\alpha \\ \theta'^{\dot\alpha} \end{pmatrix} = e^{\varepsilon^\alpha Q_\alpha + \bar{Q}_{\dot\alpha}\varepsilon^{\dot\alpha}} \begin{pmatrix} x^\mu \\ 1 \\ \theta^\alpha \\ \theta^{\dot\alpha} \end{pmatrix} = \begin{pmatrix} x^\mu + i(\varepsilon^\alpha \sigma^\mu{}_{\alpha\dot\alpha}\bar\theta^{\dot\alpha} - \theta^\alpha \sigma_{\mu\alpha\dot\alpha}\bar\varepsilon^{\dot\alpha}) \\ 1 \\ \theta^\alpha + \varepsilon^\alpha \\ \bar\theta^{\dot\alpha} - \bar\varepsilon^{\dot\alpha} \end{pmatrix}$$

Several remarks are in order here. The minus sign in the transformation of the variable $\bar\theta$ seems to be rather surprising. In addition, since the matrices (10) act on the nine-dimensional ($\mathbb{Z}_2-$graded) vector space parametrised by X, at that level, there is no need to have anticommuting variables. As a simple consequence the variable x cannot be real. All this will be corrected by the introduction of Grassmann variables in the next subsection.

2.3. Superspace

Since the supersymmetric algebra contains fermionic supercharges its seems natural to promote the space-time to a superspace by adjoining to the space-time coordinates fermionic coordinates that we take Majorana as it was done in [41]. We then consider a point in a superspace as given by $(x^\mu, \theta^\alpha, \bar\theta^{\dot\alpha})$, together with the conjugate momenta $(\partial_\mu, \frac{\partial}{\partial\theta^\alpha}, \frac{\partial}{\partial\bar\theta^{\dot\alpha}})$ which satisfy (we give the only non-vanishing graded-commutators)

$$[\partial_\mu, x^\nu] = \delta_\mu{}^\nu, \quad \left\{\frac{\partial}{\partial\theta^\alpha}, \theta^\beta\right\} = \delta_\alpha{}^\beta, \quad \left\{\frac{\partial}{\partial\bar\theta^{\dot\alpha}}, \bar\theta^{\dot\beta}\right\} = \delta_{\dot\alpha}{}^{\dot\beta}. \tag{11}$$

Differently from the previous subsection we now assume from the very beginning that the variables θ are of Grassmann type. It should be emphasise that from the properties of spinor indices we have the relation $\{\frac{\partial}{\partial\theta_\alpha}, \theta_\beta\} = -\delta^\alpha{}_\beta$. To make contact with the matrices given in (10) and the standard approach of superspace we also introduce z, ∂_z such that $[\partial_z, z] = 1$. Now, to construct a differential realisation of the supersymmetric algebra, we use several properties. Since these properties are not specific to our superalgebra, we state them to $\mathfrak{g}$ a Lie superalgebra which admits a finite dimensional representation of dimension $m + n$ (m is the dimension of the even part and n of the odd part). Denote

$$X_i = \begin{pmatrix} X_i^{00} & 0 \\ 0 & X_i^{11} \end{pmatrix}, \quad Y_a = \begin{pmatrix} 0 & Y_a^{01} \\ Y_a^{10} & 0 \end{pmatrix},$$

the matrix representation of $\mathfrak{g}$ (X are bosonic operators and Y fermionic operators), satisfying

$$[X_i, X_j] = f_{ij}{}^k X_k, \quad [X_i, Y_a] = R_{ia}{}^b Y_b, \quad \{Y_a, Y_b\} = C_{ab}{}^i X_i. \tag{12}$$

Introduce m bosonic (n fermionic) oscillators (x^A, ∂_A) $(\theta^I, \frac{\partial}{\partial\theta^I})$ satisfying the relation (11) and set

$$X = \begin{pmatrix} x^1 \\ \vdots \\ x^m \end{pmatrix}, \quad \Theta = \begin{pmatrix} \theta^1 \\ \vdots \\ \theta^n \end{pmatrix}, \quad \partial_X = (\partial_1, \cdots, \partial_m), \quad \partial_\Theta = (\partial_{\theta^1}, \cdots, \partial_{\theta^n}).$$

Assume further that the vectors $Z = \begin{pmatrix} X \\ \Theta \end{pmatrix}$ and $\partial_Z = \begin{pmatrix} \partial_X & \partial_\Theta \end{pmatrix}$ transform contravariantly (resp. covariantly) under the action of the matrices X, Y. Finally observing that the matrices $\tilde{X}_i = -X_i^t$ and $\tilde{Y}_a = iY_a^t$, with M^t the transpose of the matrix M satisfy the relations (12), one can show that the operators

$$\mathcal{X}_i = X^t \tilde{X}_i^{00} \partial_X^t + \Theta^t \tilde{X}_i^{11} \partial_\Theta^t, \quad \mathcal{Y}_a = X^t \tilde{Y}^{01} \partial_\Theta^t + \Theta^t \tilde{Y}^{10} \partial_X^t,$$

are a differential realisation of $\mathfrak{g}$, *i.e* satisfy (12).

In our special case, we define (we normalise slightly differently our operators in order that $\mathcal{P}_\mu$ leads to ∂_μ)

$$\begin{aligned}
\mathcal{L}_{\mu\nu} = X^t \tilde{L}^{00}_{\mu\nu} \partial_X^t + \Theta^t \tilde{L}^{11}_{\mu\nu} \partial_\Theta^t &= x_\mu \partial_\nu - x_\nu \partial_\mu + \theta^\alpha \sigma_{\mu\nu\alpha}{}^\beta \partial_{\theta^\beta} - \bar{\theta}^{\dot{\alpha}} \bar{\sigma}_{\mu\nu}{}^{\dot{\beta}}{}_{\dot{\alpha}} \partial_{\bar{\theta}_{\dot{\beta}}} \\
\mathcal{P}_\mu = -X^t \tilde{P}^{00}_\mu \partial_X^t - \Theta^t P^{11}_\mu \partial_\Theta^t &= z \partial_\mu \\
\mathcal{Q}_\alpha = -iX^t \tilde{Q}^{01}_\alpha \partial_\Theta^t - i\Theta^t \tilde{Q}^{10}_\alpha \partial_X^t &= z \partial_{\theta^\alpha} + i\sigma^\mu{}_{\alpha\dot{\beta}} \bar{\theta}^{\dot{\beta}} \partial_\mu \\
\bar{\mathcal{Q}}_{\dot{\alpha}} = -iX^t \tilde{\bar{Q}}^{01}_{\dot{\alpha}} \partial_\Theta^t - i\Theta^t \tilde{\bar{Q}}^{10}_{\dot{\alpha}} \partial_X^t &= -z \partial_{\bar{\theta}^{\dot{\alpha}}} - i\theta^\beta \sigma^\mu{}_{\dot{\alpha}\beta} \partial_\mu
\end{aligned} \tag{13}$$

which reduces to a differential realisation of the algebra (4), (5) and (7) for $z = 1$. From now on in order to simplify the notations $\mathcal{Q}$ and $\bar{\mathcal{Q}}$ will be denoted Q and $\bar{Q}$ respectively

$$Q_\alpha = \partial_{\theta^\alpha} + i\sigma^\mu{}_{\alpha\dot{\beta}} \bar{\theta}^{\dot{\beta}} \partial_\mu, \quad \bar{Q}_{\dot{\alpha}} = -\partial_{\bar{\theta}^{\dot{\alpha}}} - i\theta^\beta \sigma^\mu{}_{\dot{\alpha}\beta} \partial_\mu \tag{14}$$

since no confusion would be possible. For further use we are looking to some operators D_α and $\bar{D}_{\dot{\alpha}}$ which anticommute with Q_β and $\bar{Q}_{\dot{\beta}}$:

$$\{Q_\alpha, D_\beta\} = 0, \quad \{\bar{Q}_{\dot{\alpha}}, D_\beta\} = 0, \quad \left\{\bar{Q}_{\dot{\alpha}}, \bar{D}_{\dot{\beta}}\right\} = 0. \tag{15}$$

A simple calculation gives

$$\begin{aligned}
D_\alpha &= \partial_{\theta^\alpha} - i\sigma^\mu{}_{\alpha\dot{\beta}} \bar{\theta}^{\dot{\beta}} \partial_\mu, \\
\bar{D}_{\dot{\alpha}} &= \partial_{\bar{\theta}^\alpha} - i\theta^\beta \sigma^\mu{}_{\beta\dot{\alpha}} \partial_\mu.
\end{aligned} \tag{16}$$

It is easy to see that they also satisfy the supersymmetric algebra, and in particular we have

$$\left\{D_\alpha, \bar{D}_{\dot{\alpha}}\right\} = -2i\sigma^\mu{}_{\alpha\dot{\alpha}} \partial_\mu.$$

All these operators might be obtained in the more formal way of superspace. The concept of superspace was first introduced in [41]. Formally, in the same manner as we can see the Minkowski space-time as a coset of the Poincaré group by the Lorentz group, the superspace can be seen as a coset of the super-Poincaré group by the Lorentz group (see also [28]). This leads naturally to the structure of supermanifold (*i.e.* structures generalising the concept of manifold and containing Grassmann variables together with the usual "bosonic" coordinates). One can see [42] for a more formal approach (convergence problem, differentiability *etc*). Supermanifolds are intimately related to Lie supergroups as manifolds are related to Lie groups (see for instance [40]). A point in the superspace has coordinates

$$X = (x^\mu, \theta^\alpha, \bar{\theta}^{\dot{\alpha}}).$$

In the coset approach a point in the superspace is parametrised by

$$G(x, \theta, \bar{\theta}) = e^{x^\mu P_\mu + \theta^\alpha Q_\alpha + \bar{Q}_{\dot{\alpha}} \bar{\theta}^{\dot{\alpha}}},$$

and the action of a supersymmetric transformation is given by

$$G(x, \theta, \bar{\theta}) G(0, \varepsilon, \bar{\varepsilon}),$$

we consider the right action in order to have the correct sign in the supersymmetric algebra. Since we are dealing with anticommuting variables we assume further $\{\theta, \varepsilon\} = \{\theta, Q\} = \{\varepsilon, Q\} = 0$ *etc.* We recall that $[\varepsilon^\alpha Q_\alpha, \bar{Q}_{\dot{\alpha}} \bar{\theta}^{\dot{\alpha}}] = \varepsilon^\alpha \{Q_\alpha, \bar{Q}_{\dot{\alpha}}\} \bar{\theta}^{\dot{\alpha}}$ *etc.* Using the Baker-Campbell-Hausdorff formulæ which reduces to

$$e^A e^B = e^{A+B+\frac{1}{2}[A,B]},$$

since $[[A, B], A] = [[A, B], B] = 0$, we finally obtain

$$G(x, \theta, \bar{\theta}) G(0, \varepsilon, \bar{\varepsilon}) = e^{\left[x^\mu + i(\theta\sigma^\mu \bar{\varepsilon} - \varepsilon\sigma^\mu \bar{\theta})\right] P_\mu \; + \; \left[\theta^\alpha + \varepsilon^\alpha\right] Q_\alpha \; + \; \bar{Q}_{\dot{\alpha}} \left[\bar{\theta}^{\dot{\alpha}} + \bar{\varepsilon}^{\dot{\alpha}}\right]}.$$

This means that under a supersymmetry transformation we have

$$\begin{aligned} \delta_\varepsilon x^\mu &= i(\theta\sigma^\mu \bar{\varepsilon} - \varepsilon\sigma^\mu \bar{\theta}), \\ \delta_\varepsilon \theta^\alpha &= \varepsilon^\alpha, \\ \delta_\varepsilon \bar{\theta}^{\dot{\alpha}} &= \bar{\varepsilon}^{\dot{\alpha}}. \end{aligned} \tag{17}$$

Assume now that the transformations (17) are given by some supercharges

$$\delta_\varepsilon x^\mu = [\varepsilon.Q + \bar{Q}.\bar{\varepsilon}, x^\mu], \quad \delta_\varepsilon \theta^\alpha = [\varepsilon.Q + \bar{Q}.\bar{\varepsilon}, \theta^\alpha], \delta_\varepsilon \bar{\theta}^{\dot{\alpha}} = [\varepsilon.Q + \bar{Q}.\bar{\varepsilon}, \bar{\theta}^{\dot{\alpha}}],$$

gives (14) for the supercharges. We see that the benefit of introducing Grassmann variables is two-fold. Firstly the unwanted minus sign in the $\bar{\theta}$ transformation disappear. Secondly, δx is real because of (A.9).

2.4. The superfield formalism: the Wess-Zumino model

The natural objects which live in the superspace are the superfields. Superfields where firstly introduced in [43]. A superfield Φ is a field which depends on the superspace coordinates. Since the Grassmann variables are nilpotent, the superfield has a finite number of components

$$\begin{aligned} \Phi(x, \theta, \bar{\theta}) &= z(x) + \theta^\alpha \psi_\alpha(x) + \bar{\theta}_{\dot{\alpha}} \bar{\chi}^{\dot{\alpha}}(x) + \theta^\alpha \theta_\alpha n(x) + \bar{\theta}_{\dot{\alpha}} \bar{\theta}^{\dot{\alpha}} m(x) \\ &+ \theta^\alpha \sigma_{\mu\alpha\dot{\alpha}} \bar{\theta}^{\dot{\alpha}} v^\mu(x) + \theta^\alpha \theta_\alpha \bar{\theta}_{\dot{\alpha}} \bar{\lambda}^{\dot{\alpha}}(x) + \bar{\theta}_{\dot{\alpha}} \bar{\theta}^{\dot{\alpha}} \theta^\alpha \omega_\alpha(x) + \theta^\alpha \theta_\alpha \bar{\theta}_{\dot{\alpha}} \bar{\theta}^{\dot{\alpha}} d(x), \end{aligned} \tag{18}$$

with z, m, n, d complex scalar field's, v_μ a complex vector field, ψ, λ left-handed spinors and $\bar{\chi}, \bar{\omega}$ right-handed spinors. Since in all supersymmetric calculus one has to be careful with the signs, and since for Grassmann variables we have $\theta^\alpha \psi_\alpha = -\theta_\alpha \psi^\alpha$, when the summation on the spinor indices will be omitted $\theta.\psi$ means $\theta^\alpha \psi_\alpha$ and $\bar{\theta}.\bar{\psi}$ means $\bar{\theta}_{\dot{\alpha}} \bar{\psi}^{\dot{\alpha}}$. In the sequel we will intensively follow this convention. See the reference of Wess and Bagger [29] for more details and useful identities (see also Appendix Appendix A).

A scalar superfield is a superfield which transforms under a supersymmetric transformation

$$\Phi'(X') = \Phi(X).$$

At the infinitesimal level

$$\delta\Phi(X) = \Phi'(X) - \Phi(X) = (\varepsilon.Q + \bar{Q}.\bar{\varepsilon})\Phi(X). \tag{19}$$

It turns out that a scalar superfield is a reducible representation of the supersymmetric algebra. Several types of superfields constructed from Φ may be defined. The chiral superfield is defined by

$$\bar{D}_{\dot{\alpha}}\Phi = 0. \tag{20}$$

We can have also defined an antichiral superfield $\bar{\Phi}$ satisfying $D_\alpha\bar{\Phi} = 0$. The chiral superfield leads to the first supersymmetric (where supersymmetry is realised in a linear way) model in field theory in four space-time dimensions. It is known as the Zess-Zumino model [26]. In fact this model was historically constructed in components and not in the superfield language.

Observing that

$$\bar{D}_{\dot{\alpha}}y^\mu = 0, \quad y^\mu = x^\mu - i\theta\sigma^\mu\bar{\theta},$$

means that the chiral superfield Φ depends only on the variables y and θ. Developing $\Phi(y,\theta)$ using (A.12) and (A.13) we get

$$\begin{aligned}\Phi(y,\theta) &= z(y) + \sqrt{2}\theta.\psi(y) - \theta.\theta F(y) \qquad (21)\\ &= z(x) + \sqrt{2}\theta.\psi(x) - \theta.\theta\ F(x) - i\theta\sigma^\mu\bar{\theta}\ \partial_\mu z + \frac{i}{\sqrt{2}}\theta.\theta\partial_\mu\psi(x)\sigma^\mu\bar{\theta} - \frac{1}{4}\theta.\theta\bar{\theta}.\bar{\theta}\ \Box z.\end{aligned}$$

The minus sign and $\sqrt{2}$ factor are introduced such that the kinetic part of the Lagrangian is correctly normalised. A chiral superfield contains two complex scalars z and F and one left-handed Weyl spinors ψ. The adjoint of Φ is given by

$$\Phi^\dagger(y,\theta) = z^\dagger(x) + \sqrt{2}\bar{\theta}.\bar{\psi} - \bar{\theta}.\bar{\theta}F^\dagger(x) + i\theta\sigma^\mu\bar{\theta}\partial_\mu z^\dagger(x) + \frac{i}{\sqrt{2}}\bar{\theta}.\bar{\theta}\theta\sigma^\mu\partial_\mu\bar{\psi}(x) - \frac{1}{4}\theta.\theta\bar{\theta}.\theta\ \Box z^\dagger(x), \tag{22}$$

and contains two complex scalars and one right-handed Weyl spinor. It turns out that $\Phi^\dagger$ is an anti-chiral superfield. To determine the transformation of the chiral superfield under a supersymmetric transformation, we first observe that

$$Q_\alpha y^\mu = 0, \quad \bar{Q}_{\dot{\alpha}}y^\mu = -2i\theta^\alpha\sigma^\mu{}_{\alpha\dot{\alpha}}. \tag{23}$$

Since Φ only depends on y and θ, we have

$$\begin{aligned}Q_\alpha\Phi(y,\theta) &= Q_\alpha y^\mu\frac{\partial\Phi}{\partial y^\mu} + Q_\alpha\theta^\beta\frac{\partial\Phi}{\partial\theta^\beta} = \frac{\partial\Phi}{\partial\theta^\alpha}\\ \bar{Q}_{\dot{\alpha}}\Phi(y,\theta) &= \bar{Q}_{\dot{\alpha}}y^\mu\frac{\partial\Phi}{\partial y^\mu} + \bar{Q}_{\dot{\alpha}}\theta^\beta\frac{\partial\Phi}{\partial\theta^\beta} = -2i\theta^\alpha\sigma^\mu{}_{\alpha\dot{\alpha}}\frac{\partial\Phi}{\partial x^\mu}.\end{aligned} \tag{24}$$

The transformation of Φ is then given by

$$\begin{aligned} \delta_\varepsilon \Phi &= (\varepsilon.Q + \bar{Q}.\bar{\varepsilon}).\Phi \\ &= \sqrt{2}\varepsilon.\psi + \sqrt{2}\theta\left(-\sqrt{2}\varepsilon F - i\sqrt{2}\sigma^\mu\bar{\varepsilon}\partial z\right) + i\sqrt{2}\theta.\theta\partial_\mu\psi\sigma^\mu\bar{\varepsilon}. \end{aligned} \tag{25}$$

Expressed in components this gives

$$\begin{aligned} \delta_\varepsilon z &= \sqrt{2}\varepsilon.\psi = \sqrt{2}\varepsilon^\alpha\psi_\alpha, \\ \delta_\varepsilon \psi_\alpha &= -\sqrt{2}F\varepsilon_\alpha - i\sqrt{2}\sigma^\mu{}_{\alpha\dot{\alpha}}\bar{\varepsilon}^{\dot{\alpha}}\partial_\mu z, \\ \delta_\varepsilon F &= -i\sqrt{2}\partial_\mu\psi^\alpha\sigma^\mu{}_{\alpha\dot{\alpha}}\bar{\varepsilon}^{\dot{\alpha}}. \end{aligned} \tag{26}$$

Identity (A.13) has been used to simplify $-2i\sqrt{2}\theta\sigma^\mu\bar{\varepsilon}\theta\partial_\mu\psi$. These transformation laws show two points. First, they show that supersymmetry is a symmetry which maps bosons into fermions and fermions into bosons. Secondly they show that the highest component of the superfield Φ, namely F, transforms as a total derivative. This is the key point to construct invariant Lagrangians. Indeed, by definition the product of superfields is a superfield. However, the highest component of a superfield transforms as a total derivative. This means that it is a good candidate to construct invariant Lagrangians. Two types of terms may be constructed

$$\Phi^\dagger\Phi_{|\theta\theta\bar{\theta}\bar{\theta}} = \partial_\mu z^\dagger(x)\partial^\mu z(x) + \frac{i}{2}\Big(\psi(x)\sigma^\mu\partial_\mu\bar{\psi}(x) - \partial_\mu\psi(x)\sigma^\mu\bar{\psi}(x)\Big) + F^\dagger F + \text{derivative terms,} \tag{27}$$

where we have used (A.12) to simplify the z parts and (A.13) and (A.14) for the ψ part. These terms contribute to the kinetic part of the Lagrangian. The interacting part is given by the superpotential $W(\phi)$, which is a polynomial in Φ. This means that W is a holomorphic function. For renormalisation arguments (see below) W is of degree three. Using

$$\begin{aligned} \Phi_i\Phi_j &= z_iz_j + \sqrt{2}\theta\Big(z_i\psi_j + z_j\psi_i\Big) - \theta.\theta\Big(z_iF_j + z_jF_i + \psi_i.\psi_j\Big) \\ \Phi_i\Phi_j\Phi_k &= z_iz_jz_k + \sqrt{2}\theta\Big(z_iz_j\psi_k + z_iz_k\psi_j + z_jz_k\psi_i\Big) \\ &- \theta.\theta\Big(z_iz_jF_k + z_iz_kF_j + z_jz_kF_i + z_i\psi_j.\psi_k + z_j\psi_i.\psi_k + z_k\psi_i.\psi_j\Big), \end{aligned} \tag{28}$$

the most general invariant Lagrangian is given by the superpotential

$$W(\Phi) = \alpha^i\Phi_i + \frac{1}{2}m^{ij}\Phi_i\Phi_j + \frac{1}{6}\lambda^{ijk}\Phi_i\Phi_j\Phi_k. \tag{29}$$

The Wess-Zumino model is then given by

$$\begin{aligned} \mathcal{L}_{\text{W.Z.}} &= (\Phi^{\dagger i}\Phi_i)_{|\theta.\theta\bar{\theta}.\bar{\theta}} + W(\Phi)_{\theta.\theta} + W^\dagger(\Phi)_{\bar{\theta}.\bar{\theta}} \\ &= \partial_\mu z^{i\dagger}\partial^\mu z_i + \frac{i}{2}\Big(\psi_i\sigma^\mu\partial_\mu\bar{\psi}^i - \partial_\mu\psi_i\sigma^\mu\bar{\psi}^i\Big) + F^{i\dagger}F_i \\ &- \alpha^iF_i - \frac{1}{2}m^{ij}(z_iF_j + z_jF_j + \psi_i.\psi_j) + \text{c.c.} \\ &- \frac{1}{6}\lambda^{ijk}(z_iz_jF_k + z_iz_kF_j + z_jz_kF_i + z_i\psi_j.\psi_k + z_j\psi_i.\psi_k + z_k\psi_i.\psi_j) + \text{c.c.} \end{aligned} \tag{30}$$

The equations of motion for the auxiliary fields F give

$$F^{\dagger i} = \alpha^i + m^{ij} z_j + \frac{1}{2}\lambda^{ijk} z_j z_k = \frac{\partial W(z)}{\partial z_i}. \tag{31}$$

If we eliminate F from the Lagrangian and use $W(z)$, the Lagrangian reduces to

$$\begin{aligned} \mathcal{L}_{\rm W.Z.} &= \partial_\mu z^{i\dagger} \partial^\mu z_i + \frac{i}{2}\Big(\psi_i \sigma^\mu \partial_\mu \bar\psi^i - \partial_\mu \psi_i \sigma^\mu \bar\psi^i\Big) \\ &- \frac{1}{2}\sum_i \left|\frac{\partial W}{\partial z_i}\right|^2 - \frac{1}{2}\frac{\partial^2 W}{\partial z_i \partial z_j}\psi_i.\psi_i - \frac{1}{2}\frac{\partial^2 \bar W}{\partial z^{\dagger i}\partial z^{\dagger j}}\bar\psi^i.\bar\psi^i, \end{aligned} \tag{32}$$

$V_F = \frac{1}{2}\sum_i \left|\frac{\partial W}{\partial z_i}\right|^2$, the so-called $F-$terms, contributes to a scalar potential of degree four (since W is of degree three) and $\frac{1}{2}\frac{\partial^2 W}{\partial z_i \partial z_j}\psi_i.\psi_i$ to Yukawa interactions between fermions and scalars. It has to be emphasised that all the interaction terms of the Wess-Zumino model are encoded into the superpotential. Furthermore, the auxiliary fields gives an equal number of degrees of freedom (d.o.f) on-shell and off-shell:

on-shell	z	2 d.o.f.	off-shell	z	2 d.o.f.
	ψ	2 d.o.f.		ψ	4 d.o.f.
	F	0 d.o.f.		F	2 d.o.f.

To proceed further in a construction of supersymmetric models, one uses another type of superfields: the real superfield V such that $V^\dagger = V$. This superfield enables us to construct supersymmetric Yang-Mills theories [44], and to couple chiral superfields to vector superfields in an invariant way. This means that if the various fields are in some representation of the gauge group, the superpotential has to be invariant. Finally, it should be observed that models constructed along these lines are not acceptable physically, since in their spectrum they contain bosons and fermions of the same masses. Since scalar particles with the quantum numbers of known fermions (electron, *etc*) have not been observed supersymmetry has to be broken.

3. Lie algebras of order F

As it has been mentioned previously, Lie (super)algebras are binary algebras. This means that one can always multiply two elements. Indeed, the multiplication law is given by the commutators or the anticommutators. In this section we introduce higher order algebras *i.e.* defined by higher order products and try to implement them for constructing new symmetries in physics. Of course the new structures considered would have to respect the principle of Relativistic and Quantum Physics. Thus we will by-pass the no-go theorems which restrict drastically the possible extensions of the symmetry of the space-time. In this lecture, we construct some $F-$ary algebras which can be seen as a possible extension of the Lie (super)algebras. We then show that these new structures can be applied in physics in two different ways, leading to higher order non-trivial extensions of the Poincaré algebra. It is well known that $(1+2)-$dimensional space-time is exceptional. In such dimensional space-time, one can consider representations which are neither bosons nor fermions but anyons. An anyon is a particle of arbitrary spin. We take advantage of this situation to define an $F-$order non-trivial extension of the Poincaré algebra, which maps a relativistic anyon of spin $s \in \mathbb{R}$ to a relativistic anyons of spin $s \pm 1/F$. In the second application, we play with the way the Noether theorem is implemented in Quantum Field Theory, and construct a non-trivial cubic-extension of the Poincaré algebra for arbitrary dimensional space-time.

3.1. Lie algebras of order F: definition

The basic idea to construct higher order extensions of the Poincaré algebra is to define higher order extensions of Lie superalgebras. In supersymmetric theories, the extensions of the Poincaré algebra are obtained from a "square root" of the translations, "$QQ \sim P$". It is tempting to consider other alternatives where the new algebra is obtained from yet higher order roots. Basically, in such extensions, the generators of the Poincaré algebra are obtained as $F-$fold symmetric products of more fundamental generators, leading to the "$F^{\text{th}}-$root" of translation: "$Q^F \sim P$" with F a positive integer. It is important to stress that such structures are not Lie (super)algebras (even though they contain a Lie subalgebra), and as such avoid *a priori* the Coleman-Mandula [1] as well as the Haag-Lopuszanski-Sohnius no-go theorems [3]. Furthermore, as far as we know, no no-go theorem associated with such types of extensions has been considered in the literature. This can open interesting possibilities to search for a field theoretic realization of a non-trivial extension of the Poincaré algebra which *is not the supersymmetric one.* If successful, this might throw a new light on how to construct physical models. In small space-time dimensions $D \leq 2$, several authors were able to define along these lines an extension of supersymmetry called Fractional Supersymmetry [45]. In the same manner than Lie superalgebras are the underlying mathematical structure of supersymmetry, Lie algebras of order F are the mathematical structure associated to Fractional Supersymmetry. These algebras were introduced in [19, 20]. Subsequently Lie algebras of order three were studied on some formal ground. The basis of the theory of contractions and deformations in the context of Lie algebras of order three has been studied in [21], some progress in the classification of Lie algebras (of order three) were undertaken in [21, 46] and a group together with the parameters of the transformation for Lie algebras of order three were defined [47].

We give now the abstract mathematical structure which generalises the theory of Lie superalgebras and their representations. Let F be a positive integer and $q = \exp\left(\frac{2i\pi}{F}\right)$. We consider a complex vector space $\mathfrak{g}$ together with a linear map ε from $\mathfrak{g}$ into itself satisfying $\varepsilon^F = 1$. We set $\mathfrak{g}_k$ $(k = 0, \cdots, F-1)$ the eigenspace corresponding to the eigenvalue q^k of ε, so that $\mathfrak{g} = \oplus_{k=0}^{F-1} \mathfrak{g}_k$. The map ε is called the grading. If $\mathfrak{g}$ is endowed with the following structures we will say that $\mathfrak{g}$ is Lie algebra of order F [19].

Definition 3.1 *Let $F \in \mathbb{N}^*$. A $\mathbb{Z}_F$-graded $\mathbb{C}-$vector space $\mathfrak{g} = \mathfrak{g}_0 \oplus \mathfrak{g}_1 \oplus \mathfrak{g}_2 \cdots \oplus \mathfrak{g}_{F-1}$ is called a complex Lie algebra of order F if*

(i) *$\mathfrak{g}_0$ is a complex Lie algebra.*

(ii) *For all $i = 1, \ldots, F-1$, $\mathfrak{g}_i$ is a representation of $\mathfrak{g}_0$. If $X \in \mathfrak{g}_0$, $Y \in \mathfrak{g}_i$ then $[X, Y]$ denotes the action of X on Y for any $i = 1, \cdots, F-1$.*

(iii) *For all $i = 1, \ldots, F-1$, there exists an $F-$linear, $\mathfrak{g}_0-$equivariant map*

$$\{\cdots\} : \mathcal{S}^F(\mathfrak{g}_i) \to \mathfrak{g}_0,$$

where $\mathcal{S}^F(\mathfrak{g}_i)$ denotes the $F-$fold symmetric product of $\mathfrak{g}_i$, satisfying the following (Jacobi) identity

$$\sum_{j=1}^{F+1} [Y_j, \{Y_1, \ldots, Y_{j-1}, Y_{j+1}, \ldots, Y_{F+1}\}] = 0, \tag{33}$$

for all $Y_j \in \mathfrak{g}_i$, $j = 1, .., F+1$.

We would like to make now several observations. We clearly see that Definition 3.1 reduces to Lie algebras when $F = 1$ and to Lie superalgebras when $F = 2$. Thus Lie algebras of order

F constitute a possible generalisation of Lie (super)algebras. Furthermore, we have given the definition of Lie superalgebras (Definition 2.1) and of Lie algebra of order F (Definition 3.1) in a same manner in order to stress on there similitude. It should also be observed that for any $i = 1, \ldots, F-1$, the $\mathbb{Z}_F$−graded vector spaces $\mathfrak{g}_0 \oplus \mathfrak{g}_i$ is a Lie algebra of order F. We call these type of algebras *elementary Lie algebras of order F*. Finally, as for Lie superalgebras one can deduce more Jacobi identities from the definition above. Indeed, we have

$$[[X_1, X_2], X_3] + [[X_2, X_3], X_1] + [[X_3, X_1], X_2] = 0, \qquad \text{J1}$$

$$[[X_1, X_2], Y_1] + [[X_2, Y_1], X_1] + [[Y_1, X_1], X_2] = 0, \qquad \text{J2}$$

$$[X_1, \{Y_1, \ldots, Y_F\}] = \{[X_1, Y_1], \ldots, Y_F\} + \cdots + \{Y_1, \ldots, [X, Y_F]\}. \qquad \text{J3}$$

for any $X_1, X_2, X_3 \in \mathfrak{g}_0$, $Y_1, \cdots, Y_F \in \mathfrak{g}_1$. The identity J_1 is the consequence of the fact that $\mathfrak{g}_0$ is a Lie algebra, J_2 is equivalent to assume that $\mathfrak{g}_1$ is a representation of $\mathfrak{g}_0$ and J_3 is related to the $\mathfrak{g}_0$− equivariance of the mapping $\mathcal{S}^F(\mathfrak{g}_1) \to \mathfrak{g}_0$. Only (33) is an extra constraint.

A representation of an elementary Lie algebra of order F is a linear map $\rho : \mathfrak{g} = \mathfrak{g}_0 \oplus \mathfrak{g}_1 \to$ End(V), such that (for all $X_i \in \mathfrak{g}_0, Y_j \in \mathfrak{g}_1$)

$$\begin{array}{l}\rho([X_1, X_2]) = \rho(X_1)\rho(X_2) - \rho(X_2)\rho(X_1) \\ \rho([X_1, Y_2]) = \rho(X_1)\rho(Y_2) - \rho(Y_2)\rho(X_1) \\ \rho(\{Y_1.\cdots, Y_F\}) = \sum\limits_{\sigma \in S_F} \rho\left(Y_{\sigma(1)}\right) \cdots \rho\left(Y_{\sigma(F)}\right)\end{array} \qquad (34)$$

(S_F being the group of permutations of F elements). If $V = V_0 \oplus \cdots \oplus V_{F-1}$ then for all $a \in \{0, \cdots, F-1\}$, V_a is a $\mathfrak{g}_0$−module and we have

$$\rho(\mathfrak{g}_1)(V_a) \subseteq V_{a+1}. \qquad (35)$$

There is no need to impose further conditions, since the Jacobi identity (33) is just a consequence of the associativity of the product in End(V).

In Definition 3.1 we have defined complex Lie algebras of order F. Since we want to implement these new structures to extend non-trivially the Poincaré algebra and since this last algebra is a real Lie algebra, real forms of Lie algebras of order F should be defined. A real elementary Lie algebra of order F is given by a real Lie algebra $\mathfrak{g}_0$ and $\mathfrak{g}_1$ a real representation of $\mathfrak{g}_0$ which satisfy the axioms of complex elementary Lie algebras of order F. Of course when considering a real form of a complex $\mathfrak{g}$ some structures are lost such that the grading map ε. Many examples of Lie algebras of order F (complex and real) were constructed. In these lectures, since we study non-trivial higher order extensions of the Poincaré algebra we only give Lie algebras of order F which are related to the Poincaré algebra. There is two types of such extensions infinite-dimensional and finite-dimensional that will be studied below.

3.2. Lie algebras of order F in $(1+2)$*−dimensions*

In this subsection we construct explicitly an infinite dimensional extension of the Poincaré algebra in $(1+2)$−dimensions [23]. Studying the representations of this extension gives that this higher-order symmetry is a symmetry between relativistic anyons. We recall firstly some results upon the Lorentz and the Poincaré groups in three dimensions, and then construct a higher order extension of the Poincaré symmetry.

3.2.1. Discrete series of $\overline{SO}(1,2)$ Since $SO(1,2)$ is a non-compact Lie group its unitary representations are infinite dimensional. There are two types of unitary representations. The discrete and the continuous series. The former are either bounded from below or bounded from above and the latter are unbounded representations. These representations were studied by Bargmann [48] (see also [49]). The generators of the Poincaré algebra in $(1+2)-$dimensions are (P_0, P_1, P_3) and (J_0, J_1, J_2) and the algebra is given by

$$\begin{aligned} [J_\mu, J_\nu] &= -\epsilon_{\mu\nu\sigma}\eta^{\sigma\rho} J_\rho, \\ [J_\mu, P_\nu] &= -\epsilon_{\mu\nu\sigma}\eta^{\sigma\rho} P_\rho, \\ [P_\mu, P_\nu] &= 0, \end{aligned} \tag{36}$$

with $\eta_{\mu\nu} = \text{diag}(1,-1,-1)$ the three-dimensional Minkowski metric and $\varepsilon_{\mu\nu\rho}$ the Levi-Civita tensor normalised as follows $\varepsilon_{012} = \varepsilon^{012} = 1$. If we consider the complexified of the Poincaré algebra and introduce $L_0 = -iJ_0, L_\pm = -iJ_1 \mp J_2$ and $\Pi_0 = -iP_0, \Pi_\pm = -iP_1 \mp P_2$ (these unconventional notations come from our normalisation and from the fact that our structure constants are real for Lie algebras) the commutation relations reduce to[1]

$$\begin{array}{llll} & [L_-, \Pi_-] = 0 & [L_-, \Pi_0] = \Pi_- & [L_-, \Pi_+] = 2\Pi_0 \\ [L_0, L_\pm] = \pm L_\pm & [L_0, \Pi_-] = -\Pi_- & [L_0, \Pi_0] = 0 & [L_0, \Pi_+] = \Pi_+ \\ [L_+, L_-] = -2L_0 & [L_+, \Pi_-] = -2\Pi_0 & [L_+, \Pi_0] = -\Pi_+ & [L_+, \Pi_+] = 0. \end{array} \tag{37}$$

Since $\pi_1(SO(1,2)) = \mathbb{Z}$ there exists, in some universal covering group noted $\overline{SO}(1,2)$, representations with arbitrary spin $s \in \mathbb{R}$. This means that the eigenvalues of L_0 are not any more integer (bosonic) or half-integer (fermionic). There are two types of unitary representations for $\overline{SO}(1,2)$ the discrete or the continuous series [48, 49]. The only representations that are relevant in the sequel are the discrete series which are either bounded from below or bounded from above. We have

$$\mathcal{D}_s^+ : \begin{array}{l} L_{s0}|s_+, n\rangle = (s+n)|s_+, n\rangle \\ L_{s+}|s_+, n\rangle = \sqrt{(2s+n)(n+1)}|s_+, n+1\rangle \\ L_{s-}|s_+, n\rangle = \sqrt{(2s+n-1)n}|s_+, n-1\rangle, \end{array} \tag{38}$$

for representations bounded from below and

$$\mathcal{D}_s^- : \begin{array}{l} L_{s0}|s_-, n\rangle = -(s+n)|s_-, n\rangle \\ L_{s+}|s_-, n\rangle = -\sqrt{(2s+n-1)n}|s_-, n-1\rangle \\ L_{s,-}|s_- n\rangle = -\sqrt{(2s+n)(n+1)}|s_-, n+1\rangle. \end{array} \tag{39}$$

for representations bounded from above. For the first representation we have $L_{s,-}|s_+, 0\rangle = 0$ and for the second $L_{s,+}|s_-, 0\rangle = 0$. These representations are of dimensions $2|s|+1$ if $2s$ is a negative integer but in the general case we have an infinite number of states. In both cases the quadratic Casimir operator $Q = L_0^2 - \frac{1}{2}(L_+L_- + L_-L_+)$ has eigeinvalues $s(s-1)$. Furthermore, when $s < 0$ the representations are not unitary. There is a well-known way to characterise these representations, using functions of complex variables (see the first reference of [49]).

To define higher order extensions of the Poincaré algebra, taking the supersymmetric extension as a guideline, we choose representations for which $s = -1/F$ to build a non-trivial

[1] In the mathematical literature one prefers to take $K_0 = J_1$ for the Cartan subalgebra with $K_\pm = J_0 \mp J_2$. We have made of different choice – in the complexified of $\mathfrak{so}(1,2)$ – in such way that the eigenvalues of L_0 correspond to the eigenvalues of the $SO(2)$ subgroup of $SO(1,2)$ and can be identified with the spin of the state.

extension of the Poincaré algebra. If we observe the relations (38) and (39) with $s = -1/F$, we see an ambiguity in the square root of $-2/F$. So a priori we have four different representations for $s = -1/F$, (two bounded from below/above) with the two choices $\sqrt{-1} = \pm i$. We note $\mathcal{D}^{\pm}_{-1/F;\pm}$ (with obvious notations) these representations. Next, we can make the following identifications

- the dual representation of $\mathcal{D}^{+}_{-1/F;+}$ is obtained through the substitution $J_\mu \longrightarrow -\left(J_\mu\right)^t$ and is given by $\left[\mathcal{D}^{+}_{-1/F;+}\right]^* \cong \mathcal{D}^{-}_{-1/F;+}$;
- the complex conjugate representation of $\mathcal{D}^{+}_{-1/F;+}$ is defined by $J_\mu \longrightarrow \overline{J_\mu}$ (we have to be careful when we do such a transformation because we have *by definition* $L_\pm = -iJ_1 \mp J_2$ *for any representation*) is given by $\overline{\mathcal{D}^{+}_{-1/F;+}} \cong \mathcal{D}^{-}_{-1/F;-}$;
- the dual of the complex conjugate representation of $\mathcal{D}^{+}_{-1/F;+}$ is given by $\left[\overline{\mathcal{D}^{+}_{-1/F;+}}\right]^* \cong \mathcal{D}^{+}_{-1/F;-}$.

If we note $\psi_a \in \mathcal{D}^{+}_{-1/F;+}, \psi^a \in \mathcal{D}^{-}_{-1/F;+}, \bar{\psi}_{\dot{a}} \in \mathcal{D}^{-}_{-1/F;-}$ and $\bar{\psi}^{\dot{a}} \in \mathcal{D}^{+}_{-1/F;-}$ then we have the following transformation laws:

$$
\begin{aligned}
\psi'_a &= S_a{}^b \psi_b \\
\psi'^a &= \left(S^{-1}\right)_b{}^a \psi^b \\
\bar{\psi}'_{\dot{a}} &= \left(\bar{S}\right)_{\dot{a}}{}^{\dot{b}} \bar{\psi}_{\dot{b}} \\
\bar{\psi}'^{\dot{a}} &= \left((\bar{S})^{-1}\right)_{\dot{b}}{}^{\dot{a}} \bar{\psi}^{\dot{b}}.
\end{aligned} \tag{40}
$$

Furthermore, if we define

$$\psi^a = g^{a\dot{a}} \bar{\psi}_{\dot{a}}, \tag{41}$$

we can write the following scalar product

$$\varphi^a \psi_a = -\bar{\varphi}_{\dot{0}} \psi_0 + \sum_{a>0} \bar{\varphi}_{\dot{a}} \psi_a, \tag{42}$$

where the infinite matrix $g^{a\dot{a}}$ and its inverse $g_{\dot{a}a}$ are given by $\text{diag}(-1, 1, \cdots, 1, \cdots)$. The reason why we have a pseudo-hermitian scalar product is because we are dealing with a non-unitary representation of a non-compact Lie group. The invariant scalar product gives an explicit isomorphism between the two representations bounded from below (or above) $(\left(S^{-1}\right)_b{}^a = g^{a\dot{a}} \left(\bar{S}\right)_{\dot{a}}{}^{\dot{b}} g_{\dot{b}b})$. From now on, we choose $\sqrt{-2/F} = i\sqrt{2/F}$ for representations bounded from below and $\sqrt{-2/F} = -i\sqrt{2/F}$ for those bounded from above.

3.2.2. Representations of the three-dimensional Poincaré algebra Particles are classified according to the values of the Casimir operators of the Poincaré algebra. More precisely, for a mass m particle of positive/negative energy, the unitary irreducible representations are obtained by studying the little group leaving the rest-frame momentum $\Pi^\alpha = -iP^\alpha = (m, 0, 0)$ invariant. (In the same manner we consider $L_\mu = -iJ_\mu$ the angular momentum.) This stability group in $\overline{SO(1,2)}$ is simply the universal covering group $\mathbb{R}$ of the abelian subgroup of rotations $SO(2)$ (generated by L^0). As is well-known, such a group is not quantised. This means that the substitution $L^0 \to L^0 + s$ leaves the $SO(2)$ part invariant. But the remarkable property of $\overline{SO(1,2)}$ is that the concomitant transformation on the Lorentz boosts $L^i \to L^i + s\frac{\Pi^i}{\Pi^0+m}$ leaves the algebraic structure (36) unchanged. Indeed, on account of the mass-shell condition

$\Pi_0^2 - \Pi_1^2 - \Pi_2^2 = m^2$, one proves that $(L_0 + s, L_i + s\frac{\Pi_i}{\Pi_0+m})$ satisfy the same algebra as (L_0, L_i). Anyway, following the method of induced representations of groups expressible as a semi-direct product, we find that unitary irreducible representations for a massive particles are one dimensional.

The main difference between $SO(1,2)$, or more precisely the proper orthochronous Lorentz group, and $SO(3)$ is that $\Pi_0 + m$ never vanishes with $SO(1,2)$ and s does not need to be quantised. In Ref.[51, 52], a relativistic wave equation for massive anyons was given. First, notice that the two Casimir operators are the two scalars $P.P$ and $P.J$ and their eigenvalues for a spin$-s$ unitary irreducible representation are $-m^2$ and $-ms$, respectively. The equations of motion are then

$$\begin{aligned}(P^2 + m^2)\Psi &= 0 \\ (P.J + sm)\Psi &= 0.\end{aligned} \tag{43}$$

However, to obtain manifestly covariant equations one has to go beyond the mass-shell conditions (43) given by the induced representation. Therefore, we can start with a field which belongs to the appropriate spin$-s$ representation of the *full* Lorentz group instead of the little group. When s is a negative integer, or a negative half-integer, this representation is not unitary and is $2|s| + 1$ dimensional, and the solution of the relativistic wave equations reduces to the appropriate induced representation (see [50, 51] for an explicit calculation in the case $|s| = 1, 1/2$). When s is an arbitrary number, the representation is infinite dimensional and belongs to the discrete series of $\overline{SO(1,2)}$. A relativistic wave equation for an anyon in the continuous series was also considered in [52, 53]. For completeness, we recall in Appendix Appendix B, how M. Plyushchay obtained a very nice relativistic wave equation in [53].

3.2.3. Non-trivial extension of the Poincaré algebra in $(1+2)-$*dimensions* Using the representations (39–38), and with the sign ambiguity resolved, we can define two series of operators, belonging to a non-trivial representation of the Poincaré algebra. We denote now $\sqrt{-1} = i$. Note $A^+_{-1/F+n}$ those built from the representation bounded from below $\mathcal{D}^+_{-1/F;+}$ and $A^-_{-1/F+n}$ the charges of the representation bounded from above $\mathcal{D}^-_{-1/F;-}$. Recall that $\overline{(\mathcal{D}^+_{-1/F;+})} \cong \mathcal{D}^-_{-1/F;-}$. Using (39, 38) we get

$$\begin{aligned}
\left[L_0, A^+_{-1/F+n}\right] &= (n - 1/F)\ A^+_{-1/F+n} \\
\left[L_+, A^+_{-1/F+n}\right] &= \sqrt{(-2/F + n)(n+1)}\ A^+_{-1/F+n+1} \\
\left[L_-, A^+_{-1/F+n}\right] &= \sqrt{(-2/F + n - 1)n}\ A^+_{-1/F+n-1} \\
\\
\left[L_0, A^-_{-1/F+n}\right] &= -(n - 1/F)\ A^-_{-1/F+n} \\
\left[L_+, A^-_{-1/F+n}\right] &= -\left(\sqrt{(-2/F + n - 1)n}\right)^\star\ A^-_{-1/F+n-1} \\
\left[L_-, A^-_{-1/F+n}\right] &= -\left(\sqrt{(-2/F + n)(n+1)}\right)^\star\ A^-_{-1/F+n+1}.
\end{aligned} \tag{44}$$

We want to combine this algebra (44) in a non-trivial way with the Poincaré algebra (36). With such a choice, $A^+_{\frac{-1}{F}}$ (resp. $A^-_{\frac{-1}{F}}$) has helicity $h = -\frac{1}{F}$ ($\frac{1}{F}$ resp.). With the above choices for the square roots of the negative numbers, we know that the representations are conjugate to each other, *i.e.* $\left(A^{+}_{-1/F+n}\right)^{\dagger} \equiv A^{-}_{-1/F+n}$.

By Definition 3.1, to associate a Lie algebra of order F with $\mathcal{D}^{\pm}_{-1/F,\pm}$, we have to relate $\mathcal{S}^{F}(\mathcal{D}^{\pm}_{-1/F,\pm})$ to $\mathcal{D}_{-1}$. But here some care has to be taken, since $\mathcal{D}^{\pm}_{-1/F,\pm}$ are infinite-dimensional representations of $\mathfrak{so}(1,2)$, although $\mathcal{D}_{-1}$ is finite-dimensional. Indeed, using (44) we have

$$\left[L_{\mp},(A^{\pm}_{-\frac{1}{F},\pm})^{F}\right]=0,\quad \left[L_{0},(A^{\pm}_{-\frac{1}{F},\pm})^{F}\right]=\mp(A^{\pm}_{-\frac{1}{F}},\pm)^{F}.$$

This means that $(A^{\pm}_{-\frac{1}{F},\pm})^{F}$ might be identified with a primitive vector for the vector representation. Consider hence $\left\langle(A^{\pm}_{-\frac{1}{F}})^{F}\right\rangle \subset \mathcal{S}^{F}\left(\mathcal{D}^{\pm}_{-1/F,\pm}\right)$, the representation built from the primitive vector $(A^{\pm}_{-\frac{1}{F}})^{F}$. For this representation, *a priori* we have

$$\left[L_{\pm},\left[L_{\pm},\left[L_{\pm},(A^{\pm}_{-\frac{1}{F},\pm})^{F}\right]\right]\right]\neq 0, \tag{45}$$

but using (37) we can easily see that

$$\left[L_{\mp},\left[L_{\pm},\left[L_{\pm},\left[L_{\pm},(A^{\pm}_{-\frac{1}{F},\pm})^{F}\right]\right]\right]\right]=0. \tag{46}$$

This can be represented by means of a diagram. We denote now $[X,Y]=\mathrm{ad}(X).Y$, $(A^{\pm}_{-\frac{1}{F},\pm})^{F}=V_{\mp}$, $\mathrm{ad}(L_{\pm})^{n}.(A^{\pm}_{-\frac{1}{F},\pm})^{F}=V_{\pm(n-1)}$

$$\begin{array}{ccccccccccccc} V_{\mp} & \underset{L_{\mp}}{\overset{L_{\pm}}{\rightleftarrows}} & V_{0} & \underset{L_{\mp}}{\overset{L_{\pm}}{\rightleftarrows}} & V_{\pm} & \overset{L_{\pm}}{\longrightarrow} & V_{\pm 2} & \underset{L_{\mp}}{\overset{L_{\pm}}{\rightleftarrows}} & V_{\pm 3} & \underset{L_{\mp}}{\overset{L_{\pm}}{\rightleftarrows}} \cdots & \underset{L_{\mp}}{\overset{L_{\pm}}{\rightleftarrows}} & V_{\pm n} & \underset{L_{\mp}}{\overset{L_{\pm}}{\rightleftarrows}} \cdots \\ \swarrow{\scriptstyle L_{\mp}} & & & & & \swarrow{\scriptstyle L_{\mp}} & & & & & & & \\ 0 & & & & 0 & & & & & & & & \end{array} \tag{47}$$

This means that we can safely impose $V_n = 0$ for $n = \pm 2, \pm 3, \cdots$. This can be presented on a more formal ground. See Appendix Appendix C for more details. To construct a non-trivial extension of the Poincaré algebra we then proceed as follows. We identify $(A^{\pm}_{-\frac{1}{F},\pm})^{F}$ with $\Pi_{\mp}$ and we impose the relation $\mathrm{ad}^{3}(L_{\pm}).(A^{\pm}_{-\frac{1}{F},\pm})^{F}=0$. Thus we have the following isomorphism

$$\mathcal{D}_{-1}=\langle\Pi_0,\Pi_+,\Pi_-\rangle=\left\{(A^{\pm}_{-\frac{1}{F},\pm})^{F},\left[L_{\pm},(A^{\pm}_{-\frac{1}{F},\pm})^{F}\right],\left[L_{\pm},\left[L_{\pm},(A^{\pm}_{-\frac{1}{F},\pm})^{F}\right]\right]\right\}$$

which leads to the following brackets

$$\begin{aligned}
&\frac{1}{F!}\left\{A^{\pm}_{-\frac{1}{F}},\ldots,A^{\pm}_{-\frac{1}{F}}\right\} = \Pi_{\mp}\\
&\frac{1}{F!}\left\{A^{\pm}_{-\frac{1}{F}},\ldots,A^{\pm}_{-\frac{1}{F}},A^{\pm}_{1-\frac{1}{F}}\right\} = \pm i\sqrt{\frac{2}{F}}\Pi_0\\
&-\frac{(F-1)}{F!}\left\{A^{\pm}_{-\frac{1}{F}},\ldots,A^{\pm}_{-\frac{1}{F}},A^{\pm}_{1-\frac{1}{F}},A^{\pm}_{1-\frac{1}{F}}\right\} \pm i\frac{\sqrt{F-2}}{F!}\left\{A^{\pm}_{-\frac{1}{F}},\ldots,A^{\pm}_{-\frac{1}{F}},A^{\pm}_{2-\frac{1}{F}}\right\}_F = \Pi_{\pm}\\
&\left[L_{\pm},\left[L_{\pm},\left[L_{\pm},\left(A^{\pm}_{-\frac{1}{F}}\right)^F\right]\right]\right] = 0\\
&\vdots
\end{aligned} \tag{48}$$

How, can we address the question of the remaining brackets? In fact, it is impossible to find a decomposition

$$\mathcal{S}^F\left(\mathcal{D}^{\pm}_{-1/F}\right) = \mathcal{D}_{-1} \oplus \mathcal{D}^{\perp}_{-1}, \tag{49}$$

where $\mathcal{D}^{\perp}_{-1}$ is stable under $SO(1,2)$. Indeed, if there were such a decomposition, there would exist a $SO(1,2)$ equivariant projection

$$\pi:\ \mathcal{S}^F\left(\mathcal{D}^{\pm}_{-1/F}\right) \longrightarrow \mathcal{D}_{-1}. \tag{50}$$

But then $X^{\pm} = \pi\left(\mathcal{S}^F\left(A^{\pm}_{-1/F},\cdots,A^{\pm}_{-1/F},A^{\pm}_{3-1/F}\right)\right) \in \mathcal{D}_{-1}$ would satisfy (see 48)

$$\left[L_{\mp},\left[L_{\mp},\left[L_{\mp},X^{\pm}\right]\right]\right] = iF!\sqrt{2/F}\sqrt{2(1-2/F)}\sqrt{3(2-2/F)}\Pi_{\mp} \neq 0,$$

and this is impossible because in the vector representation $\mathcal{D}_{-1}$, $\mathrm{ad}^3(L_-)$ acts as zero. This means that the algebra given by (48) is not a Lie algebra of order three. There are in fact two ways to define a Lie algebra of order three. Either to embed $\mathcal{D}_{-1}$ into an infinite dimensional (but indecomposable) representation of $\mathfrak{so}(1,2)$ or to embed the algebra $\mathfrak{so}(1,2)$ into a infinite dimensional algebra, which here turns out to be the De Witt algebra (the centerless Virasoro algebra) [19, 54]. Since this part is technical and not relevant for the sequel, it will be developed in Appendix Appendix C.

3.2.4. Representations Before studying the representations of the algebra given by the quadratic relations (36) and (44) and the higher order brackets (48), let us draw some general feature. First, since P^2 commutes with all the generators, all states in an irreducible representation have the same mass. Second, define an anyonic-number operator (this is in fact the grading map of the definition of Lie algebras of order F) $\exp(2i\pi\mathcal{N}_A)$ which gives the phase $e^{2i\pi s}$ on a spin$-s$ anyon and assume we have a finite dimensional representation of our algebra. We have $\mathrm{tr}\exp(2i\pi\mathcal{N}_A) = 0$ showing that in each irreducible representation there are F possible states with helicity $(s, s \pm \frac{1}{F}, \ldots, s \pm \frac{F-1}{F})$, where s will be specified later) and the dimension of the space with a given helicity is always the same. This can be checked proving by that $(\exp(2i\pi\mathcal{N}_A)A_s = e^{2i\pi s}A_s\exp(2i\pi\mathcal{N}_A))$ and using cyclicity of the trace

$$
\begin{aligned}
&\operatorname{tr}\left(\exp(2i\pi\mathcal{N}_A)\left\{A^+_{-\frac{1}{F}},\ldots,A^+_{-\frac{1}{F}},A^+_{1-\frac{1}{F}}\right\}_F\right)\\
=&\ (F-1)!\times\operatorname{tr}\left(\sum_{a=0}^{F-1}e^{2i\pi\mathcal{N}_A}\left(A^+_{-\frac{1}{F}}\right)^a\left(A^+_{1-\frac{1}{F}}\right)\left(A^+_{-\frac{1}{F}}\right)^{F-a-1}\right)\\
=&\ (F-1)!\times\left(\sum_{a=0}^{F-1}e^{-\frac{2i\pi a}{F}}\right)\operatorname{tr}\left(\left(A^+_{-\frac{1}{F}}\right)^{F-1}e^{2i\pi\mathcal{N}_A}\left(A^+_{1-\frac{1}{F}}\right)\right)=0.
\end{aligned}
$$

To study the representations of the higher order extension of the Poincaré algebra, we follow the method of induced representations of Wigner. We restrict ourselves to massive representations. The massive representations are characterised by $P^\mu P_\mu = -m^2$ (recall that P_μ are antihermitian). Going in the rest-frame where $P^\mu = (im, 0, 0)$ gives that the only non-vanishing bracket are the ones involving the generator $A^\pm_{-1/F}, A^\pm_{1-1/F}$. Thus we assume that $A^\pm_s = 0$ with $s \neq -1/F, 1-1/F$ *i.e.* that the "active" charges are $A^\pm_{-1/F}$ and $A^\pm_{1-1/F}$, and the little algebra is generated by $1, L_0$ and $A^\pm_{-1/F}, A^\pm_{1-1/F}$. After appropriate rescaling of the generators, the higher order brackets are given by

$$
\begin{aligned}
(F-1)! &= \left\{A^\pm_{-1/F},\cdots,A^\pm_{-1/F},A^\pm_{1-1/F}\right\} \\
0 &= \left\{A^\pm_{s_1},\cdots,A^\pm_{s_{F-1}},A^\pm_{s_F}\right\}, s_1,\cdots,s_F=-1/F,1-1/F,\ s_1+\cdots s_F\neq 0.
\end{aligned}
\tag{51}
$$

This algebra, called the Clifford algebra of polynomial's is studied in Appendix Appendix D. A relevant representation is given by

$$
A^+_{-\frac{1}{F}} = \begin{pmatrix}
0 & 0 & 0 & \ldots & 0 & 0\\
\sqrt{1(F-1)} & 0 & 0 & \ldots & 0 & 0\\
0 & \sqrt{2(F-2)} & 0 & \ldots & 0 & 0\\
\vdots & \vdots & & \ddots & \ddots & \vdots\\
0 & 0 & \ldots & 0 & \sqrt{(F-1)1} & 0
\end{pmatrix},\quad
A^+_{1-\frac{1}{F}} = \begin{pmatrix}
0 & 0 & 0 & \ldots & 0 & 1/(F-1)!\\
0 & 0 & 0 & \ldots & 0 & 0\\
0 & 0 & 0 & \ldots & 0 & 0\\
\vdots & \vdots & & \ddots & & \vdots\\
0 & 0 & 0 & \ldots & 0 & 0
\end{pmatrix}.
\tag{52}
$$

In [23] another equivalent representation was exhibited. From the basic conjugation we obtain

$$
A^-_{-1/F} = (A^+_{-1/F})^\dagger,\quad A^-_{1-1/F} = (A^+_{1-1/F})^\dagger.
$$

As a direct consequence, the generators satisfy the additional relations

$$
A^+_{1-1/F} = \frac{1}{(F-1)!^2}(A^-_{-1/F})^{F-1},
$$

together with the quadratic relations

$$
\begin{aligned}
\left[A^-_{-1/F},A^+_{-1/F}\right] &= 2N = 2\mathrm{diag}(\frac{F-1}{2},\frac{F-3}{2},\cdots,\frac{1-F}{2}) \\
\left[N,A^\pm_{-1/F}\right] &= \mp A^\pm_{-1/F}.
\end{aligned}
\tag{53}
$$

This means that $iN, \frac{i}{2}(A^+_{-1/F}+A^-_{-1/F}), \frac{1}{2}(A^+_{-1/F}-A^-_{-1/F})$ are antihermitian and generate the finite dimensional unitary irreducible representation of $\mathfrak{su}(2)$. Thus the representation of the little algebra build with $A^{\pm}_{-1/F}$ will be unitary. In fact, we obtain an F−dimensional representation of $\mathfrak{su}(2)$. To identify the helicity content of the representation, we assume that we are starting from a vacuum state $\Omega^+_\lambda = \Omega_\lambda$ in the spin−λ representation of $SO(1,2)$ which is annihilated by $A^-_{-1/F}$. Thus acting on the vacuum state with the operator $A^+_{-1/F}$ we identify the helicity content of the representation. Indeed the states $(A^+_{-1/F})^n\Omega_\lambda$, $n = 0, \cdots, F-1$ are of helicity $h = \lambda - n/F$. We assume further that these states are of positive energy. Since the representation has to be CPT invariant, we consider also a conjugated vacuum $\Omega^-_{-\lambda} = \bar{\Omega}_\lambda$ of helicity $h = -\lambda$, negative energy and annihilated by $A^+_{-1/F}$. Acting with $A^-_{-1/F}$ on the conjugated vacuum, we obtain the states of helicity $n/F - \lambda$. This can be summarised in the following table:

states	helicity	states	helicity
Ω^+_λ	λ	$\Omega^-_{-\lambda}$	$-\lambda$
$A^+_{-1/F}\Omega^+_\lambda$	$\lambda - 1/F$	$A^-_{-1/F}\Omega^-_{-\lambda}$	$-\lambda + 1/F$
$\vdots$			$\vdots$
$\left(A^+_{-1/F}\right)^a\Omega^+_\lambda$	$\lambda - a/F$	$\left(A^-_{-1/F}\right)^a\Omega^-_{-\lambda}$	$-\lambda + a/F$
$\vdots$			$\vdots$
$\left(A^+_{-1/F}\right)^{F-1}\Omega^+_\lambda$	$\lambda - (F-1)/F$	$\left(A^-_{-1/F}\right)^{F-1}\Omega^-_{-\lambda}$	$-\lambda + (F-1)/F$

The states of positive energy and helicity $(\lambda, \lambda - \frac{1}{F}, \ldots, \lambda - \frac{F-1}{F})$ are CP- conjugated to the states of negative negative energy and helicity $(-\lambda, -\lambda + \frac{1}{F}, \ldots, -\lambda + \frac{F-1}{F})$. Having the states on shell, the states of positive energy are boosted and belong to the representations bounded from below $\mathcal{D}^+_{\lambda - n/F}$ while those of negative energy correspond to the representations bounded from above $\mathcal{D}^-_{\lambda - n/F}$. It should be interesting to have a explicit Lagrangian where the higher order extensions of the Poincaré algebra are realised, but up to now there are no results in these direction. As a final remark let us mention that if we had considered only one series of operator A, say $A^+_{n-1/F}$, the representation would not have been unitary.

3.3. Cubic extensions of the Poincaré algebra in any dimensions

Now, we would like to extend the ideas developed previously in any space-time dimensions, namely to construct higher order extensions of the Poincaré algebra in D−dimensions. We have shown that considering any semi-simple Lie algebra $\mathfrak{g}$ and any representation $\mathcal{D}$, along the lines of Appendix Appendix C, one is able to define a Lie algebra of order F starting from $\mathfrak{g} \oplus \mathcal{D}$ [19]. In particular this applies to $\mathfrak{so}(1, D-1) \oplus \mathcal{D}_v$, with $\mathcal{D}_v$ the vector representation of $\mathfrak{so}(1, D-1)$. However, this construction leads to infinite dimensional representations of the Lorentz group which are not exponentiable, and as a consequence is unacceptable for realistic physical models. This obstruction is valid as soon as the space-time dimension is greater than four. Indeed, even in four dimensions, where the situation is somewhat exceptional, we cannot construct cubic extensions along the lines of infinite dimensional algebras. In this case, the little group for massless particles is $E(2)$, the group of affine transformations in the plane. Representing the translational part by zero gives rise to $SO(2)$, which is not quantised. So we could have expected to consider, in this case, some massless states with fractional helicity. In [55] we defined a relativistic wave equation along the lines of Appendix Appendix B, considering infinite-dimensional representations corresponding formally to the massless states with fractional (real) helicity. But the solutions of the relativistic equation, however, break down "spontaneously" the $(1+3)D$ Poincaré invariance to the $(1+2)D$ Poincaré invariance and induce a compactification on a circle that produces a consistent theory for massive anyons in $D = 1+2$.

This seems to be a serious obstacle to define higher order extensions of the Poincaré algebra in the framework of Lie algebras of order F when the space-time dimension is greater than four. Therefore, if we expect to apply higher order extensions of the Poincaré algebra, we have to follow new lines. Subsequently, it was realised that finite-dimensional Lie algebras of order F could be defined. Indeed induction theorems enable us to define Lie algebras of order F from Lie algebras or Lie superalgebras [20]. Thus, this opens the way to obtain higher order extensions of the Poincaré algebra. However, we should understand in that context, how these structures would not be in conflict with the Noether and spin-statistics theorems. Since we continue to assume that bosonic fields are quantised by commutation relations although fermionic fields are quantised by anticommutation relations, a fresh look to the Noether theorem should be given in order to implement Lie algebras of order F.

3.3.1. Noether theorem in higher order algebras To understand in which way the Noether theorem can be implemented in higher order symmetries, we first recall some well-known results and give some general features in the conventional case *i.e.* when the symmetries correspond to a Lie (super)algebra. According to Noether theorem, to all the symmetries there correspond conserved currents. The symmetries are then generated by charges which are expressed in terms of the fields. We actually start from the classical field theory case, then go to the quantum case through the usual canonical quantisation procedure. Starting from a general Lagrangian $\mathfrak{L}$ at the classical level, invariant under some Lie (super)algebra $\mathfrak{g}$, one constructs through the standard procedure the Noether charges $\hat{T}_a$

$$\varepsilon^a \hat{T}_a = \int d^3x \frac{\partial \mathfrak{L}}{\partial \partial_0 \Psi} \delta_\varepsilon \Psi \tag{54}$$

associated with the transformation

$$\delta_\varepsilon \Psi = \varepsilon^a T_a \Psi \tag{55}$$

where T_a generate the Lie (super)algebra in some appropriate (matrix) representation. Upon use of eq.(55) in eq.(54) one gets

$$\hat{T}_a = \int d^3x \Pi(x) T_a \Psi(x) \tag{56}$$

where $\Pi(x) = \frac{\partial \mathfrak{L}}{\partial \partial_0 \Psi}$ is the conjugate momentum. Equation (56) is the general relation between $\hat{T}_a$ and T_a. At the quantum level, $\hat{T}_a$ and Ψ are operators acting in some Hilbert space and we have

$$[\varepsilon^a \hat{T}_a, \Psi(x)] = \delta_\varepsilon \Psi. \tag{57}$$

Thus the conserved charges $\hat{T}_a$ from Noether's currents, realise the algebra in the sense that

$$[\delta_a, \delta_b]\Psi \equiv \delta_a(\delta_b \Psi) - \delta_b(\delta_a \Psi) = f_{ab}{}^c \delta_c \Psi \tag{58}$$

where Ψ is a field operator and δ is *by definition* given by

$$\delta_a \Psi \equiv [\hat{T}_a, \Psi]. \tag{59}$$

Equation (58) reads then equivalently

$$[\hat{T}_a, [\hat{T}_b, \Psi]] - [\hat{T}_b, [\hat{T}_a, \Psi]] = f_{ab}{}^c [\hat{T}_c, \Psi]. \tag{60}$$

Now due to Jacobi identity ($[A,[B,C]]+\text{cyclic}=0$), one can recast the left-hand side of Eq.(60) in the form $[[\hat{T}_a,\hat{T}_b],\Psi]$ to get

$$[[\hat{T}_a,\hat{T}_b],\Psi]=f_{ab}{}^c[\hat{T}_c,\Psi] \qquad (\text{for any } \Psi) \tag{61}$$

meaning that

$$[\hat{T}_a,\hat{T}_b]=f_{ab}{}^c\hat{T}_c \tag{62}$$

at least on some (sub-)space of field operators Ψ.

Now consider a Lie algebra of order three where the cubic brackets are given by

$$\{Y_i,Y_j,Y_k\}=Q_{ijk}{}^aX_a, \tag{63}$$

all the results above remain true (Eqs.(54) and (59)) but the analogy with the steps described above stops at Eq.(60). Now the algebra is realised (with $\hat{Y}_i,\hat{X}_a$ the conserved charges associated to (63))

$$(\delta_i.\delta_j.\delta_k+\text{perm})\,\Psi \;=\left[\hat{Y}_i,\left[\hat{Y}_j,\left[\hat{Y}_k,\Psi\right]\right]\right]+\text{perm}=Q_{ijk}{}^a\left[\hat{X}_a,\Psi\right], \tag{64}$$

but, the (generalised) Jacobi identities (see Eq. (33)), do not allow to obtain

$$[\left\{\hat{Y}_i,\hat{Y}_j,\hat{Y}_k\right\},\Psi]=Q_{ijk}{}^a[\hat{X}_a,\Psi] \tag{65}$$

which would have been the analog of Eq.(61) above. In other words, due to the lack of some appropriate Jabobi identities, the quantised version of the Noether charges algebra is just (64) and cannot be cast simply in a Ψ independent form. We should stress at this level that the difference with the conventional algebras and algebras of the type (63) we are pointing out, does not mean the absence of a realisation of this algebra in terms of Noether charges, as we will see latter on.

We thus define the adjoint representation where the generators of the Lie algebra of order three act on some given operator Φ which belongs to some endomorphism space End(V) of some vector space V. In Quantum Mechanics, V reduces to some Hilbert space. The adjoint representation of an elementary Lie algebra of order three is a linear map

$$\begin{aligned}\text{ad}:\mathfrak{g}=\mathfrak{g}_0\oplus\mathfrak{g}_1 &\;\to\; \mathfrak{L}(\text{End}(V),\text{End}(V))\\ g &\;\mapsto\; \text{ad}(g)\end{aligned} \tag{66}$$

$$\tag{67}$$

such that for all $\Phi\in\text{En}(V)$ we have $\text{ad}(g).\Phi=[\text{ad}(g),\Phi]$. In the adjoint representation, the algebra is realised through multiple-commutators. Furthermore, it is a matter of calculation to check that the Jacobi identities (33) are satisfied.

Another difference with usual quadratic algebras lies on the tensor product of representations (this has implications on $N-$particles states at the quantum level). Assume that we have two representations ρ_1 of and ρ_2 of (63), then $\rho_1\,\hat{\otimes}\,1+1\,\hat{\otimes}\,\rho_2$ is a representation if $\hat{\otimes}$ is the twisted tensor product, such that for homogeneous elements G_1,G_2,G_3,G_4 of $\mathfrak{g}$ we have $(G_1\hat{\otimes}G_2)(G_3\hat{\otimes}G_4)=q^{|G_2||G_3|}G_1G_3\hat{\otimes}G_2G_4$, where $|G|$ denotes the $\mathbb{Z}_3-$grading of G and q is a cubic primitive root of the unity. Indeed, if we calculate

$$\{\rho_1(Y_i)\,\otimes\,1+1\,\hat{\otimes}\,\rho_2(Y_i),\rho_1(Y_j)\,\otimes\,1+1\,\hat{\otimes}\,\rho_2(Y_j),\rho_1(Y_k)\,\otimes\,1+1\,\hat{\otimes}\,\rho_2(Y_k)\} \tag{68}$$

(we assume here for simplicity a matrix representation), the mixed terms like

$$\{\rho_1(Y_i) \otimes 1, \rho_1(Y_j) \otimes 1, 1 \hat{\otimes} \rho_2(Y_k)\}$$

vanish because we have $1+q+q^2=0$ and (68) reduces to

$$\{\rho_1(Y_i),\rho_1(Y_j),\rho_1(Y_k)\} \hat{\otimes} 1 + 1 \hat{\otimes} \{\rho_2(Y_i),\rho_2(Y_j),\rho_2(Y_k)\} = Q_{ijk}{}^a\Big(\rho_1(X_a) \hat{\otimes} 1 + 1 \hat{\otimes} \rho_2(X_a)\Big).$$

This result has been established in a more formal way, defining a coproduct for Lie algebras of order three [47]. This twisted tensor product is in fact a general feature of $n-$ary algebras as Clifford algebras of polynomial's (see Appendix Appendix D).

To end this subsection, let us mention that we have shown that it is possible to endow a Lie algebra of order three with the structure of a Hopf algebra [47]. The benefit of this construction is two-fold. Firstly, the parameters of the transformation associated to the graded part of an elementary Lie algebra of order three have been identified, and turn out to be related to the three-exterior algebra (see Eq.[D.10], Appendix Appendix D). Secondly, a group associated to a Lie algebra of order three was constructed. This latter being simply some matrix group where the matrix elements belong to the three-exterior algebra, in a straight analogy with Lie supergroups associated to Lie superalgebras [47].

3.3.2. Lie algebras of order three in any space-time dimensions As mentioned previously, it seems rather difficult to define higher order extensions of the Poincaré algebra in the framework of infinite dimensional Lie algebras of order F when the space-time dimension is bigger than three. A major progress, in this direction, was done when it was realised that it was possible to construct *finite-dimensional* Lie algebras of order F from the usual Lie algebras and Lie superalgebras [20], by an induction theorem.

Theorem 3.2 *(M. Rausch de Traubenberg, M. J. Slupinski, [20]).*
Let $\mathfrak{g}_0$ be a Lie algebra and $\mathfrak{g}_1$ be a $\mathfrak{g}_0$-module such that:
(i) $\mathfrak{g} = \mathfrak{g}_0 \oplus \mathfrak{g}_1$ is a Lie algebra of order $F_1 > 1$;
(ii) $\mathfrak{g}_1$ admits a $\mathfrak{g}_0$-equivariant symmetric form of order $F_2 \geq 2$.
Then $\mathfrak{g} = \mathfrak{g}_0 \oplus \mathfrak{g}_1$ inherits the structure of a Lie algebra of order $F_1 + F_2$.

The proof of this theorem is not very difficult, the only delicate point is to prove the Jacobi identities (33). Indeed one can show that these last identities result on some factorisation property [20]. This theorem can be extended to include the $F_1 = 1$ cases, and as a simple consequence, one can define elementary Lie algebras of order three from any Lie algebras [2]. Let $\mathfrak{g}_0$ be any Lie algebra and let $\mathfrak{g}_1$ be its adjoint representation. Introduce $\{J_a, a=1,\cdots,\dim \mathfrak{g}_0\}$ a basis of $\mathfrak{g}_0$, $\{A_a, a=1,\cdots, \dim \mathfrak{g}_0\}$ the corresponding basis of $\mathfrak{g}_1$. The invariant forms are given by the Casimir operators. In particular, for the Killing form, set $g_{ab} = Tr(A_a A_b)$ and denote $f_{ab}{}^c$ the structure constants. Then one can endow $\mathfrak{g} = \mathfrak{g}_0 \oplus \mathfrak{g}_1$ with a Lie algebra of order 3 structure given by

$$\begin{aligned} [J_a, J_b] &= f_{ab}{}^c J_c, \\ [J_a, A_b] &= f_{ab}{}^c A_c, \\ \{A_a, A_b, A_c\} &= g_{ab} J_c + g_{ac} J_b + g_{bc} J_a. \end{aligned} \tag{69}$$

The algebra (69) enables us to define a cubic extension of the Poincaré algebra in any space-time dimension. Consider the real Lie algebra $\mathfrak{g}_0 = \mathfrak{so}(1,D)$ (we could have equally chosen $\mathfrak{g}_0 =$

[2] It is also possible to construct Lie algebras of order F from Lie superalgebras. For instance, in [20] a quadratic extension of the Poincaré algebra has been obtained from the orthosymplectic Lie superalgebra.

$\mathfrak{so}(2,D-1))$ and define $\mathfrak{g} = \mathfrak{so}(1,D)\oplus\text{ad}(\mathfrak{so}(1,D)) = \langle L_{MN} = -L_{NM}, 0 \le M < N \le D+1 >\rangle\oplus$ $\langle A_{MN} = -A_{NM}, 0 \le M < N \le D+1\rangle$. The algebra is given by

$$\begin{aligned}
[L_{MN}, L_{PQ}] &= \eta_{NQ}L_{PM} - \eta_{MQ}L_{PN} + \eta_{NP}L_{MQ} - \eta_{MP}L_{NQ}, \\
[L_{MN}, A_{PQ}] &= \eta_{NQ}A_{PM} - \eta_{MQ}A_{PN} + \eta_{NP}A_{MQ} - \eta_{MP}A_{NQ}, \\
\{A_{MN}, A_{PQ}, A_{RS}\} &= (\eta_{MP}\eta_{NQ} - \eta_{MQ}\eta_{NP})L_{RS} + (\eta_{MR}\eta_{NS} - \eta_{MS}\eta_{NR})L_{PQ} + (\eta_{PR}\eta_{QS} - \eta_{PS}\eta_{QR})L_{MN}
\end{aligned} \tag{70}$$

with $\eta_{MN} = \text{diag}(1,-1,\cdots,-1)$. Using vector indices of $\mathfrak{so}(1,D-1)$ coming from the inclusion $\mathfrak{so}(1,D-1)\subset\mathfrak{so}(1,D)$, $\mathfrak{g}_0$ is generated by $L_{\mu\nu}, L_{\mu D}$, with $\mu,\nu = 0,\cdots,D-1$ and the graded part by $A_{\mu\nu}, A_{\mu D}$. Letting $R\to\infty$ after the Inönü-Wigner contraction [20],

$$\begin{array}{ll}
J_{\mu\nu} = L_{\mu\nu}, & P_\mu = \frac{1}{R}L_{\mu D} \\
T_{\mu\nu} = \frac{1}{\sqrt[3]{R}}A_{\mu\nu}, & V_\mu = \frac{1}{\sqrt[3]{R}}A_{\mu D},
\end{array} \tag{71}$$

one sees that $L_{\mu\nu}$ and P_μ generate the Poincaré algebra in $D-$dimensions and that $T_{\mu\nu}, V_\mu$ are in respectively the adjoint and vector representations of $\mathfrak{so}(1,D-1)$. This Lie algebra of order three is therefore a non-trivial extension of the Poincaré algebra where translations are cubes of more fundamental generators. The subspace generated by $L_{\mu\nu}, P_\mu, V_\mu$ is also a Lie algebra of order three extending the Poincaré algebra. The trilinear symmetric brackets have the simple form

$$\{V_\mu, V_\nu, V_\rho\} = \eta_{\mu\nu}P_\rho + \eta_{\mu\rho}P_\nu + \eta_{\rho\nu}P_\mu, \tag{72}$$

where $\eta_{\mu\nu} = \text{diag}(1,-1,\cdots,-1)$ is the Minkowski metric.

This algebra has been firstly studied in four-space time dimensions [56, 57, 58] and then in any dimensions [24], and it has been realised that it induces naturally a symmetry on generalised gauge fields or $p-$forms. An analysis of the implementation of the Noether theorem in relation with the spin-statistics theorem was given in [58] in four dimensions. Furthermore, the cubic algebra (72) is not the only non-trivial extension of the Poincaré algebra one can construct. In fact in [21] a classification of all cubic extensions of the Poincaré algebra in four space-time dimensions was given. Furthermore in [46], along the lines of the classification of Bacry and Lévy-Leblond (where a classification of kinematical algebras was undertaken [59]), we have classified all kinematical Lie algebras of order three in four dimensions and showed that they are related through generalised Inonü-Wigner contractions of the algebra (70) [46].

3.3.3. Representations In this section we construct some multiplets of the cubic extension of the Poincaré algebra (72). Since, the generators V_μ are in the vector representation of the Lorentz algebra, a multiplet contains states of the same statistics. This is rather different from the supersymmetric extension of the Poincaré algebra. To proceed further, from the matrix representations given in (D.8) or (D.9), one can construct representations of the algebra (72). The representations are specified by the representation of the vacuum. If the vacuum is in the trivial representation of the Lorentz group the multiplet consists of three spinors

$$\mathbf{\Psi} = \begin{pmatrix}\Psi_1\\ \Psi_2\\ \Psi_3\end{pmatrix},$$

transforming like $\delta_\varepsilon\mathbf{\Psi} = \varepsilon^\mu V_\mu\mathbf{\Psi}$. An invariant Lagrangian with three spinors was constructed in [56, 57]. However, there is more interesting multiplets obtained when the vacuum is in the spinor representation of the Lorentz group [24]. In this case, a multiplet contains $p-$forms. Generalised gauge fields or $p-$forms which are fully antisymmetric tensors are generalisations of the usual electromagnetic gauge field. The revival of interest for the p–forms is mainly due to there appearance in supergravity or string theory. Furthermore, the pth antisymmetric gauge fields naturally couple to $(p-1)$–dimensional extended objects. However, the p–forms are known to be relatively rigid in the sense that there is a few number of consistent interactions for them [60]. Despite these restrictions p–forms may present interesting symmetry properties, as it is the case, for instance, with the duality transformations between types IIA and IIB string theories [61]. In this section we investigate a new possible symmetry among p–forms.

Since the properties of spinors and $p-$forms depends on the space-time dimension, from now on, we restrict to the case where $D = 4n$. The other cases are analogous. We thus concentrate on the case where the vacua are in the spinor representations of the Lorentz algebra. We take two copies $\mathbf{\Psi}_\pm, \mathbf{\Lambda}_\pm$ transforming with $V_\pm$, and two copies of the vacuum in the spinor representation $\Omega_\pm$, $\omega_\pm$. Since $\Psi_\pm$ transforms with (D.9) we have

$$\mathbf{\Psi}_\pm = \begin{pmatrix} \Psi_{1\pm} \\ \Psi_{2\mp} \\ \Psi_{3\pm} \end{pmatrix},$$

with Ψ_{i+} (resp. Ψ_{i-}) left-handed (resp. right-handed) spinors. Thus for instance we have

$$\Psi_+ \otimes \Omega_+ = \begin{pmatrix} \Psi_{1+} \otimes \Omega_+ \\ \Psi_{2-} \otimes \Omega_+ \\ \Psi_{3+} \otimes \Omega_+ \end{pmatrix}.$$

Consequently, from the decomposition of the product of spinors (E.7) one gets the four possible multiplets (See Appendix Appendix E for notations)

$$\begin{aligned}
\Xi_{++} &= \mathbf{\Psi}_+ \otimes \Omega_+ = \begin{pmatrix} \Xi_{1++} \\ \tilde{\Xi}_{2-+} \\ \tilde{\tilde{\Xi}}_{3++} \end{pmatrix} = \begin{pmatrix} A_{[0]} \oplus A_{[2]} \oplus \cdots \oplus A_{[2n]+} \\ \tilde{A}_{[1]} \oplus \tilde{A}_{[3]} \oplus \cdots \oplus \tilde{A}_{[2n-1]} \\ \tilde{\tilde{A}}_{[0]} \oplus \tilde{\tilde{A}}_{[2]} \oplus \cdots \oplus \tilde{\tilde{A}}_{[2n]+} \end{pmatrix}, \\
\Xi_{--} &= \mathbf{\Psi}_- \otimes \Omega_- = \begin{pmatrix} \Xi_{1--} \\ \tilde{\Xi}_{2+-} \\ \tilde{\tilde{\Xi}}_{3--} \end{pmatrix} = \begin{pmatrix} A'_{[0]} \oplus A'_{[2]} \oplus \cdots \oplus A'_{[2n]-} \\ \tilde{A}'_{[1]} \oplus \tilde{A}'_{[3]} \oplus \cdots \oplus \tilde{A}'_{[2n-1]} \\ \tilde{\tilde{A}}'_{[0]} \oplus \tilde{\tilde{A}}'_{[2]} \oplus \cdots \oplus \tilde{\tilde{A}}'_{[2n]-} \end{pmatrix}, \\
\Xi_{-+} &= \mathbf{\Lambda}_- \otimes \omega_+ = \begin{pmatrix} \xi_{1-+} \\ \tilde{\xi}_{2++} \\ \tilde{\tilde{\xi}}_{3-+} \end{pmatrix} = \begin{pmatrix} A_{[1]} \oplus A_{[3]} \oplus \cdots \oplus A_{[2n-1]} \\ \tilde{A}_{[0]} \oplus \tilde{A}_{[2]} \oplus \cdots, \tilde{A}_{[2n]+} \\ \tilde{\tilde{A}}_{[1]} \oplus \tilde{\tilde{A}}_{[3]} \oplus \cdots \oplus \tilde{\tilde{A}}_{[2n-1]} \end{pmatrix}, \\
\Xi_{+-} &= \mathbf{\Lambda}_+ \otimes \omega_- = \begin{pmatrix} \xi_{1+-} \\ \tilde{\xi}_{2--} \\ \tilde{\tilde{\xi}}_{3+-} \end{pmatrix} = \begin{pmatrix} A'_{[1]} \oplus A'_{[3]} \oplus \cdots \oplus A'_{[2n-1]} \\ \tilde{A}'_{[0]} \oplus \tilde{A}'_{[2]} \oplus \cdots \oplus \tilde{A}'_{[2n]-} \\ \tilde{\tilde{A}}'_{[1]} \oplus \tilde{\tilde{A}}'_{[3]} \oplus \cdots \oplus \tilde{\tilde{A}}'_{[2n-1]} \end{pmatrix}.
\end{aligned} \tag{73}$$

From the general property of representations of Lie algebras of order three (35), one can assume that, for instance, for the multiplet Ξ_{++}, the fields A are in the $(-1)-$graded sector, $\tilde{A}$ in the $0-$graded sector and $\tilde{\tilde{A}}$ in the $1-$graded sector. The same classification also holds for the other multiplets Since anti-self dual $2n-$forms are complex (see (E.11)), we take all the $p-$forms above as being complex. But in $4n-$dimensional space-time a left-handed spinor is complex conjugate to a right-handed spinor, this gives us the opportunity to take the multiplets above as being conjugated $\Xi^*_{++} = \Xi_{--}$ and $\Xi^*_{+-} = \Xi_{-+}$. This gives for the multiplets Ξ_{++} and Ξ_{--}

$$
\begin{array}{ll}
A^{\star}_{[2p]} = A'_{[2p]}\,, & A^{\star}_{[2n]+} = A'_{[2n]-}\,, \\
\tilde{\tilde{A}}^{\star}_{[2p]} = \tilde{\tilde{A}}'_{[2p]}\,, & \tilde{\tilde{A}}^{\star}_{[2n]+} = \tilde{\tilde{A}}'_{[2n]-}\,, \\
\tilde{A}^{\star}_{[2p+1]} = A'_{[2p+1]}\,. &
\end{array}
\tag{74}
$$

We now calculate the transformation law of the different fields. Since the results are similar for all the multiplets (73), we only give the results for the multiplet Ξ_{++}. From the transformation law $\delta_\varepsilon \Xi_{++} = (\varepsilon^\mu V_\mu^+ \Psi_+) \otimes \Omega_+$, we obtain

$$
\begin{aligned}
\delta_\varepsilon \Xi_{1++} &= \varepsilon^\mu \Sigma_\mu \bar{\Xi}_{2-+}\,, \\
\delta_\varepsilon \bar{\Xi}_{2-+} &= \varepsilon^\mu \tilde{\Sigma}_\mu \Xi_{3++}\,, \\
\delta_\varepsilon \Xi_{3++} &= \varepsilon^\mu \partial_\mu \Xi_{1++}\,.
\end{aligned}
\tag{75}
$$

To calculate (75), we use the identity (E.5). We proceed differently for Ξ_{1++} and $\bar{\Xi}_{2-+}$, in order to avoid the presence of $\Sigma^{(\ell)}$ or $\tilde{\Sigma}^{(\ell)}$ with $\ell > 2n$. Starting from $\delta_\varepsilon \bar{\Xi}_{2-+} = \varepsilon^\mu \tilde{\Sigma}_\mu \Xi_{3++}$, and using (E.8) we first have $\delta_\varepsilon \tilde{A}_{[2p+1]} = \varepsilon^\mu 2^{-n} \mathrm{Tr}\left(\tilde{\Sigma}_\mu \Xi_{3++} \mathcal{C}_+ \Sigma^{(2p+1)}\right) = \varepsilon^\mu 2^{-n} \mathrm{Tr}\left(\Sigma^{(2p+1)} \tilde{\Sigma}_\mu \Xi_{3++} \mathcal{C}_+\right)$. Using (E.5), we calculate $\Sigma^{(2p+1)}\tilde{\Sigma}_\mu$. Then, from the trace identities for the Σ matrices (E.4), one obtains the transformations laws for $\tilde{A}_{[2p+1]}$. In order to calculate $\delta_\varepsilon \Xi_{1++} = \varepsilon^\mu \Sigma_\mu \bar{\Xi}_{2-+}$, we proceed in the reverse order. Firstly, using (E.7) and the identity (E.5), we compute the product $\Sigma_\mu \bar{\Xi}_{2-+}$. Then, using the trace formulae, we get the transformation laws of $A_{[2p]}$. The last case $\delta_\varepsilon \Xi_{3++}$ is not difficult to handle. For instance, this procedure gives for $\delta \Xi_{1++}$ (when $k < n$)

$$
\begin{aligned}
\delta A_{[2k]}{}^{\nu_1 \cdots \nu_{2k}} &= \frac{1}{2^n} \mathrm{Tr}\left(\sum_{p=0}^{n} \frac{1}{(2p+1)!} \left[\tilde{\Sigma}^{\nu_1 \cdots \nu_{2k}} \Sigma_{\mu \mu_{2p+1} \cdots \mu_1} + (2p+1) \eta_{\mu \mu_{2p+1}} \tilde{\Sigma}^{\nu_1 \cdots \nu_{2k}} \Sigma_{\mu_{2p} \cdots \mu_1} \right] \right) \\
&\times \ \varepsilon^\mu \tilde{A}_{[2p+1]}{}^{\mu_1 \cdots \mu_{2p+1}}.
\end{aligned}
$$

Using the trace formulæ (E.4), the first term gives $\tilde{A}_{[2k-1]} \wedge \varepsilon$ and the second $i_\varepsilon \tilde{A}_{[2k+1]}$. Finally we get [24]

$$
\begin{array}{llll}
\delta_\varepsilon A_{[0]} = i_\varepsilon \tilde{A}_{[1]} & & & \\
& & \delta_\varepsilon \tilde{A}_{[1]} &= i_\varepsilon \tilde{\tilde{A}}_{[2]} + \tilde{\tilde{A}}_{[0]} \wedge \varepsilon \\
\vdots & & \vdots & \\
\delta_\varepsilon A_{[2p]} = i_\varepsilon \tilde{A}_{[2p+1]} + \tilde{A}_{[2p-1]} \wedge \varepsilon & & & \\
& & \delta_\varepsilon \tilde{A}_{[2p+1]} &= i_\varepsilon \tilde{\tilde{A}}_{[2p+2]} + \tilde{\tilde{A}}_{[2p]} \wedge \varepsilon \\
\vdots & & \vdots & \\
& & \delta_\varepsilon \tilde{A}_{[2n-1]} &= i_\varepsilon \tilde{\tilde{A}}_{[2n]_+} + \tilde{\tilde{A}}_{[2n-2]} \wedge \varepsilon \\
\delta_\varepsilon A_{[2n]_+} = \tilde{A}_{[2n-1]} \wedge \varepsilon - i\, {}^\star\!\left(\tilde{A}_{[2n-1]} \wedge \varepsilon\right) & & &
\end{array}
\tag{76}
$$

$$
\delta_\varepsilon \tilde{\tilde{A}}_{[0]} = \varepsilon.\partial A_{[0]}, \qquad \ldots \qquad \delta_\varepsilon \tilde{\tilde{A}}_{[2n]_+} = \varepsilon.\partial A_{[2n]_+},
$$

the second term in $\delta_\varepsilon A_{[2n]+}$ ensure its self-duality. For the definition and conventions see (E.13) and (E.15). It is interesting to observe that the transformation (76) have a geometrical interpretation in terms of the natural operations on $p-$forms.

Journal of Physics: Conference Series **175** (2009) 012003 doi:10.1088/1742-6596/175/1/012003

3.3.4. Invariant Lagrangian The transformations (76) suggest that an invariant Lagrangian should be obtained coupling the field A with the fields $\tilde{\tilde{A}}$ and the field $\tilde{A}$ with themselves. In other words to couple (-1)–graded sector with 1–graded sector, and grade 0–graded sector with itself. Furthermore, if we consider for example the Ξ_{++} multiplet, in order to have a real Lagrangian, one has also to take into consideration the conjugate multiplet Ξ_{--} (see (74)). For the Ξ_{++} and Ξ_{--} multiplets, the Lagrangian writes [24]

$$\begin{aligned}
\mathfrak{L} &= \mathfrak{L}(\Xi_{++}) + \mathfrak{L}(\Xi_{--}) = \mathfrak{L}_{[0]} + \ldots + \mathfrak{L}_{[2n]} + \mathfrak{L}'_{[0]} \ + \ldots + \mathfrak{L}'_{[2n]} = \\
&= dA_{[0]} d\tilde{\tilde{A}}_{[0]} + \ldots + \\
&\quad -\frac{1}{2}\frac{1}{(2p+2)!}\, d\tilde{A}_{[2p+1]}\, d\tilde{A}_{[2p+1]} - \frac{1}{2}\frac{1}{(2p)!}\, d^\dagger \tilde{A}_{[2p+1]}\, d^\dagger \tilde{A}_{[2p+1]} \\
&\quad +\frac{1}{(2p+3)!}\, dA_{[2p+2]}\, d\tilde{\tilde{A}}_{[2p+2]} + \frac{1}{(2p+1)!}\, d^\dagger A_{[2p+2]}\, d^\dagger \tilde{\tilde{A}}_{[2p+2]} \\
&\quad +\ldots+ \\
&\quad +\frac{1}{2}\frac{1}{(2n+1)!}\, dA_{[2n]_+}\, d\tilde{\tilde{A}}_{[2n]_+} + \frac{1}{2}\frac{1}{(2n-1)!}\, d^\dagger A_{[2n]_+}\, d^\dagger \tilde{\tilde{A}}_{[2n]_+} \\
&\quad +dA'_{[0]}\, d\tilde{\tilde{A}}'_{[0]} + \ldots + \qquad\qquad (77) \\
&\quad -\frac{1}{2}\frac{1}{(2p+2)!}\, d\tilde{A}'_{[2p+1]}\, d\tilde{A}'_{[2p+1]} - \frac{1}{2}\frac{1}{(2p)!}\, d^\dagger \tilde{A}'_{[2p+1]}\, d^\dagger \tilde{A}'_{[2p+1]} \\
&\quad +\frac{1}{(2p+3)!}\, dA'_{[2p+2]}\, d\tilde{\tilde{A}}'_{[2p+2]} + \frac{1}{(2p+1)!}\, d^\dagger A'_{[2p+2]}\, d^\dagger \tilde{\tilde{A}}'_{[2p+2]} \\
&\quad +\ldots+ \\
&\quad +\frac{1}{2}\frac{1}{(2n+1)!}\, dA'_{[2n]_-}\, d\tilde{\tilde{A}}'_{[2n]_-} + \frac{1}{2}\frac{1}{(2n-1)!}\, d^\dagger A'_{[2n]_-}\, d^\dagger \tilde{\tilde{A}}'_{[2n]_-}\,.
\end{aligned}$$

Here $\omega_{[p]}\omega'_{[p]}$ stands for $\omega_{[p]\mu_1\cdots\mu_p}\ \omega'_{[p]}{}^{\mu_1\cdots\mu_p}$, where $\omega_{[p]}$ and $\omega'_{[p]}$ are two p–forms. For the definition of the exterior derivative d and its adjoint $d^\dagger$ see (E.17) and (E.18). To prove that (77) is invariant under (76), we firstly note that $\delta_\varepsilon\mathfrak{L}(\Xi_{++})$ and $\delta_\varepsilon\mathfrak{L}(\Xi_{--})$ do not mix. It is thus sufficient to check separately their invariance, which we do here only for $\mathfrak{L}(\Xi_{++})$ as an illustration. Starting from a specific normalisation for $\mathfrak{L}_{[0]}$, its variation fixes the normalisation for $\mathfrak{L}_{[1]}$. By a step-by-step process, the normalisations for $\mathfrak{L}_{[p]}, 0 \le p \le 2n$ are also fixed. At the very end, all the terms of $\delta_\varepsilon\mathfrak{L}$ compensate each other, up to a total derivative. The key observations in this compensation process, is to remark that we have

$$(i_\varepsilon . A_{[p+1]})A_{[p]} = A_{[p+1]}(A_{[p]} \wedge \varepsilon),$$

together with the fact that the Lagrangian can be rewritten in a Fermi-like form (see below (84)). This shows, the Lagrangian (77) is invariant.

If one considers the terms involving the (anti–)self–dual $2n$–form one can have further simplifications. Indeed, for the self–dual $2n$–forms we have

$$\begin{aligned}
\mathfrak{L}_{[2n]} &= \tfrac{1}{2}\tfrac{1}{(2n+1)!}\, dA_{[2n]_+}\, d\tilde{\tilde{A}}_{[2n]_+} + \tfrac{1}{2}\tfrac{1}{(2n-1)!}\, d^\dagger A_{[2n]_+}\, d^\dagger \tilde{\tilde{A}}_{[2n]_+} = \\
&= \tfrac{1}{(2n+1)!}\, dA_{[2n]_+}\, d\tilde{\tilde{A}}_{[2n]_+}
\end{aligned}$$

because of the self–duality condition ${}^\star A_{[2n]_+} = iA_{[2n]_+}$. A more interesting way of regrouping the terms involving the self–dual and the anti–self–dual $2n$–forms is to introduce the real $2n$–forms

$$A_{1\,[2n]} = \frac{1}{\sqrt{2}}\left(A_{[2n]_+} + A'_{[2n]_-}\right), \tilde{\tilde{A}}_{1\,[2n]} = \frac{1}{\sqrt{2}}\left(\tilde{\tilde{A}}_{[2n]_+} + \tilde{\tilde{A}}'_{[2n]_-}\right), \tag{78}$$

such that

$$\mathcal{L}_{[2n]} + \mathcal{L}'_{[2n]} = \frac{1}{(2n+1)!} dA_{1\,[2n]} d\tilde{\tilde{A}}_{1\,[2n]} + \frac{1}{(2n-1)!} d^\dagger A_{1\,[2n]} d^\dagger \tilde{\tilde{A}}_{1\,[2n]} \tag{79}$$

since $A_{[2n]_+} A'_{[2n]_-} = 0$ when $D = 1 + (4n-1)$, for a self–dual and an anti–self–dual $2n$–form. The final real $2n$–forms are neither self–dual nor anti–self–dual, which is in agreement with the representation theory of the Poincaré algebra. In the same manner, we introduce the real fields

$$\begin{aligned}
A_{1\,[2p]} &= \tfrac{1}{\sqrt{2}}\left(A_{[2p]} + A'_{[2p]}\right), & A_{2\,[2p]} &= \tfrac{i}{\sqrt{2}}\left(A_{[2p]} - A'_{[2p]}\right), \\
\tilde{A}_{1\,[2p+1]} &= \tfrac{1}{\sqrt{2}}\left(\tilde{A}_{[2p+1]} + \tilde{A}'_{[2p+1]}\right), & \tilde{A}_{2\,[2p+1]} &= \tfrac{i}{\sqrt{2}}\left(\tilde{A}_{[2p+1]} - \tilde{A}'_{[2p+1]}\right), \\
\tilde{\tilde{A}}_{1\,[2p]} &= \tfrac{1}{\sqrt{2}}\left(\tilde{\tilde{A}}_{[2p]} + \tilde{\tilde{A}}'_{[2p]}\right), & \tilde{\tilde{A}}_{2\,[2p]} &= \tfrac{i}{\sqrt{2}}\left(\tilde{\tilde{A}}_{[2p]} - \tilde{\tilde{A}}'_{[2p]}\right).
\end{aligned} \tag{80}$$

With this new fields the Lagrangian is not diagonal, introducing the fields

$$\begin{aligned}
\hat{A}_{1\,[2p]} &= \tfrac{1}{\sqrt{2}}\left(A_{1\,[2p]} + \tilde{\tilde{A}}_{1\,[2p]}\right), & \hat{\hat{A}}_{1\,[2p]} &= \tfrac{1}{\sqrt{2}}\left(A_{1\,[2p]} - \tilde{\tilde{A}}_{1\,[2p]}\right), & p &= 0, \ldots, n, \\
\hat{A}_{2\,[2p]} &= \tfrac{1}{\sqrt{2}}\left(A_{2\,[2p]} + \tilde{\tilde{A}}_{2\,[2p]}\right), & \hat{\hat{A}}_{2\,[2p]} &= \tfrac{1}{\sqrt{2}}\left(A_{2\,[2p]} - \tilde{\tilde{A}}_{2\,[2p]}\right), & p &= 0, \ldots, n-1,
\end{aligned} \tag{81}$$

we have ($2p \neq 2n$ since for $\mathcal{L}_{2n}$ we have only one term)

$$\begin{aligned}
\mathcal{L}_{2p+1} + \mathcal{L}'_{2p+1} = & - \frac{1}{2}\frac{1}{(2p+2)!} d\tilde{A}_{1\,[2p+1]} d\tilde{A}_{1\,[2p+1]} - \frac{1}{2}\frac{1}{(2p)!} d^\dagger \tilde{A}_{1\,[2p+1]} d^\dagger \tilde{A}_{1\,[2p+1]} \\
& + \frac{1}{2}\frac{1}{(2p+2)!} d\tilde{A}_{2\,[2p+1]} d\tilde{A}_{2\,[2p+1]} + \frac{1}{2}\frac{1}{(2p)!} d^\dagger \tilde{A}_{2\,[2p+1]} d^\dagger \tilde{A}_{2\,[2p+1]} \\
\mathcal{L}_{2p} + \mathcal{L}'_{2p} = & \ \frac{1}{2}\frac{1}{(2p+1)!} d\hat{A}_{1\,[2p]} d\hat{A}_{1\,[2p]} + \frac{1}{2}\frac{1}{(2p-1)!} d^\dagger \hat{A}_{1\,[2p]} d^\dagger \hat{A}_{1\,[2p]} \\
& - \frac{1}{2}\frac{1}{(2p+1)!} d\hat{\hat{A}}_{1\,[2p]} d\hat{\hat{A}}_{1\,[2p]} - \frac{1}{2}\frac{1}{(2p-1)!} d^\dagger \hat{\hat{A}}_{1\,[2p]} d^\dagger \hat{\hat{A}}_{1\,[2p]} \\
& - \frac{1}{2}\frac{1}{(2p+1)!} d\hat{A}_{2\,[2p]} d\hat{A}_{2\,[2p]} - \frac{1}{2}\frac{1}{(2p-1)!} d^\dagger \hat{A}_{2\,[2p]} d^\dagger \hat{A}_{2\,[2p]} \\
& + \frac{1}{2}\frac{1}{(2p+1)!} d\hat{\hat{A}}_{2\,[2p]} d\hat{\hat{A}}_{2\,[2p]} + \frac{1}{2}\frac{1}{(2p-1)!} d^\dagger \hat{\hat{A}}_{2\,[2p]} d^\dagger \hat{\hat{A}}_{2\,[2p]}
\end{aligned} \tag{82}$$

Usually, the kinetic term for a p–form $\omega_{[p]}$ writes, with our conventions for the metric, $(-1)^p \frac{1}{2(p+1)!}\, d\omega_{[p]}\, d\omega_{[p]}$. However, in the Lagrangian written as sum of terms above (79) and (82) some of the kinetic terms have the wrong sign. This implies that these fields have an energy density not bounded from bellow. We propose a possible way to construct a Lagrangian with correct signs for the various kinetic terms, based on a special choice for the physical fields. The main idea is related to Hodge duality. However, the duality transformation will act here on the $p-$forms with respect to the Lorentz group $SO(1, D-1)$. This should be contrasted with the case of the usual duality transformations (generalising the electric-magnetic duality) which act on the field strengths with respect to $SO(1, D-1)$, or equivalently on the potentials themselves but with respect to the little group $SO(D-2)$. Firstly, the decomposition (E.6) is purely conventional, and we could also have decomposed this product on the set of $p-$forms with $p \geq 2n$. Moreover, looking at the Lagrangians (79) and (82) there is a kind of duality between the kinetic term and the gauge fixing term. This means that for the field with the wrong signs, we define the Hodge dual

$$\begin{array}{rcl} \hat{\hat{A}}_{1\,[2p]} & \rightarrow & \hat{\hat{B}}_{1\,[4n-2p]} = {}^\star\hat{\hat{A}}_{1\,[2p]}\,, \\ \hat{A}_{2\,[2p]} & \rightarrow & \hat{B}_{2\,[4n-2p]} = {}^\star\hat{A}_{2\,[2p]}\,, \\ \tilde{A}_{2\,[2p+1]} & \rightarrow & \tilde{B}_{2\,[4n-2p-1]} = {}^\star\tilde{A}_{2\,[2p+1]}\,, \\ \hat{\hat{A}}_{1\,[2n]} & \rightarrow & \hat{\hat{B}}_{1\,[D-2n]} = {}^\star\hat{\hat{A}}_{1\,[2n]}, \end{array} \tag{83}$$

with $p = 0, \ldots, n-1$. Thus, starting from the Lagrangian (77) and performing the field redefinitions (80), (81) and (83) we get a new Lagrangian with the correct signs (that we do not write but is easy to obtain), because of relation (E.19). This Lagrangian is given in [24]. In this transformation the kinetic term of A becomes the gauge fixing term (see below Section 3.3.5) of B and *vice versa*. Using (E.16), one can obtain the transformations law of the new fields. The field content is then one one–form, one three–form, ..., one $(4n-1)$–form and two zero–forms, two two–forms, ... and two $4n$–form, all the p–forms have a kinetic term and a gauge fixing term; the only exceptions are the zero–forms, which have only kinetic terms, and the $4n$–forms, which have only gauge fixing terms. Let us emphasise that the substitutions (83) are done with respect to the gauge fields. This is quite different from the usual duality transformations (generalising the electric–magnetic duality) where the duality transformations are done with respect to the field strengths.

With the "duality" transformations (83) the number of degree of freedom is not the same for $\omega_{[p]}$ and $\rho_{[D-p]} = {}^\star\omega_{[p]}$, which is not the case for the usual duality transformations. Hence, the two Lagrangians describe inequivalent theories.

Note finally that the sum of the kinetic terms and the gauge fixing terms gives rises to Fermi–like terms. For instance

$$\frac{1}{2}\frac{1}{(p+1)!}dA_{[p]}dA_{[p]} + \frac{1}{2}\frac{1}{(p-1)!}d^\dagger A_{[p]}d^\dagger A_{[p]} = \frac{1}{2}\frac{1}{p!}\partial_\mu A_{[p]}\partial^\mu A_{[p]}, \tag{84}$$

and note, in particular that it is more easy to check the invariance of the Lagrangian (77) when it is written as a sum of Fermi-like terms.

3.3.5. Gauge invariance Let us analyse gauge invariance. Gauge transformations for $p-$forms, are given by

$$A_{[p]} \rightarrow A_{[p]} + d\chi_{[p-1]}, \; p > 0, \tag{85}$$

where $\chi_{[p-1]}$ is a $(p-1)$-form. We observe that (77) (or the new Lagrangian that we do not have given) is not invariant. Indeed, the terms $d^\dagger A_{[p]} d^\dagger A_{[p]}$ in the Lagrangian (77) fixes partially the gauge, and the Lagrangian is invariant provided $\chi_{[p-1]}$ satisfies an additional constraint

$$d^\dagger d\chi_{[p-1]} = 0. \tag{86}$$

This means that this last term is a gauge fixing term, analogous to the Feynman gauge fixing term for the electromagnetic field. The peculiar form of our Lagrangian leads also to another gauge invariance. Since it contains a kinetic term and a gauge fixing term, there is some kind of duality between these two terms. It was in fact this duality which enabled us to make the field redefinition (83). At the level of gauge invariance, this duality translates in the invariance of the Lagrangian under the transformation

$$A_{[p]} \to A_{[p]} + d^\dagger \chi_{[p+1]},\ p < 4n, \tag{87}$$

such that $dd^\dagger\chi_{[p+1]} = 0$. However, from the Poincaré theorem (since we are in $\mathbb{R}^{4n}$ there is no topological obstruction), $dd^\dagger\chi_{[p+1]} = 0$ implies that there exists a $(p-1)$–form $\lambda_{[p-1]}$ such that $d^\dagger\chi_{[p+1]} = d\lambda_{[p-1]}$. Thus (87) is not a new symmetry. While in the case of gauge theories the gauge invariance guarantees that the physical (on-shell) quantities are gauge fixing independent, in our case $(d^\dagger A_{[p]})^2$ cannot be traded for any other gauge-fixing function since it is imposed by the invariance under (76), and is thus expected to affect the physical degrees of freedom. This shows that the effective degrees of freedom of $A_{[p]}$ are dictated only by the gauge freedom eq.(85), supplemented by $d^\dagger d\chi_{[p-1]} = 0$. An immediate consequence of the latter constraint is that the usual Lorentz condition $d^\dagger A_{[p]} = 0$ cannot be imposed in general to eliminate the unphysical components. This means that the way one should eliminate the unphysical components cannot be handled in a usual manner [62]. On top of that, such a condition is not stable under our transformation laws (76). For instance, if we put $\partial_\mu A^\mu_{[1]} = 0$, then $\partial_\mu \delta_\varepsilon A^\mu_{[1]} = 0$ gives $\varepsilon^\mu \partial_\mu A_{[0]} = 0$, which is obviously too strong.

The gauge invariance (85) and the field equations imply (for a p–form $A_{[p]}$, with $p \leq D-2$) $P^\mu A_{[p]\mu\mu_2\cdots\mu_p} = 0$ and $P^2 = 0$ (with P^μ the momentum), thus $A_{[p]}$ gives rise to a massless state in the p–order antisymmetric representation of the little group $SO(4n-2)$. But, in our decomposition, there are also appearing p–forms with $p = 4n-1, 4n$. Of course these p–forms do not propagate. It is interesting to note that similar phenomena are well-known in the context of type IIA, IIB string theory [63] in 10 space-time dimensions where 9– and 10–forms appear. Actually, subsequent to the early works on two-forms in [64, 65], several authors studied the classical and quantum properties of the non-propagating 3- and 4-forms [66, 67] in four dimensions. In particular, it was pointed out in [67] that the gauge fixing term for a 4-form takes the form of a kinetic term for a scalar field, in exact analogy with our results. The gauge condition for p–forms is crucial in order to eliminate the unphysical degrees of freedom such that massless p–form $(p < D-1)$ has $\begin{pmatrix} p \\ D-1 \end{pmatrix}$ degrees of freedom off–shell and $\begin{pmatrix} p \\ D-2 \end{pmatrix}$ on–shell. In our case, relation (86) is not enough to eliminate the unphysical degrees of freedom. However, here the situation is not as simple, since in our case there is some mixing between gauge invariance and (76). This means that (76) should itself have some role in the elimination of the superfluous degree of freedom. For a discussion of the compatibility of the transformation (76) and gauge transformation see [57].

The Lagrangian constructed so far is a free theory. It should be interesting to construct interacting theories invariant under (76). It has been proved by a brute force method, that when $D = 4$ and we consider only multiplets of the types (73), no interacting terms are allowed [57]. In the general case, it is still an open question to know whether or not interacting terms compatible with (76) exist.

Let us mention to conclude this section that invariant Lagrangians including mass terms were also constructed [24].

4. Conclusion

There are many mathematical structures generalising Lie superalgebras which can be defined. In particular one of this structure called Lie algebras of order F can be used to implement higher order extensions of the Poincaré algebra. In these lectures we have studied explicitly two examples of non-trivial higher order extensions of the Poincaré algebra based on Lie algebras of order F, which are not the supersymmetric ones. In the former, we have constructed higher order symmetries in three space-time dimensions which act on relativistic anyons. In the latter, cubic extensions of the Poincaré algebra in any space-time dimensions are obtained. These new symmetries have a natural geometrical interpretation on generalised gauge fields or $p-$forms. Invariant Lagrangians in this last cases have revealed some problems to be solved: construction of interacting Lagrangian, relation between the cubic symmetries and gauge invariance.

In order to understand deeper these structures a program to study Lie algebras of order F on a formal way has been investigated. The basis of the theory of contractions and deformations in the Gerstenhaber sense together with a classification of Lie algebras associated to $\mathfrak{sl}(2)$ and the four dimensional-Poincaré algebra have been initiated [21]. A classification of kinematical algebras of order three have been investigated in [46]. Since Lie algebras of order three correspond to transformations at the infinitesimal level, group associated to ternary algebras has been defined in the context of Hopf algebras, and the parameters of the transformation have been identified [47]. However some points remains to be studied. Simple Lie algebras of order three have been defined in [20], however there is no general classification of simple complex Lie algebras of order F (and even of order three) analogous to the classification of the Lie (super)algebras. It should be also interesting to have a analogous of a Coleman-Mandula theorem in that context. Finally, in order to construct interacting Lagrangians some adapted superspace will certainly be relevant.

Acknowledgments

I would like to acknowledge the organising committee, and in particular R. Campoamor-Stursberg to offer me the opportunity to give these lectures, and for the kind invitation to Madrid.

Appendix A. Convention and useful identities for spinor calculus in four dimensions

In this appendix, we collect useful relations and conventions for the four dimensional spinor calculus. The metric is taken to be

$$\eta_{\mu\nu} = \text{diag}(1, -1, -1, -1) \tag{A.1}$$

In the $\mathfrak{so}(1,2) = \mathfrak{sl}(2,\mathbb{C})$ notations of dotted and undotted indices for two-dimensional spinors, the spinor conventions to raise/lower indices are as follows (we have minor differences as compared to the notations of Wess and Bagger [29]): $\psi_\alpha = \varepsilon_{\alpha\beta}\psi^\beta$, $\psi^\alpha = \varepsilon^{\alpha\beta}\psi_\beta$, $\bar\psi_{\dot\alpha} = \varepsilon_{\dot\alpha\dot\beta}\bar\psi^{\dot\beta}$, $\bar\psi^{\dot\alpha} = \varepsilon^{\dot\alpha\dot\beta}\bar\psi_{\dot\beta}$ with $(\psi_\alpha)^* = \bar\psi_{\dot\alpha}$, $\varepsilon_{12} = \varepsilon_{\dot1\dot2} = -1$, $\varepsilon^{12} = \varepsilon^{\dot1\dot2} = 1$.
The $4D$ Dirac matrices, in the Weyl representation, are

$$\Gamma_\mu = \begin{pmatrix} 0 & \sigma_\mu \\ \bar\sigma_\mu & 0 \end{pmatrix}, \tag{A.2}$$

with

$$\sigma_\mu = (1, \sigma_i), \bar{\sigma}_\mu = (1, -\sigma_i), \tag{A.3}$$

the σ_i $(i = 1, 2, 3)$ denoting the Pauli matrices. The index structure of the σ_μ−matrices is as follows: $\sigma_\mu \rightarrow \sigma_{\mu\alpha\dot{\alpha}}$, $\bar{\sigma}_\mu \rightarrow \bar{\sigma}_\mu{}^{\dot{\alpha}\alpha}$. The following relation holds,

$$\bar{\sigma}_\mu{}^{\dot{\alpha}\alpha} = \sigma_{\mu\,\beta\dot{\beta}}\varepsilon^{\alpha\beta}\varepsilon^{\dot{\alpha}\dot{\beta}}, \tag{A.4}$$

and (A.3) leads to

$$\mathrm{Tr}(\sigma^\mu \bar{\sigma}^\nu) = 2\eta^{\mu\nu}. \tag{A.5}$$

Furthermore, the Lorentz generators for the spinor representations are given by

$$\begin{aligned} \sigma_{\mu\nu\alpha}{}^\beta &= \frac{1}{4}\left(\sigma_{\mu\alpha\dot{\alpha}}\bar{\sigma}_\nu{}^{\dot{\alpha}\beta} - \sigma_{\nu\alpha\dot{\alpha}}\bar{\sigma}_\mu{}^{\dot{\alpha}\beta}\right) \\ \bar{\sigma}_{\mu\nu}{}^{\dot{\alpha}}{}_{\dot{\beta}} &= \frac{1}{4}\left(\bar{\sigma}_\mu{}^{\dot{\alpha}\alpha}\sigma_{\nu\alpha\dot{\beta}} - \bar{\sigma}_\nu{}^{\dot{\alpha}\alpha}\sigma_{\mu\alpha\dot{\beta}}\right) \end{aligned} \tag{A.6}$$

We adopt the usual spinor summation convention

$$\psi.\lambda = \psi^\alpha\lambda_\alpha = -\psi_\alpha\lambda^\alpha, \quad \bar{\psi}.\bar{\lambda} = \bar{\psi}_{\dot{\alpha}}\bar{\lambda}^{\dot{\alpha}}. \tag{A.7}$$

Since $\psi_\alpha^\dagger = \bar{\psi}_{\dot{\alpha}}$ and $\sigma^{\dagger\mu} = \sigma^\mu$, we have

$$(\psi.\lambda)^\dagger = \bar{\lambda}.\bar{\psi}, \tag{A.8}$$

$$\left(\psi\sigma^\mu\bar{\lambda}\right)^\dagger = \lambda\sigma^\mu\bar{\psi}. \tag{A.9}$$

For anticommuting Grassmann spinors we have

$$\theta^\alpha\theta^\beta = -\frac{1}{2}\varepsilon^{\alpha\beta}\theta.\theta, \tag{A.10}$$

$$\bar{\theta}^{\dot{\alpha}}\bar{\theta}^{\dot{\beta}} = \frac{1}{2}\bar{\varepsilon}^{\dot{\alpha}\dot{\beta}}\bar{\theta}.\bar{\theta}, \tag{A.11}$$

$$(\theta\sigma^\mu\bar{\theta})\,(\theta\sigma^\nu\bar{\theta}) = \frac{1}{2}\theta.\theta\;\bar{\theta}.\bar{\theta}\;\eta^{\mu\nu}, \tag{A.12}$$

$$(\theta\sigma^\mu\bar{\theta})\;\theta.\lambda = -\frac{1}{2}\theta.\theta\;\lambda\sigma^\mu\bar{\theta}, \tag{A.13}$$

$$(\theta\sigma^\mu\bar{\theta})\;\bar{\theta}.\bar{\lambda} = -\frac{1}{2}\bar{\theta}.\bar{\theta}\;\theta\sigma^\mu\bar{\lambda}. \tag{A.14}$$

Journal of Physics: Conference Series **175** (2009) 012003 doi:10.1088/1742-6596/175/1/012003

Appendix B. Relativistic wave equations for anyons

We give now a relativistic wave equation for anyons. Following M. Plyushchay it is based on the R−deformed Heisenberg algebra with reflection. This algebra is defined by generators and relations. Consider the generators $a^\pm, \mathcal{R}$ that satisfy [68]

$$[a^-, a^+] = 1 + \nu\mathcal{R}, \quad \left\{R, a^\pm\right\} = 0, \quad \mathcal{R}^2 = 1, \tag{B.1}$$

with $\nu \in \mathbb{R}$ the deformation parameter. The operators $a^\pm$ together with the quadratic operators generate the superalgebra $\mathfrak{osp}(1|2)$ whose bosonic part is $\mathfrak{sl}(2, \mathbb{R}) \cong \mathfrak{so}(1, 2)$ and fermionic part the two-dimensional (Majorana) spinor representation

$$\begin{aligned}\left\{a^\pm, a^\pm\right\} = 4L_\pm, &\quad \{a^+, a^-\} = 4L_0,\\ \left[L_\pm, a^\mp\right] = \mp a^\mp, &\quad [L_0, a^\pm] = \pm\tfrac{1}{2}a^\pm,\\ [L_+, L_-] = -2L_0, &\quad [L_0, L_\pm] = -L_\pm.\end{aligned} \tag{B.2}$$

This is the generalisation of the well-known realisation of the algebra $\mathfrak{sl}(2, \mathbb{R}) \cong \mathfrak{so}(1, 2)$ by the usual harmonic oscillator. When $\nu > -1$, the algebra (B.1) admits an infinite dimensional unitary representation (when $\nu = -(2k + 1)$ the algebra admits finite dimensional representations) [53]

$$\begin{aligned} a^+ \left|n\right\rangle &= \sqrt{n + 1 + \frac{\nu}{2}(1 - (-1)^{n+1})}\ \left|n+1\right\rangle,\\ a^- \left|n\right\rangle &= \sqrt{n + \frac{\nu}{2}(1 - (-1)^{n})}\ \left|n-1\right\rangle,\end{aligned} \tag{B.3}$$

this representation is bounded from below since we have $a^- \left|0\right\rangle = 0$, unitary $(a^-)^\dagger = a^+$ and the vectors $|n>$ are orthonormal $< n|m >= \delta_{mn}$. We denote $\mathcal{R}_\nu = \{\left|n\right\rangle,\ n \in \mathbb{N}\}$ this representation. We also have $\mathcal{R}\left|n\right\rangle = (-1)^n \left|n\right\rangle$. Furthermore, because of relations (B.2) the representation of (B.3) is an irreducible representation of $\mathfrak{osp}(1|2)$ but a reducible representation of $\mathfrak{sl}(2, \mathbb{R})$. Indeed, it decomposes on the direct sum of the two irreducible representations

$$\mathcal{R}_\nu = \mathcal{D}^+_{\frac{1+\nu}{4}} \oplus \mathcal{D}^+_{\frac{3+\nu}{4}} \tag{B.4}$$

with

$$\begin{aligned}\mathcal{D}^+_{\frac{1+\nu}{4}} &= \frac{1}{2}(1 + \mathcal{R})\mathcal{R}_\nu = \Big\{\left|2n\right\rangle,\ n \in \mathbb{N}\Big\},\\ \mathcal{D}^+_{\frac{3+\nu}{4}} &= \frac{1}{2}(1 - \mathcal{R})\mathcal{R}_\nu = \Big\{\left|2n+1\right\rangle,\ n \in \mathbb{N}\Big\}.\end{aligned} \tag{B.5}$$

In order to obtain a relativistic wave equation for relativistic anyons, and to make contact with the literature [53], in this appendix we are working with physical quantities. This means that differently from the rest of the text we are directly working with the tri-momentum and the angular momentum. In particular, this means that they are given by hermitian quantities (although in the rest of the text they are antihermitian). In order to avoid confusion, we denote here $\Pi_\mu = -iP_\mu, L_\mu = -iJ_\mu$, the tri-momentum and angular momentum. In this basis the Lie algebra $\mathfrak{so}(1,2)$ writes $[L_\mu, L_\nu] = i\varepsilon_{\mu\nu\rho}\eta^{\rho\sigma}L_\sigma$ and $L_\pm = L_1 \mp iL_2$. We also define $\Pi_\pm = \Pi_1 \mp i\Pi_2$. The Dirac $\Gamma-$matrices in the Majorana representation are taken to be $\gamma^0 = -i\sigma^2, \gamma^1 = i\sigma^2, \gamma^2 = -i\sigma^3$ with spinor matrix elements $\gamma^\mu{}_\alpha{}^\beta$, and the spinor indices can be raised and lowed as in Appendix Appendix A. We also define the hermitian spinor operators

$$L_1 = \frac{1}{\sqrt{2}}(a^+ + a^-), \quad L_2 = \frac{i}{\sqrt{2}}(a^+ - a^-).$$

A direct calculation using (B.1) and (B.2) gives

$$\begin{aligned} [L_\alpha, L_\beta] &= -i\epsilon_{\alpha\beta}(1+\nu\mathcal{R}), \\ \{L_\alpha, L_\beta\} &= -4i(L.\gamma)_{\alpha\beta} = \begin{pmatrix} L_0+L_1 & L_2 \\ L_2 & L_0-L_1 \end{pmatrix}. \end{aligned} \tag{B.6}$$

Finally we introduce the spinor operator

$$D_\alpha = (\Pi.\gamma)_\alpha{}^\beta L_\beta + \varepsilon m L_\alpha,$$

with $\varepsilon = \pm 1$, and a little algebra gives

$$\begin{aligned} L^\alpha L_\alpha &= -i(1+\nu\mathcal{R}), \\ D^\alpha D_\alpha &= -i(\Pi^2 - m^2)(1+\nu\mathcal{R}), \\ L^\alpha D_\alpha &= 4i(\Pi.L - m\varepsilon\frac{1}{4}(1+\nu\mathcal{R})). \end{aligned} \tag{B.7}$$

Now we have all the material and identities to define the spinor set of equations. Taking $|\psi\rangle \in \mathcal{R}_\mu$ we have

$$|\psi\rangle = |\psi_+\rangle + |\psi_-\rangle = \sum_{n=0}^{+\infty}\psi_{2n}(x)\,|2n\rangle + \sum_{n=0}^{+\infty}\psi_{2n+1}(x)\,|2n+1\rangle \tag{B.8}$$

and we assume the relativistic wave equations

$$D_\alpha\,|\psi\rangle = 0. \tag{B.9}$$

The covariance of (B.9) under the $(1+2)-$dimensional Poincaré group is checked in [53]. Using the identities (B.7) a direct calculation gives

$$\begin{array}{ll} (\Pi^2 - m^2)\,|\psi_+\rangle = 0, & (\Pi.L - m\varepsilon\frac{1+\nu}{4})\,|\psi_+\rangle = 0, \\ (\Pi^2 - m^2)\,|\psi_-\rangle = 0, & (\Pi.L - m\varepsilon\frac{1-\nu}{4})\,|\psi_-\rangle = 0. \end{array} \tag{B.10}$$

This means that $|\psi_\pm\rangle$ describe a particle of mass m and spin $\frac{1\pm 1}{4}$. Now if we solve equation (B.9) in the rest frame we obtain that $|\psi_+\rangle = \psi_0\,|0\rangle$, which describes a anyons of mass m, energy εm and helicity $\frac{1+\nu}{4}$ although we have $|\psi_-\rangle = 0$. Solving the equation in any frame gives

$$|\psi_+\rangle = \sum_{n=0}^{+\infty}\sqrt{\frac{\Gamma(2s+n)}{\Gamma(2s)\Gamma(m+1)}}\left(\frac{\Pi_1 + i\Pi_2}{\Pi_0 + \varepsilon m}\right)^n \psi_0\,|2n\rangle\,. \tag{B.11}$$

This is proven by induction. Assuming

$$\psi_{2n} = \sqrt{\frac{\Gamma(2s+n)}{\Gamma(2s)\Gamma(m+1)}} \left(\frac{\Pi_1 + i\Pi_2}{\Pi_0 + \varepsilon m}\right)^n \psi_0,$$

using

$$\frac{\Pi_0 - \varepsilon m}{\Pi_1 - i\Pi_2} = \frac{\Pi_1 + i\Pi_2}{\Pi_0 + \varepsilon m},$$

gives the correct value for ψ_{2n+2}. Thus equations (B.9) describe a relativistic anyon of mass m, helicity $\frac{1+\nu}{4}$ and energy of sign ε. It has one degree of freedom as it should.

Appendix C. Lie algebras of order F associated with anyons

In this appendix we give an abstract and an explicit construction to define a Lie algebra of order F for relativistic anyons. Recall that the vector representation may be obtained in the Verma module language formalism [69]. Consider $\mathfrak{U}(\mathfrak{so}(1,2))$ the universal enveloping algebra. The Poincaré-Birkhoff-Witt theorem gives

$$\mathfrak{U}(\mathfrak{so}(1,2)) = \Big\{L_+^m L_-^n L_0^p,\ m,n,n \in \mathbb{N}\Big\}.$$

Consider now the two-sided ideal generated by $L_-, L_0 + \mathbf{I}$ (where $\mathbf{I}$ denotes the identity of $\mathfrak{U}(\mathfrak{so}(1,2))$) and set

$$\mathcal{V}_{-1} = \mathfrak{U}(\mathfrak{so}(1,2))/I.$$

We have, in $\mathcal{V}_{-1}$

$$L_0.\mathbf{I} = -\mathbf{I}, \quad L_-\mathbf{I} = 0.$$

Thus $\mathbf{I}$ is the highest weight representation of $\mathcal{V}_{-1}$, and we have

$$\mathcal{V}_{-1} = \Big\{L_+^n, n \in \mathbb{N}\Big\}.$$

But as seen in (46) ($L_- L_+^3 = 0$) and (47), $M_{-1} = \langle L_+^n, n > 2\rangle$ is an invariant subspace of $\mathcal{V}_{-1}$, and the standard finite dimensional vector representation is given by

$$\mathcal{D}_{-1} = \mathcal{V}_{-1}/M_{-1}.$$

In the same vain, we define the Verma module associated with the representation of spin $-1/F$. Recall that among the four representations of Section 3.2.1, there are two inequivalent representations, one bounded from below and one bounded from above. Since the construction works equally well on both representations, in the discussion, we only consider the representation bounded from below. In this language, we have

$$\mathcal{D}^+_{-1/F} = \mathcal{V}^+_{-1/F} = \mathfrak{U}(\mathfrak{so}(1,2))/ < L_-, L_0 + \frac{1}{F}\mathbf{I} > .$$

In $\mathcal{D}^+_{-1/F}$ we have $L_0.\mathbf{I} = -1/F\mathbf{I}$, $L_-\mathbf{I} = 0$, this means that in $\mathcal{S}^F(\mathcal{D}^+_{-1/F})$, we have $\mathbf{I}\otimes\cdots\otimes\mathbf{I}$ ($F-$times) is such that $L_-.\mathbf{I}\otimes\cdots\otimes\mathbf{I} = 0$, $L_0.\mathbf{I}\otimes\cdots\otimes\mathbf{I} = -\mathbf{I}\otimes\cdots\otimes\mathbf{I}$. Thus $\mathbf{I}\otimes\cdots\otimes\mathbf{I}$ is a primitive vector of $\mathcal{V}_{-1}$ meaning that there is an $\mathfrak{so}(1,2)-$equivariant map

$$i : \mathcal{V}_{-1} \hookrightarrow \mathcal{S}^F(\mathcal{D}^+_{-1/F}).$$

Taking the coset by M_{-1} we obtain an $\mathfrak{so}(1,2)-$equivariant inclusion

$$\mathcal{D}_{-1} \hookrightarrow \mathcal{S}^F(\mathcal{D}^+_{-1/F})/i(M_{-1}).$$

If we denote by $\mathcal{S}^F(\mathcal{D}^+_{-1/F})_{\text{red}} = \mathcal{S}^F(\mathcal{D}^+_{-1/F})/i(M_{-1})$ we have

$$\Big(\mathfrak{so}(1,2) \oplus \mathcal{S}^F(\mathcal{D}^+_{-1/F})_{\text{red}}\Big) \oplus \mathcal{S}^F(\mathcal{D}^+_{-1/F}),$$

is a Lie algebra of order three.. This can be summarised in the following diagram

$$\begin{array}{ccccc} V_{-1} & \xrightarrow{i\cong} & \left\langle \mathbf{I}^{\otimes^F}_{-\frac{1}{F}} \right\rangle & \hookrightarrow & \mathcal{S}^F(\mathcal{D}_{-\frac{1}{F}}) \\ \downarrow & & \downarrow & & \downarrow \\ \mathcal{D}_{-1} = V_{-1}/M_{-1} & \xrightarrow{i\cong} & \left\langle \mathbf{I}^{\otimes^F}_{-\frac{1}{F}} \right\rangle/i(M_{-1}) & \hookrightarrow & \mathcal{S}^F_{\text{red}}(\mathcal{D}_{-\frac{1}{F}}). \end{array}$$

This commutative diagram also shows that there is no mapping from $\mathcal{S}^F(D_{-1/F})$ into $\mathcal{D}_{-1}$ and thus no Lie algebras of order F associated with $(\mathfrak{so}(1,2) \oplus \mathcal{D}_{-1}) \oplus \mathcal{D}_{-1/F}$. A Lie of algebra of order F can however be defined if $\mathcal{D}_{-1}$ is extended into an infinite dimensional representation.

To examine some properties of $\mathcal{S}^F(\mathcal{D}^+_{-1/F})_{\text{red}}$ (indecomposability), and to clarify the abstract construction above, we obtain the same result in an explicit way. It is well known that $\mathfrak{so}(1,2)$ can be realised by differential operators

$$L_- = x\partial_y, L_+ = -y\partial_x, L_0 = \frac{1}{2}(y\partial_y - x\partial_x).$$

Consider now $\mathcal{F}$ the vector space of functions on $\mathbb{R}^{2*}_+ = \big\{(x,y) \in \mathbb{R}^2, x, y > 0\big\}$. The following subspaces of $\mathcal{F}$

$$\begin{aligned} \mathcal{D}_{-n} &= \Big\langle\, x^{2n}, x^{2n-1}y, \cdots, xy^{2n-1}, y^{2n} \Big\rangle, \quad n \in \mathbb{N}/2, \\ \mathcal{D}^+_{-\lambda} &= \Big\langle\, x^{2\lambda}\left(\frac{y}{x}\right)^m, m \in \mathbb{N} \Big\rangle, \quad \lambda \in \mathbb{R} \setminus \mathbb{N}/2, \\ \mathcal{D}^-_{-\lambda} &= \Big\langle\, y^{2\lambda}\left(\frac{x}{y}\right)^m, m \in \mathbb{N} \Big\rangle, \quad \lambda \in \mathbb{R} \setminus \mathbb{N}/2, \end{aligned} \tag{C.1}$$

are representations of $\mathfrak{so}(1,2)$. The representation $\mathcal{D}_{-n}$ is the $(2n+1)$−dimensional irreducible representation and the representations $\mathcal{D}^\pm_{-\lambda}$ are infinite dimensional representations, bounded from below and above respectively. It is important to emphasise that the representations given in (C.1) do not have the normalisations conventionally taken.

To define $S^F(\mathcal{D}^+_{-1/F})_{\text{red}}$, we consider the multiplication map $m_F : \mathcal{F} \times \cdots \times \mathcal{F} \to \mathcal{F}$ given by

$$m_F(f_1, \cdots, f_F) = f_1 \cdots f_F \tag{C.2}$$

which is multilinear and totally symmetric and hence induces a map μ_F from $\mathcal{S}^F(\mathcal{F})$ into $\mathcal{F}$. Restricting to $\mathcal{S}^F\left(D^\pm_{-1/F}\right)$ one sees that

$$\begin{aligned} S^F(\mathcal{D}^+_{-1/F})_{\text{red}} &\stackrel{\text{def}}{=} \mu_F\Big(S^F(D^+_{-1/F})\Big) = \left\langle x^2\left(\frac{y}{x}\right)^m, \ m \in \mathbb{N} \right\rangle \supset \mathcal{D}_{-1} \\ S^F(\mathcal{D}^-_{-1/F})_{\text{red}} &\stackrel{\text{def}}{=} \mu_F\Big(S^F(D^-_{-1/F})\Big) = \left\langle y^2\left(\frac{x}{y}\right)^m, \ m \in \mathbb{N} \right\rangle \supset \mathcal{D}_{-1}. \end{aligned} \tag{C.3}$$

Journal of Physics: Conference Series **175** (2009) 012003 doi:10.1088/1742-6596/175/1/012003

If we construct a diagram analogous to (47) (for $\mathcal{D}^+_{-1/F}$), we have

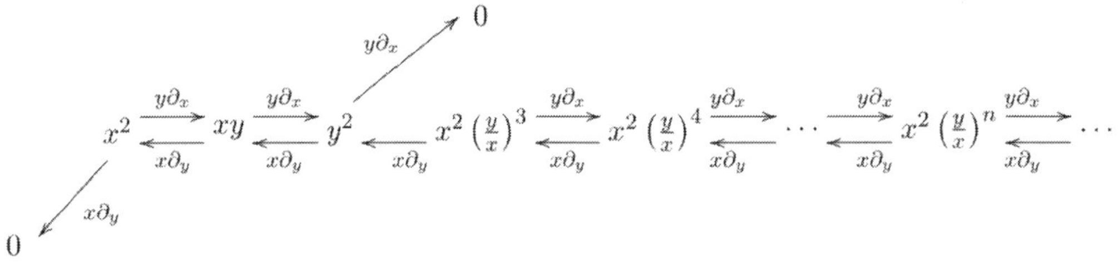

Looking at the representations defined in (C.3) *i.e.* $S^F\left(\mathcal{D}^{\pm}_{-1/F}\right)_{\text{red}}$, one sees that, even though $\mathcal{D}_{-1}$ is a subspace stable under $\mathfrak{so}(1,2)$ there is no complement stable under $\mathfrak{so}(1,2)$ [23]. Indeed, these representations cannot be built from a primitive vector. This is due to the fact that $L^3_+\left(x^2\right) = 0$ and consequently we cannot reach $x^{-1}y^3$ from x^2 but conversely $L^3_-(x^{-1}y^3) = 6x^2$. This is the reason why there is no $F-$Lie algebra structure on $\mathfrak{so}(1,2) \oplus \mathcal{D}_{-1}$. With the normalisations (C.1) and (C.3), denoting $\mathcal{D}^+_{-1/F} = \{A_{-1/F+n},\ n \in \mathbb{N}\}$ and $\mathcal{S}^F_{\text{red}}(\mathcal{D}^+_{-1/F}) = \{P_{-1+n},\ n \in \mathbb{N}\}$, the trilinear brackets are given by

$$\left\{A^+_{-1/F+n_1}, \cdots, A^+_{-1/F+n_F}\right\} = P_{-1+n_1+\cdots n_F}.$$

Appendix D. Clifford algebras of polynomial

Clifford algebras of a polynomial's is a direct generalisation of usual Clifford algebras for higher degree polynomials. Consider p a polynomial of degree n with k variables. Since degree n polynomials are isomorphic to symmetric tensors of order n, we can write

$$p(x^1, \cdots, x^k) = x^{i_1}\cdots x^{i_n} g_{i_1\cdots i_n}. \tag{D.1}$$

The Clifford algebra of the polynomial p denoted $\mathcal{C}_p$ is the algebra generated by k primitive elements $g_1, \cdots, g_k$ such that

$$\{g_{i_1}, \cdots, g_{i_n}\} = n! g_{i_1\cdots i_n}. \tag{D.2}$$

The algebra defined by relations (D.2) can be real if the tensor g is real, or complex. It appears as an example of a Lie algebras of order n where $\mathfrak{g}_0 = \mathbb{R}$ or $\mathbb{C}$. The relations (D.2) means that if we consider P in $\mathcal{C}_p$ defined by $P = x^k g_k$, we have

$$(x^1 g_1 + \cdots x^k g_k)^n = p(x_1, \cdots, x_k).1\ , \tag{D.3}$$

where 1 denotes the unity of $\mathcal{C}_p$ that we omit from now on. This algebra has been introduced in a formal way by N. Roby [70]. It is a natural generalisation of the usual Clifford algebra, but this algebra is very different from the usual Clifford algebra. Since it is defined through n-th order relations, the number of independent monomials increases with polynomial's degree (for instance, $(g_1g_2)^k, k \geq 0$ are all independent). This means that we do not have enough constraints among the generators to order them in some fixed way and, as a consequence, $\mathcal{C}_p$ turns out to be an infinite-dimensional algebra. Thus any finite dimensional representation are non-faithful. Several properties on representations was then established such that the dimension of a representation of Clifford algebras is a multiple of the degree of the polynomial [71], but the first systematic way to obtain a matrix representation was given in [72]. Subsequently, an extensive study of the representations of Clifford algebras of cubic polynomials was undertaken by Revoy [73] and a family of inequivalent representations can be obtained. See also Ref.[12] for many references on Clifford algebras of polynomial's.

For a self contained presentation, we give an algorithm to construct a matrix representation of the Clifford algebra of a given polynomial, for more details and comments see [72, 12]. In this process, two basic polynomials will be considered. The sum polynomial $S(x) = (x^1)^n + \cdots + (x^k)^n$ and the product polynomial $\pi(x) = x^1 \cdots x^n$.

For the linearisation of the sum polynomial, consider the $2r+1$ following matrices (with $r = [k/2]$)

$$\begin{array}{ll}
\Pi_1 = \pi_1 \otimes I^{\otimes^{(r-1)}} & \Pi_2 = \pi_2 \otimes I^{\otimes^{(r-1)}} \\
\vdots & \vdots \\
\Pi_{2\ell-1} = \pi_3^{\otimes^{(\ell-1)}} \otimes \pi_1 \otimes I^{\otimes^{(r-\ell-1)}} & \Pi_{2\ell} = \pi_3^{\otimes^{(\ell-1)}} \otimes \pi_2 \otimes I^{\otimes^{(r-\ell-1)}} \\
\vdots & \vdots \\
\Pi_{2r-1} = \pi_3^{\otimes^{(r-1)}} \otimes \pi_1 & \Pi_{2r} = \pi_3^{\otimes^{(r-1)}} \otimes \pi_2 \\
\end{array}
\quad \Pi_{2r+1} = \pi_3^{\otimes^{r}} . \tag{D.4}$$

with the matrices π_1, π_2, π_3 being defined by

$$\pi_1 = \begin{pmatrix} 0 & 1 & 0 & \cdots & 0 \\ 0 & 0 & 1 & \cdots & 0 \\ \vdots & & \ddots & \ddots & \vdots \\ 0 & 0 & \cdots & 0 & 1 \\ 1 & 0 & 0 & \cdots & 0 \end{pmatrix}, \quad \pi_3 = \begin{pmatrix} 1 & 0 & \cdots & 0 \\ 0 & q & \cdots & 0 \\ \vdots & & \ddots & \\ 0 & 0 & \cdots & q^{n-1} \end{pmatrix}, \quad \pi_2 = (\sqrt{q})\pi_3\pi_1, \tag{D.5}$$

$\sqrt{q}$ being there only when n is even and $q = e^{2i\frac{\pi}{n}}$. Many authors have considered this set of matrices, see [12] for references. It is not difficult to see that the $\Pi-$matrices satisfy the relation

$$\Pi_i \Pi_j = q \Pi_j \Pi_i, \quad \Pi_i^n = 1, \quad i < j \tag{D.6}$$

and as a consequence of the relation root-coefficients that we have $(x^1\Pi_1 + \cdots x^k\Pi_k)^n = (x^1)^n + \cdots (x^k)^n$. The algebra generated by elements satisfying (D.6) generate the generalised Clifford algebra. This algebra together with its representations have been classified by Morris [74].

The product polynomial is linearised by the matrices H_{ii+1} (with H_{ij} the canonical matrices with a one at the intersection of the $i-$th line and $j-$th column and zero elsewhere). Indeed an easy calculation gives $(x^i H_{12} + \cdots x^n H_{n1})^n = x^1 \cdots x^n$.

The matrices Π and H allow to linearise any polynomial, since an arbitrary polynomial is a sum of monomials

$$p(x) = \sum_{\ell=1}^{q} m_\ell(x).$$

Each monomial is a particular case of the polynomial $(x_1)^{a_1}(x_2)^{a_2} \cdots (x_p)^{a_p}$ which can be linearised by the $n \times n$ $H-$matrices

$$(x_1)^{a_1}(x_2)^{a_2}\cdots(x_p)^{a_p} =$$
$$\left[x_1\left(\sum_{i=1}^{a_1} H_{ii+1}\right) + x_2\left(\sum_{i=1+a_1}^{a_1+a_2} H_{ii+1}\right) + \cdots + x_p\left(\sum_{i=1+a_1+\cdots+a_{p-1}}^{n} H_{ii+1}\right)\right]^n .$$

Thus any monomial of the polynomial p can be linearised as above and we have $p(x) = \sum_{\ell=1}^q M_\ell^n$, with M_ℓ, q $n\times n$ matrices. Introducing the commuting matrices $\tilde{M}_\ell = I^{\otimes^{\ell-1}} \otimes M_\ell \otimes I^{\otimes^{q-\ell-1}}$ we have using the linearisation of the sum polynomial by the $\Pi-$matrices

$$p(x) = \sum_{\ell=1}^q \tilde{M}_\ell^n = (\sum_{\ell=1}^q (\Pi_\ell \otimes \tilde{M}_\ell)^n.$$

Since $\Pi_\ell \otimes \tilde{M}_\ell$ are linear in x^k we have $\sum_{\ell=1}^q \Pi_\ell \otimes \tilde{M}_\ell = x^1 G_1 + \cdots x^k G_k$. This ends end the process of linearisation. Of course this process is far from being unique and many different matrices of different size can be obtained. See [72, 12] for examples and comments.

As an illustration of this process consider the little algebra (51) of Section 3.2.4. Equation (51) say that the generators $A_{-1/F}$ and $A_{1-1/F}$ generate the Clifford algebra of the polynomial $x^{F-1}y$. The process above allows to find a linearisation of the polynomial by the $n\times n$ matrices

$$A^+_{-\frac{1}{F}} = \begin{pmatrix} 0 & 0 & 0 & \cdots & 0 & 0 \\ a_1 & 0 & 0 & \cdots & 0 & 0 \\ 0 & a_2 & 0 & \cdots & 0 & 0 \\ & & & & & \\ \vdots & \vdots & & \ddots & \ddots & \vdots \\ 0 & 0 & \cdots & 0 & a_{F-1} & 0 \end{pmatrix}, \quad A^+_{1-\frac{1}{F}} = \begin{pmatrix} 0 & 0 & 0 & \cdots & 0 & \frac{1}{a_1\cdots a_{F-1}} \\ 0 & 0 & 0 & \cdots & 0 & 0 \\ 0 & 0 & 0 & \cdots & 0 & 0 \\ & & & & & \\ \vdots & \vdots & & \ddots & & \vdots \\ 0 & 0 & 0 & \cdots & 0 & 0 \end{pmatrix}. \quad \text{(D.7)}$$

(For physical reasons the normalisation of these matrices is not the same of the normalisation which would have been obtained by the algorithm above *i.e* that the non-zero matrix elements are not equal to one (see Section 3.2.4).) The matrix $A^+_{-1/F}$ can be obtained in a different way. Indeed, (51) implies that the first power of $A^+_{-\frac{1}{F}}$ which is equal to zero is F (in other words the rank of $A^+_{-\frac{1}{F}}$ is $F-1$). Writing $A^+_{-1/F}$ in its Jordan form using the relations (51) gives a solution for $A^+_{1-1/F}$ of the type above. However, it is known for $F=3$ that there exists other solutions for the matrix $A^+_{1-1/F}$ (see [70, 12]). But these matrices would not respect the grading, see (35). Indeed, if some of the matrix elements which are equal to zero in (D.7) are different from zero, the matrices would not be consistent with the Poincaré algebra *i.e.* we obtain equations where both sides do not have the same helicity. Thus we take the matrices as in (D.7).

In the same way, the representation of the cubic extension of the Poincaré algebra in any space-time dimensions (see Section 3.3) are related to Clifford algebras of polynomial's. Indeed, writing the R.H.S. of (72) as $g_{\mu\nu\rho} = \frac{1}{6}(\eta_{\mu\nu}\delta_\rho{}^\sigma + \eta_{\nu\rho}\delta_\mu{}^\sigma + \eta_{\rho\mu}\delta_\nu{}^\sigma)P_\sigma$, we define $p(x) = g_{\mu\nu\rho}x^\mu x^\nu x^\rho = \frac{1}{2}(x.x)(x.P)$. In [56, 57], along the algorithm above, we have found two different representations of the Clifford algebra of the polynomial $\frac{1}{2}(x.x)(x.P)$, but only one was compatible with the Poincaré algebra (see [56]). For the second ones, writing

$$\frac{1}{2}(x.x)(x.p) = \frac{1}{2}(x^\mu \Gamma_\mu)^2(x.p) = \frac{1}{2}\begin{pmatrix} 0 & \Lambda^{\frac{1}{3}} x^\mu \Gamma_\mu & 0 \\ 0 & 0 & \Lambda^{\frac{1}{3}} x^\mu \Gamma_\mu \\ \Lambda^{-\frac{2}{3}} x^\mu P_\mu & 0 & 0 \end{pmatrix}^3$$

we obtain

$$V_\mu = \frac{1}{\sqrt[3]{2}}\begin{pmatrix} 0 & \Lambda^{\frac{1}{3}}\Gamma_\mu & 0 \\ 0 & 0 & \Lambda^{\frac{1}{3}}\Gamma_\mu \\ \Lambda^{-\frac{2}{3}} P_\mu & 0 & 0 \end{pmatrix}, \tag{D.8}$$

where Λ is a parameter with dimension of mass (that we take normalised to one) and Γ_μ are the Dirac $\Gamma-$matrices in $D-$space-time dimensions. We also renormalise the matrices V in such a way that the factor $\sqrt[3]{2}$ cancels. If D is even, the representation (D.8) is reducible. Taking the Dirac $\Gamma-$matrices in the chiral representation

$$\Gamma_\mu = \begin{pmatrix} 0 & \Sigma_\mu \\ \tilde{\Sigma}_\mu & 0 \end{pmatrix},$$

(where $\Sigma_0 = \tilde{\Sigma}_0 = 1$ and $\tilde{\Sigma}_i = -\Sigma_i$ are the generators of the Clifford algebra of $SO(D-1)$) we obtain

$$V_\mu^+ = \begin{pmatrix} 0 & \Sigma_\mu & 0 \\ 0 & 0 & \tilde{\Sigma}_\mu \\ P_\mu & 0 & 0 \end{pmatrix}, \quad V_\mu^- = \begin{pmatrix} 0 & \tilde{\Sigma}_\mu & 0 \\ 0 & 0 & \Sigma_\mu \\ P_\mu & 0 & 0 \end{pmatrix}, \tag{D.9}$$

two inequivalent representations.

A peculiar Clifford algebra of polynomial is given when $p = 0$. In this case, the algebra (D.2) reduces to

$$\{e_{i_1}, \cdots, e_{i_n}\} = 0. \tag{D.10}$$

This algebra has been defined in [75] and called the n-exterior algebra. An explicit basis of this infinite dimensional algebra was exhibited. This is clearly a generalisation of the Grassmann algebra.

Appendix E. Some properties on spinors and $p-$forms in $(1 + (4n-1))-$dimensions
In this section, we give a collection of useful identities on $p-$forms and spinors. The Minkowski metric is taken to be $\eta_{\mu\nu} = \text{diag}(1, -1, \cdots, -1)$, and the Levi-Civita tensor is normalised as follows $\varepsilon_{01\cdots D-1} = -\varepsilon^{01\cdots D-1} = 1$. For latter convenience, we denote

$$\begin{aligned} \delta^{(\mu)_\ell}_{(\nu)_\ell} &= \delta^{\mu_1\cdots\mu_\ell}_{\nu_1\cdots\nu_\ell} = \begin{vmatrix} \delta^{\mu_1}_{\nu_1} & \cdots & \delta^{\mu_\ell}_{\nu_1} \\ \vdots & & \vdots \\ \delta^{\mu_1}_{\nu_\ell} & \cdots & \delta^{\mu_\ell}_{\nu_\ell} \end{vmatrix} \\ \varepsilon_{(\mu)(\nu)} &= \varepsilon_{\mu_1\cdots\mu_{2n}\nu_1\cdots\nu_{2n}} \\ \eta_{(\mu)(\nu)} &= \eta_{\mu_1\nu_1}\cdots\eta_{\mu_{2n}\nu_{2n}}. \end{aligned} \tag{E.1}$$

All the properties below come, for instance, from an explicit matrix realisation of the Dirac matrices and from the properties of spinors (see *e.g.* [76]).

Appendix E.1. Dirac matrices

The Dirac Γ−matrices in the chiral representation are given by

$$\Gamma_\mu = \begin{pmatrix} 0 & \Sigma_\mu \\ \tilde{\Sigma}_\mu & 0 \end{pmatrix},$$

(where $\Sigma_0 = \tilde{\Sigma}_0 = 1$ and $\tilde{\Sigma}_i = -\Sigma_i$ are the generators of the Clifford algebra of $SO(D-1)$). We introduce further the fully antisymmetric matrices:

$$\Gamma^{(\ell)} : \Gamma_{\mu_1 \cdots \mu_\ell} = \frac{1}{\ell!} \sum_\sigma \Gamma_{\mu_{\sigma(1)}} \cdots \Gamma_{\mu_{\sigma(\ell)}} \tag{E.2}$$

which write

$$\begin{aligned} \Gamma_{\mu_1 \cdots \mu_{2\ell}} &= \frac{1}{(2\ell)!} \begin{pmatrix} \Sigma_{\mu_1} \tilde{\Sigma}_{\mu_2} \cdots \tilde{\Sigma}_{\mu_{2\ell}} + \text{ perm} & 0 \\ 0 & \tilde{\Sigma}_{\mu_1} \Sigma_{\mu_2} \cdots \Sigma_{\mu_{2\ell}} + \text{ perm} \end{pmatrix} \\ &= \begin{pmatrix} \Sigma_{\mu_1 \cdots \mu_{2\ell}} & 0 \\ 0 & \tilde{\Sigma}_{\mu_1 \cdots \mu_{2\ell}} \end{pmatrix} \\ \Gamma_{\mu_1 \cdots \mu_{2\ell+1}} &= \frac{1}{(2\ell+1)!} \begin{pmatrix} 0 & \Sigma_{\mu_1} \tilde{\Sigma}_{\mu_2} \cdots \Sigma_{\mu_{2\ell+1}} + \text{ perm} \\ \tilde{\Sigma}_{\mu_1} \Sigma_{\mu_2} \cdots \tilde{\Sigma}_{\mu_{2\ell+1}} + \text{ perm} & \end{pmatrix} \\ &= \begin{pmatrix} 0 & \Sigma_{\mu_1 \cdots \mu_{2\ell+1}} \\ \tilde{\Sigma}_{\mu_1 \cdots \mu_{2\ell+1}} & 0 \end{pmatrix} \end{aligned} \tag{E.3}$$

These matrices are generically denoted by

$$\Gamma^{(\mu)_\ell} = \Gamma^{\mu_1 \cdots \mu_\ell}, \quad \Gamma_{(\mu)_\ell} = \Gamma_{\mu_1 \cdots \mu_\ell}$$

and we have the trace formulæ

$$\begin{aligned} \frac{1}{2^n} \text{Tr} \left(\Sigma^{(\mu)_{2a}} \Sigma_{(\nu)_{2b}} \right) &= \delta_{ab} \left(\delta^{(\mu)_{2a}}_{(\nu)_{2a}} - i \delta_{an} \varepsilon^{(\mu)(\mu')} \eta_{(\mu')(\nu)} \right) \\ \frac{1}{2^n} \text{Tr} \left(\tilde{\Sigma}^{(\mu)_{2a}} \tilde{\Sigma}_{(\nu)_{2b}} \right) &= \delta_{ab} \left(\delta^{(\mu)_{2a}}_{(\nu)_{2a}} + i \delta_{an} \varepsilon^{(\mu)(\mu')} \eta_{(\mu')(\nu)} \right) \\ \frac{1}{2^n} \text{Tr} \left(\Sigma^{(\mu)_{2a+1}} \tilde{\Sigma}_{(\nu)_{2b+1}} \right) &= \delta_{ab} \delta^{(\mu)_{2a+1}}_{(\nu)_{2a+1}}. \end{aligned} \tag{E.4}$$

We have assumed here that $\Sigma^{(\mu)_{2n}}$ projects onto the self-dual $2n$−forms. Correspondingly we assume that $\tilde{\Sigma}^{(\mu)_{2n}}$ projects onto the anti-self-dual $2n$−forms. This convention fixes the second term in the first equation in (E.4). Moreover, we have the following properties

$$\Gamma_{\mu_1 \cdots \mu_\ell} \Gamma_\mu = \Gamma_{\mu_1 \cdots \mu_\ell \mu} + \eta_{\mu_\ell \mu} \Gamma_{\mu_1 \cdots \mu_{\ell-1}} + \cdots + (-1)^{\ell-1} \eta_{\mu_1 \mu} \Gamma_{\mu_2 \cdots \mu_\ell}. \tag{E.5}$$

Appendix E.2. Spinors and p−forms

The Dirac matrices act naturally on spinors. When the dimension is even, the spinor space decomposes into left-handed and right-handed spinors $\mathcal{S} = \mathcal{S}_+ \oplus \mathcal{S}_-$. Furthermore, when $D = 4n$, left-handed spinors are in the complex conjugate representation of right-handed spinors. In a straight analogy with Appendix Appendix A, we denote $\Psi_L \in \mathcal{S}_+$ and $\Psi_R \in \mathcal{S}_-$ by there components $\Psi_L \to \Psi_A$ and $\Psi_R \to \bar{\Psi}^{\dot{A}}$. This leads to the following index structure for the $\Gamma-$matrices: $\Sigma_\mu \to \Sigma_{\mu A\dot{B}}, \tilde{\Sigma}_\mu \to \tilde{\Sigma}_\mu{}^{\dot{A}B}$. The charge conjugation matrix is given by

$$\mathcal{C} = \begin{pmatrix} C_+{}^{AB} & 0 \\ 0 & C_-{}_{\dot{A}\dot{B}} \end{pmatrix}, \quad \mathcal{C}^{-1} = \begin{pmatrix} C^{-1}{}_{+AB} & 0 \\ 0 & C^{-1}{}_-{}^{\dot{A}\dot{B}} \end{pmatrix},$$

and allows to raise and lower the indices $\Psi^A = \Psi_B C_+{}^{AB}, \Psi_A = \Psi^B C^{-1}{}_{+AB}$ *etc.*

The tensor product of two spinors decomposes on the set of $p-$forms ($[p]$ denotes $p-$forms and $[2n]_\pm$ (anti-)self-dual $2n-$form see (E.11))

$$\begin{aligned} \mathcal{S} \otimes \mathcal{S} &= [0] \oplus [1] \oplus \cdots [4n], \\ \mathcal{S}_+ \otimes \mathcal{S}_+ &= [0] \oplus [2] \oplus \cdots [2n]_+ \\ \mathcal{S}_- \otimes \mathcal{S}_- &= [0] \oplus [2] \oplus \cdots [2n]_- \\ \mathcal{S}_+ \otimes \mathcal{S}_- &= [1] \oplus [3] \oplus \cdots [2n-1]. \end{aligned} \tag{E.6}$$

Introducing, $A_{[p]} \in [p], 0 \le p \le 2n-1$ and $A_{[2n]_\pm} \in [2n]_\pm$ (E.6) give (in the sequel $A_{[p]}\Gamma^{(p)}$ alway means $A_{[p]\mu_1\cdots\mu_p}\Gamma^{\mu_p\cdots\mu_1}$):

$$\begin{aligned} \Xi_{++} = \Psi_+ \otimes \Psi'_+ &= \sum_{p=0}^{n-1} \frac{1}{(2p)!} A_{[2p]}\Sigma^{(2p)}\mathcal{C}_+^{-1} + \frac{1}{2}\frac{1}{(2n)!} A_{[2n]_+}\Sigma^{(2n)}\mathcal{C}_+^{-1} \\ \Xi_{--} = \Psi_- \otimes \Psi'_- &= \sum_{p=0}^{n-1} \frac{1}{(2p)!} A'_{[2p]}\tilde{\Sigma}^{(2p)}\mathcal{C}_-^{-1} + \frac{1}{2}\frac{1}{(2n)!} A'_{[2n]_-}\tilde{\Sigma}^{(2n)}\mathcal{C}_-^{-1} \\ \Xi_{+-} = \Psi_+ \otimes \Psi'_- &= \sum_{p=0}^{n-1} \frac{1}{(2p+1)!} A_{[2p+1]}\Sigma^{(2p+1)}\mathcal{C}_-^{-1} \\ \Xi_{-+} = \Psi_- \otimes \Psi'_+ &= \sum_{p=0}^{n-1} \frac{1}{(2p+1)!} A'_{[2p+1]}\tilde{\Sigma}^{(2p+1)}\mathcal{C}_+^{-1}. \end{aligned} \tag{E.7}$$

In the first relation in (E.7) the $2n-$form is self-dual, and in the second the $2n-$form is anti-self-dual. Using the trace relations (E.4), one gets

$$\begin{aligned} A_{[2p]} &= \frac{1}{2^n}\mathrm{Tr}\left(\Xi_{++}\mathcal{C}_+\Sigma^{(2p)}\right), \quad A_{[2n]_+} = \frac{1}{2^n}\mathrm{Tr}\left(\Xi_{++}\mathcal{C}_+\Sigma^{(2n)}\right), \\ A'_{[2p]} &= \frac{1}{2^n}\mathrm{Tr}\left(\Xi_{--}\mathcal{C}_-\tilde{\Sigma}^{(2p)}\right), \quad A'_{[2n]_-} = \frac{1}{2^n}\mathrm{Tr}\left(\Xi_{--}\mathcal{C}_-\tilde{\Sigma}^{2n}\right), \\ A_{[2p+1]} &= \frac{1}{2^n}\mathrm{Tr}\left(\Xi_{+-}\mathcal{C}_-\tilde{\Sigma}^{(2p+1)}\right), \\ A'_{[2p+1]} &= \frac{1}{2^n}\mathrm{Tr}\left(\Xi_{-+}\mathcal{C}_+\Sigma^{(2p+1)}\right). \end{aligned} \tag{E.8}$$

where $\mathrm{Tr}\left(\Xi_{+-}\mathcal{C}_-\tilde{\Sigma}^{(2p+1)}\right) = \mathrm{Tr}\left(\Xi_{+-}\mathcal{C}_-\tilde{\Sigma}_{\mu_1\cdots\mu_{2p+1}}\right)$ to simplify notations.

Finally, we introduce some operations on $p-$forms. The Hodge duality is a linear map $\star : [p] \longrightarrow [D-p]$. If $A_{[p]} \in [p], {}^\star A_{[p]} = B_{[D-p]} \in [D-p]$ is given by

$$B_{[D-p]\mu_1 \cdots \mu_{D-p}} = \frac{1}{p!} \varepsilon_{\mu_1 \cdots \mu_{D-p} \nu_1 \cdots \nu_p} A_{[p]}^{\nu_1 \cdots \nu_p}, \tag{E.9}$$

and it is easy to prove that

$$^{\star\star} A_{[p]} = (-1)^{(D-1)(p-1)} A_{[p]}. \tag{E.10}$$

When the dimension is even, one can define a (anti-)self-dual $(D/2)-$form:

$$^\star A_{[D/2]} = \begin{cases} \pm A_{[D/2]} & \text{when } D/2 \text{ is an odd number } (\star^2 = 1) \\ \pm i A_{[D/2]} & \text{when } D/2 \text{ is an even number } (\star^2 = -1) \end{cases} \tag{E.11}$$

This means that (anti-)self-dual $2n-$forms are complex representations of the Lorentz group when $D = 4n$.

Next, introducing $\varepsilon \in [1]$, one defines

- The inner product

$$\begin{array}{rccc} i_\varepsilon : & [p] & \longrightarrow & [p-1] \\ & A_{[p]} & \longmapsto & i_\varepsilon A_{[p]}, \end{array} \tag{E.12}$$

  in components we have

$$(i_\varepsilon A_{[p]})_{\mu_1 \cdots \mu_{p-1}} = A_{[p]\mu_1 \cdots \mu_p} \varepsilon^{\mu_p}. \tag{E.13}$$

  Notice the difference of convention, useful for our purpose: the summation is done on the last index instead of the first one.

- The exterior product

$$\begin{array}{rccc} \wedge : & [p] & \longrightarrow & [p+1] \\ & A_{[p]} & \longmapsto & A_{[p]} \wedge \varepsilon, \end{array} \tag{E.14}$$

  in components reads

$$(A_{[p]} \wedge \varepsilon)_{\mu_1 \cdots \mu_{p+1}} = \frac{1}{p!} \delta^{\nu_1 \cdots \nu_{p+1}}_{\mu_1 \cdots \mu_{p+1}} A_{[p]\nu_1 \cdots \nu_p} \varepsilon_{\nu_{p+1}}. \tag{E.15}$$

- Now, with $A_{[p]} \in [p]$ and ${}^\star A_{[p]} = B_{[D-p]} \in [D-p]$, we have that

$$\begin{aligned} {}^\star\left(A_{[p]} \wedge \varepsilon\right) &= (-1)^p i_\varepsilon B_{[D-p]}, \\ {}^\star\left(i_\varepsilon A_{[p]}\right) &= (-1)^{D(p-D)} B_{[D-p]} \wedge \varepsilon. \end{aligned} \tag{E.16}$$

Then, one defines the exterior derivative d which maps $[p] \to [p+1]$ by

$$(dA_{[p]})_{\mu_1\cdots\mu_{p+1}} = \frac{1}{p!}\delta^{\nu_1\cdots\nu_{p+1}}_{\mu_1\cdots\mu_{p+1}}\partial_{\nu_1}A_{[p]\nu_2\cdots\nu_{p+1}} \tag{E.17}$$

and its adjoint $d^\dagger$ which maps $[p] \to [p-1]$ is defined by $d^\dagger = (-1)^{pD+D}\star d\star$, and gives

$$(d^\dagger A_{[p]})_{\mu_2\cdots\mu_p} = \partial^{\mu_1}A_{[p]\mu_1\nu_2\cdots\nu_p}. \tag{E.18}$$

It can the be shown that we have the following relation

$$\begin{aligned} &\frac{1}{(p+1)!}dA_{[p]}dA_{[p]} + \frac{1}{(p-1)!}d^\dagger A_{[p]}d^\dagger A_{[p]} \\ = \ &(-1)^{D-1}\left(\frac{1}{(D-p-1)!}d^\dagger B_{[D-p]}d^\dagger B_{[D-p]} + \frac{1}{(D-p+1)!}dB_{[D-p]}dB_{[D-p]}\right) \end{aligned} \tag{E.19}$$

with $B_{[D-p]} = {}^\star A_{[p]}$.

References

[1] Greenberg O. W. *Phys. Rev.* **135** (1964) 1447-1450; O'Raifeartaigh L *Phys. Rev* **139** (1965), B 1052-1062; Coleman S and Mandula J *Phys. Rev.* **159** (1967) B 1251-1256.

[2] Georgi H and Glashow S. L. *Phys. Rev. Lett.* **32** (1974) 438; Fritzsch H. and Minkowski P. *Annals Phys.* **93** (1975) 193. Gursey F, Ramond P. and Sikivie P. *Phys. Lett.* **B 60** (1976) 177; Slansky R *Phys. Rept.* **79** (1981) 1; Ross G. G. *Grand Unified Theories*, The Benjanin/Cummings Publishing Company, INC, (1985).

[3] Haag R., Lopuszanski J. T. and Sohnius M. F. *Nucl. Phys.* **B88** (1975) 257-274.

[4] Weinberg S. *The quantum theory of fields. Vol. 3: Supersymmetry,* Cambridge, UK: Univ. Pr. (2000).

[5] Nambu Y. *Phys. Rev.* **D7** (1973) 2405.

[6] Bars I. and Günaydin M. *J. Math. Phys.* **20** (1979) 1977-1993.

[7] Bars I and Günaydin M. *Phys. Rev.* **D22** (1980) 1403-1413.

[8] Vainerman L. and Kerner R. *J. Math. Phys* **37** (1995) 2553-2565.

[9] Kerner R. *Class. and Quantum Grav.* **14 (1A)** (1997) A203-A225.

[10] Borowiec A. Bazunova N. and Kerner R. *Lett. Math. Phys.* **67** (2004) 195-206.

[11] de Azcárraga J. A. and Pérez Bueno J. C. *Commun. Math. Phys.* **184** (1997), 669. [arXiv:hep-th/9605213].

[12] Rausch de Traubenberg M arXiv:hep-th/9802141 (Habilitation Thesis).

[13] Filippov V. T. *Sibirsk. Math. Zh* **26** (1985) 126-140.

[14] Gnedbaye A. V. *C. R. Acad. Sci. Paris Sér. I Math.* **321** (1995) 147-152.

[15] Michor P. W. and Vinogradov A. M. *Rend. Sem. Mat. Univ. Politec. Torino* **54** (1996) 373-392.

[16] Goze N. and Remm E. arXiv:0803.0553[math.RA].

[17] Bagger J. and Lambert N *Phys. Rev.* **D75**, (2007) 045020 [arXiv:hep-th/0611108].

[18] De Medeiros P., Figueroa-O'Farrill J. M. and Méndez-Escobar E. *JHEP* **0807** (2008) 111 [arXiv:0805.4363 [hep-th]];

[19] Rausch de Traubenberg M. and Slupinski M. J. *J. Math. Phys.* **41**, (2000), 4556–4571 [arXiv:hep-th/9904126].

[20] Rausch de Traubenberg M. and Slupinski M. J. *J. Math. Phys.* **43**, (2002), 5145–5160 [arXiv:hep-th/0205113].

[21] Goze M., Rausch de Traubenberg M. and Tanasa A. *J. Math. Phys.* **48** (2007) 093507, [arXiv:math-ph/0603008].

[22] Kerner R. *J. Math. Phys.* **33** (1992) 403; Kerner R. *Class. Quantum Grav.* **9** (1992) S137; Wills Toro L. A. *J. Math. Phys.* **42** (2001) 3915; Wills Toro L. A. Sánchez L. A., Osorio J. M. and Jaramillo D. E. *J. Math. Phys.* **42** (2001) 3935; Wills Toro L. A. *J. Math. Phys.* **42** (2001) 3947; Wills Toro L. A., Sánchez L. A. and Leleu X., *Int. J. Theor. Phys.* **42** (2003) 57; Wills Toro L. A., Sánchez L. A. , and Bleecker D. *Int. J. Theor. Phys.* **42** (2003) 73; Beckers J. and Debergh N. *Int. J. Mod. Phys.* **A8** (1993) 5041; Beckers J., Debergh N. and Nikitin A. G. *Fortsch. Phys.* **43** (1995) 67-80; Beckers J., Debergh N., Nikitin A. G. *Fortsch. Phys.* **43** (1995) 81-96; Nikitin A. G. and Tretynyk V. V. *J. Phys.* **A 28**, (1995) 1655; Niederle J. and Nikitin A. G. *J. Phys.* **A 32**, (1999) 5141.

[23] Rausch de Traubenberg M. and Slupinski M. J., *Mod. Phys. Lett.* **A12** (1997) 3051 [hep-th/9609203].
[24] Moultaka G., Rausch de Traubenberg M. and Tanasa A. Proceedings of the XIth International Conference Symmetry Methods in Physics, Prague 21-24 June 2004, arXiv:hep-th/0407168.
[25] Gol'fand T. A. and Likhtman E. P., *JETP Lett.* **13** (1971) 32; Ramond P. *Phys. Rev.* **D 3** (1971) 2415; Neveu A. and Schwarz J. *Nucl. Phys.* **B 31** (1971) 86; Volkov D. V. and Akulov V. P. *Phys. Lett.* **B 46** (1973) 109.
[26] Wess J. and Zumino B. *Nucl. Phys.* **B 70** (1974) 39-50.
[27] Kac V. G. *Functional Anal. Appl.* **9** (1975), no. 3, 263–265; Kac V. G. *Comm. Math. Phys.* **53** (1977) 31; Nahm W. and Scheunert M. *J. Math. Phys.* **17** (1976) 1626; Scheunert M. *The Theory of Lie Superalgebras: an Introduction*, Lecture Notes in Mathematics 716 (Springer, Berlin); Kaplansky I. *Pacific J. Math* **86** (1980) 93; Freund P. O. G. and Kaplansky I. *J. Math. Phys.* **17** (1976) 228.
[28] West P. *Introduction to supersymmetry and supergravity* (Singapore, Singapore: World Scientific 1990).
[29] Wess J and Bagger J *Supersymmetry and Supergravity* (Princeton University Press, 1983).
[30] Sohnius M. F. *Phys. Rept.* **128** (1985), 39-204.
[31] Freund P. G. O. *Introduction To Supersymmetry,* (Cambridge Univ. Press, 1983).
[32] Terning J. *Modern Supersymmetry*, (Oxford, UK: Clarendon 2006).
[33] Nilles H. P. *Phys. Rept.* **110** (1984) 1-162.
[34] Martin S. P. arXiv:hep-ph/9709356 (v4, June 2006).
[35] Aitchison I. J. R. *Supersymmetry in particle physics: An elementary introduction*, (Cambridge, UK: Univ. Pr. 2007).
[36] Dondi P. H. and Sohnius M. *Nucl. Phys.* **B81** (1974) 317.
[37] Salam A. and Strathdee J. A. *Nucl. Phys.* **B80** (1974) 499-505.
[38] Strathdee J. A. *Int. J. Mod. Phys.* **A2** (1987) 273.
[39] Freedman D.Z, van Nieuwenhuizen P. and Ferrara S. *Phys. Rev.* **D13** (1976) 3214-3218; Deser S. and Zumino B. *Phys. Lett.* **B62** (1976) 335.
[40] Cornwell J. F. *Group Theory in Physics, Vol 3: Supersymmetries and Infinite Dimensional Algebras*, (London, UK: Academic 1989 (Techniques of physics, 10)).
[41] Salam A. and Strathdee J. A. *Nucl. Phys.* **B76** (1974) 477-482.
[42] DeWitt B. *Supermanifolds*, (Cambridge University Press, Cambridge 1992 −second edition−).
[43] Salam A. and Strathdee J. A. *Phys. Rev.* **D11** (1975) 1521-1535.
[44] Wess J. and Zumino B. *Nucl. Phys.* **B78** (1974) 1; Salam A. and Strathdee J. A. *Phys. Lett.* **B51** (1974) 353; Ferrara S. and Zumino B. *Nucl. Phys.* **B79** (1974) 413-421.
[45] Ahn C., Bernard D. and Leclair A. *Nucl. Phys.* **B346** (1990) 409; Matheus-Valle J. L. and Monteiro M. A. R *Mod. Phys. Lett.* **A7** (1992) 3023; Durand S. *Mod. Phys. Lett* **A8** (1993) 2323 [hep-th/9305130]; Debergh N. *J. Phys.* **A26** (1993) 7219; Mohammedi N. *Mod. Phys. Lett.* **A10** (1995) 1287 [hep-th/9412133]; Colatto L. P. and Matheus-Valle J. L. *J. Math. Phys.* **37** (1996) 6121 [hep-th/9504101]; de Azcàrraga J. A. and Macfarlane A. J. *J. Math .Phys.* **37** (1996) 1115 [hep-th/9506177]; Fleury N. and Rausch de Traubenberg M. *Mod. Phys. Lett.* **A11** (1996) 899 [hep-th/9510108]; Dunne R. S., Macfarlane A. J., de Azcárraga J. A. and Pérez Bueno J. C. *Int. J. Mod. Phys.* **A 12** (1997) 3275; [hep-th/9610087]; Daoud M. and Kibler M. arXiv:math-ph/9912024; Aghababaei Samani K. and Mostafazadeh A. *Nucl. Phys.* **B595** (2001) 467 [hep-th/0007008]; Perez A., Rausch de Traubenberg M. and Simon P. *Nucl. Phys.* **B482** (1996) 325 [hep-th/9603149]; Rausch de Traubenberg M. and Simon P. *Nucl. Phys.* **B517** (1998) 485 [hep-th/9606188]; Matheus-Valle J. L. and Monteiro M. A. R. *Phys. Lett.* **B300** (1993) 66; Saidi E. H., Sedra M. B. and Zerouaoui J *Class. and Quantum Grav.* **12** (1995) 1567; Saidi E. H., Sedra M. B. and Zerouaoui J. *Class.Quant.Grav.* **12** (1995) 2705; Ahmedov H. and Dayi O. F. *Mod. Phys. Lett.* **A15** (2000) 1801 [math.qa/9905164]; Ahmedov H. and Dayi O. F. *J. Phys.* **A32** (1999) 6247 [math.qa/9903093]; Kheirandish F. and Khorrami M. *Eur. Phys. J.* **C18** (2001) 795 [hep-th/0007013]; Kheirandish F. and Khorrami M. *Int. J . Mod. Phys.* **A 16** (2001) 2165 [hep-th/0004154]; Durand S. *Mod. Phys. Lett.* **A7** (1992) 2905 [hep-th/9205086]; Filippov A.T., Isaev A. P. and Kurdikov A. B. *Int. J. Mod. Phys.* **A8** (1993) 4973 [hep-th/9212157].
[46] Campoamor-Stursberg R. and Rausch de Traubenberg M. *J. Math. Phys.* **49** (2008) 063506 [arXiv:0801.2630 [hep-th]].
[47] Goze M. and Rausch de Traubenberg M. arXiv:0809.4212 [math-ph].
[48] Bargmann V. *Ann. of Math* **48** (1947) 568-640.
[49] Gel'fand I., Graev M. and Vilenkin N., *Generalized functions* (Academic, New-York, 1966) Vol 5; Lang S. $SL_2\mathbb{R}$, (Springer, Berlin, 1985).
[50] Binegar B. *J. Math. Phys.* **23** (1982) 1511.
[51] Jackiw R. and Nair V. P. *Phys. Rev.* **D43** (1991) 1933.
[52] Plyushchay M. S. *Phys. Lett.* **B262** (1991) 71; Plyushchay M. S. *Nucl. Phys.* **B362** (1991) 54; Plyushchay

M. S. *Phys. Lett.* **B273** (1991) 250; Cortes J. L. and Plyushchay M. S. *J. Math. Phys.* **35** (1994) 6049 [arXiv:hep-th/9405193].

[53] Plyushchay M. S. *Mod. Phys. Lett.* **A12** (1997) 1153 [arXiv:hep-th/9705034].

[54] Rausch de Traubenberg M. *Nucl. Phys. Proc. Suppl.* **102** (2001) 256 [arXiv:hep-th/0109106].

[55] Klishevich S. M., Plyushchay M. S. and Rausch de Traubenberg M. *Nucl. Phys.* **B616** (2001) 419 [arXiv:hep-th/0101190].

[56] Mohammedi N., Moultaka G. and Rausch de Traubenberg M. *Int. J. Mod. Phys.* **A19**, (2004), 5585–5608 [arXiv:hep-th/0305172].

[57] Moultaka G., Rausch de Traubenberg M. and Tanasa A. *Int. J. Mod. Phys.* **A20**, (2005), 5779–5806 [arXiv:hep-th/0411198].

[58] Rausch de Traubenberg M. *Pr. Inst. Mat. Nats. Akad. Nauk Ukr. Mat. Zastos.*, 50, Part 1, 2, 3, *Natsional. Akad. Nauk Ukrani, Ĭnst. Mat., Kiev*, 2004, pp. 578-585, [arXiv:hep-th/0312066].

[59] Bacry H. and Lévy-Leblond J. *J. Math. Phys.* **9** (1968) 1605-1614; Cariñena J. F. , del Olmo M. A. and Santander M. *J. Phys. A: Math. Gen.* **14** (1981) 1-14.

[60] Henneaux M. and Knaepen B. *Phys. Rev.* **D 56** (1997) 6076, [arXiv:hep-th/9706119]; Henneaux M. and Knaepen B. *Int. J. Mod. Phys.* **A15** (2000) 3535-3548 [arXiv:hep-th/9912052].

[61] Meessen P. and Ortin T. *Nucl. Phys.* **B541**, 195 (1999) [arXiv:hep-th/9806120].

[62] Kaul R. K. *Phys. Rev.* **D 18** (1978) 1127.

[63] Polchinski J. *Phys. Rev. Lett.* **75** (1995) 4724 [`arXiv:hep-th/9510017`]; Polchinski J. and Cai Y. *Nucl. Phys.* **B296** (1988) 91. Townsend P. K. *Phys. Lett.* **B88** (1979) 97; Namazie M. A. and Storey D. *J. Phys.* **A13** (1980) L161; Kimura T. *Prog. Theor. Phys.* **65** (1981) 338; Siegel W. *Phys. Lett.* **B 93** (1980) 170; Hata H., Kugo T. and Ohta N., *Nucl. Phys.* **B 178** (1981) 527; Thierry-Mieg J. *Nucl. Phys.* **B 335** (1990) 334.

[64] Cremmer E. and Scherk J. *Nucl. Phys.* **B 72** (1974) 117-124.

[65] Deser S. *Phys. Rev.* **187** (1969) 1931-1969; Kalb M. and Ramond P. *Phys. Rev.* **D9** (1974) 2273; Nambu Y. *Phys. Rept.* **23** (1976) 250.

[66] Sezgin E. and van Nieuwenhuizen P. *Phys. Rev.* **D 22**, 301 (1980); Townsend P. K. CERN-TH-3067; *Lecture given at 18th Winter School of Theoretical Physics, Karpacz, Poland, Feb 18 - Mar 3, 1981.*

[67] Siegel W. *Phys. Lett.* **B103** (1981) 107.

[68] Ohnuki Y and Kamefuchi S. *Quantum Field Theory and Parastatistics* (Springer Verlag, 1982); Macfarlane A. J. in Proc. *International Workshop on Symmetry Methods in Physics*, Eds. Sissakian A. N., Pogosyan G. S. and Vinitsky S. I. (JINR, 1994) p. 319; Plyushchay M. S. *Nucl. Phys.* **B491**, 619 (1997) [arXiv:hep-th/9701091].

[69] Kac V. G. and Raina A. K. *Bombay Lectures on Highest Weight Representations of Infinite Dimensional Lie Algebras*, World Scientific (Singapore, New Jersey, Hong Kong, 1987).

[70] Roby N. *C. R. Acad. Sc. Paris* **268** (1969) A484–A486; Revoy Ph. *C. R. Acad. Sc. Paris* **284** (1977) A985–A988.

[71] Haile D. H. and Tesser S. *J. Alg.* **116** (1988) 372.

[72] Fleury N. and Rausch de Traubenberg M. *J. Math. Phys.* **33** (1992) 3356; Fleury N. and Rausch de Traubenberg M. *Adv. Apl. Clif. Alg.* **4** (1994) 123.

[73] Revoy Ph. *Adv. Appl. Cliff. Alg.* **3** (1993) 39-55.

[74] Morris A. O. *Quart. J. Math., Oxford* **18** (1967) 7–12; Morris A. O. *Quart. J. Math., Oxford* **19** (1968) 289–299.

[75] Roby N. *Bull. Sc. Math.* **94** (1970) 49-57.

[76] Rausch de Traubenberg M. to appear in Proceedings of the 7th International Conference on Clifford algebras and their applications. (ICCA), May 2005, Toulouse, arXiv:hep-th/0506011.

Workshop on Higher Symmetries in Physics IOP Publishing
Journal of Physics: Conference Series **175** (2009) 012004 doi:10.1088/1742-6596/175/1/012004

(Super)integrability from coalgebra symmetry: formalism and applications

A Ballesteros[1], A Blasco[1], F J Herranz[1], F Musso[1] and O Ragnisco[2]

[1] Departamento de Física, Universidad de Burgos, E-09001 Burgos, Spain
[2] Dipartimento di Fisica, Università di Roma Tre and Instituto Nazionale di Fisica Nucleare sezione di Roma Tre, Via Vasca Navale 84, I-00146 Roma, Italy

E-mail: angelb@ubu.es, ablasco@ubu.es, fjherranz@ubu.es, fmusso@ubu.es, ragnisco@fis.uniroma3.it

Abstract. The coalgebra approach to the construction of classical integrable systems from Poisson coalgebras is reviewed, and the essential role played by symplectic realizations in this framework is emphasized. Many examples of Hamiltonians with either undeformed or q-deformed coalgebra symmetry are given, and their Liouville superintegrability is discussed. Among them, (quasi-maximally) superintegrable systems on N-dimensional curved spaces of nonconstant curvature are analysed in detail. Further generalizations of the coalgebra approach that make use of comodule and loop algebras are presented. The generalization of such a coalgebra symmetry framework to quantum mechanical systems is straightforward.

1. Introduction

Poisson coalgebras are Poisson algebras endowed with a compatible coproduct structure. The aim of this contribution is to provide a self-contained review of a recently introduced symmetry approach in which Poisson coalgebras play an essential role as 'hidden' dynamical symmetries underlying the (super)integrability properties of a wide class of N-dimensional (ND) classical Hamiltonian systems. Within this construction, once a symplectic realization of the coalgebra is given, their generators play the role of dynamical symmetries of the Hamiltonian –which is written as a function of them– while the coproduct map (coalgebra structure) is used to 'propagate' the integrability to any arbitrary dimension.

From this so-called coalgebra approach, many well-known (super)integrable systems have been recovered, and some integrable deformations for them as well as new ND integrable systems have also been obtained (see [1]–[8] and references therein). As a remarkable application, this framework has been recently used to introduce integrable Hamiltonians describing geodesic flows on spaces with either constant or nonconstant curvature, and (super)integrable potential terms preserving the coalgebra symmetry can also be considered on such spaces [9]–[16].

Although in this contribution we shall concentrate on classical mechanical systems obtained from (commutative) Poisson coalgebras, we stress that all the (super)integrability properties of the coalgebra symmetric systems that we are going to describe are preserved at the quantum mechanical level by using the corresponding (noncommutative) operator coalgebras. In this way, the coalgebra approach has also been used, for instance, to solve in [17, 18] the quantum Calogero–Gaudin system [19], as well as supersymmetric generalizations of this model and of its

q-deformations, which have been constructed by starting from the underlying supersymmetric coalgebra structures [20, 21].

The paper is organized as follows. In section 2 we present in a complete and self-contained way the general coalgebra approach to integrable Hamiltonians, that will be illustrated in section 3 by applying it to three relevant Poisson coalgebras, namely $\mathfrak{sl}(2,\mathbb{R})$, its non-standard quantum deformation, $\mathfrak{sl}_z(2,\mathbb{R})$ and the two-photon Poisson-coalgebra h_6. Many physically interesting Hamiltonian systems will be explicitly constructed on these and other coalgebras in sections 4, 5, 6 and 7. Finally, some possible generalizations of the coalgebra formalism will be sketched in section 8.

2. Hamiltonian systems on Poisson coalgebras

First of all, let us fix the terminology concerning (Liouville) integrability and superintegrability (see, for instance, [22]). An ND Hamiltonian $H^{(N)}$ is called *completely integrable* if there exists a set of $(N-1)$ globally defined and functionally independent constants of the motion in involution that Poisson-commute with $H^{(N)}$. The Hamiltonian will be called *maximally superintegrable* (MS) if there exists a set of $(2N-2)$ globally defined functionally independent constants of the motion Poisson-commuting with $H^{(N)}$; among them, at least two different subsets of $(N-1)$ constants in involution can be found. Finally, a Hamiltonian system will be called *quasi-maximally superintegrable* (QMS) if it has $(2N-3)$ independent integrals with the abovementioned properties.

We also recall that a coalgebra (A,Δ) is a (unital, associative) algebra A endowed with a coproduct map [23, 24]:

$$\Delta : A \to A \otimes A$$

which is coassociative

$$(\Delta \otimes \mathrm{id}) \circ \Delta = (\mathrm{id} \otimes \Delta) \circ \Delta$$

that is, the following diagram is commutative:

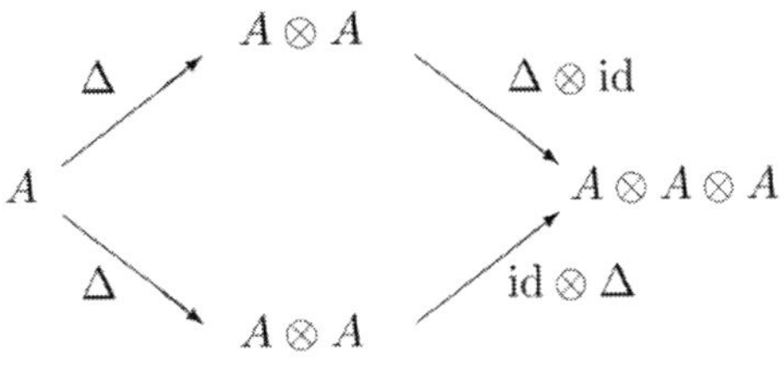

Due to the coassociativity property, the comultiplication Δ provides a 'two-fold way' for the definition of the objects on $A\otimes A\otimes A$ that, as we shall explain in section 2.2, will be deeply connected with *superintegrability* properties.

Let us now summarize the general construction of [2]. Let (A,Δ) be a Poisson coalgebra with l generators X_i $(i=1,\ldots,l)$, and r functionally independent Casimir functions $\mathcal{C}_j(X_1,\ldots,X_l)$ (with $j=1,\ldots,r$). The coassociative coproduct $\Delta \equiv \Delta^{(2)}$ is a Poisson map with respect to the usual Poisson bracket on $A\otimes A$:

$$\{X_i \otimes X_j, X_r \otimes X_s\}_{A\otimes A} = \{X_i, X_r\}_A \otimes X_j X_s + X_i X_r \otimes \{X_j, X_s\}_A.$$

Then, the m-th coproduct map $\Delta^{(m)}(X_i)$

$$\Delta^{(m)} : A \to A \otimes A \otimes \ldots^{m)} \otimes A \tag{1}$$

Table 1. Functions obtained by applying the coproduct map.

X_1	X_2	$\dots$	X_l	$\mathcal{C}_1$	$\mathcal{C}_2$	$\dots$	$\mathcal{C}_r$
$\Delta^{(2)}(X_1)$	$\Delta^{(2)}(X_2)$	$\dots$	$\Delta^{(2)}(X_l)$	$\Delta^{(2)}(\mathcal{C}_1)$	$\Delta^{(2)}(\mathcal{C}_2)$	$\dots$	$\Delta^{(2)}(\mathcal{C}_r)$
$\Delta^{(3)}(X_1)$	$\Delta^{(3)}(X_2)$	$\dots$	$\Delta^{(3)}(X_l)$	$\Delta^{(3)}(\mathcal{C}_1)$	$\Delta^{(3)}(\mathcal{C}_2)$	$\dots$	$\Delta^{(3)}(\mathcal{C}_r)$
$\vdots$	$\vdots$		$\vdots$	$\vdots$	$\vdots$		$\vdots$
$\Delta^{(m)}(X_1)$	$\Delta^{(m)}(X_2)$	$\dots$	$\Delta^{(m)}(X_l)$	$\Delta^{(m)}(\mathcal{C}_1)$	$\Delta^{(m)}(\mathcal{C}_2)$	$\dots$	$\Delta^{(m)}(\mathcal{C}_r)$
$\vdots$	$\vdots$		$\vdots$	$\vdots$	$\vdots$		$\vdots$
$\Delta^{(N)}(X_1)$	$\Delta^{(N)}(X_2)$	$\dots$	$\Delta^{(N)}(X_l)$	$\Delta^{(N)}(\mathcal{C}_1)$	$\Delta^{(N)}(\mathcal{C}_2)$	$\dots$	$\Delta^{(N)}(\mathcal{C}_r)$

can be defined by applying recursively the coproduct $\Delta^{(2)}$ in the form

$$\Delta^{(m)} := (\mathrm{id} \otimes \mathrm{id} \otimes \dots^{m-2)} \otimes \mathrm{id} \otimes \Delta^{(2)}) \circ \Delta^{(m-1)}. \tag{2}$$

Such an induction ensures that $\Delta^{(m)}$ is also a Poisson map.

In this way, we can construct the set of functions shown in table 1. From them, given a smooth function $\mathcal{H}(X_1, \dots, X_l)$, the N-sites Hamiltonian is defined as the N-th coproduct of $\mathcal{H}$:

$$H^{(N)} := \Delta^{(N)}(\mathcal{H}(X_1, \dots, X_l)) = \mathcal{H}(\Delta^{(N)}(X_1), \dots, \Delta^{(N)}(X_l)). \tag{3}$$

From [2] it can be proven that the set of $r \cdot N$ functions ($m = 1, \dots, N; j = 1, \dots, r$)

$$C_j^{(m)} := \Delta^{(m)}(\mathcal{C}_j(X_1, \dots, X_l)) = \mathcal{C}_j(\Delta^{(m)}(X_1), \dots, \Delta^{(m)}(X_l)), \tag{4}$$

such that $C_j^{(1)} = \mathcal{C}_j$, Poisson-commute with the Hamiltonian

$$\left\{C_j^{(m)}, H^{(N)}\right\}_{A\otimes A\otimes \dots^{N)}\otimes A} = 0 \tag{5}$$

and all these constants are, by construction, in involution:

$$\left\{C_i^{(m)}, C_j^{(n)}\right\}_{A\otimes A\otimes \dots^{N)}\otimes A} = 0 \qquad m, n = 1, \dots, N \quad i, j = 1, \dots, r. \tag{6}$$

This construction can be applied to *any* Poisson coalgebra. But there are two relevant families of Poisson coalgebras that will constitute the core of the integrable systems presented in this paper:

- *Lie–Poisson algebras.* The Poisson analogue of any Lie algebra with generators X_i ($i = 1, \dots, l$) is a coalgebra when endowed with the (primitive) coproduct map [25]

$$\Delta(X_i) = X_i \otimes 1 + 1 \otimes X_i \qquad \Delta(1) = 1 \otimes 1. \tag{7}$$

  Equivalently, this means that Lie algebras are cocommutative Hopf algebras.
- *Poisson analogues of quantum algebras and groups* [23, 24, 25]. These are also (deformed) coalgebras (A_z, Δ_z), where z is the quantum deformation parameter ($q = \mathrm{e}^z$), and the deformed coproduct gives rise to a noncocommutative Hopf algebra. The 'classical' limit $z \to 0$ (or, equivalently, $q \to 1$) provides the corresponding Lie–Poisson coalgebra. In fact, these q-Poisson coalgebras can be written as quadratic Poisson structures on certain (dual) Lie–Poisson groups, that are invariant with respect to the group multiplication (which is nothing but a coproduct map on the dual, see [6]).

Some remarks are in order.

- In general, we cannot say *a priori* that $H^{(N)}$ is a completely integrable Hamiltonian system because [8]: (i) We have to determine the number of degrees of freedom of $H^{(N)}$ by choosing a explicit symplectic realization of the coalgebra (A, Δ). (ii) Once this number is fixed, we have to check whether the number of independent invariants extracted from $C_j^{(m)}$ (under such a specific symplectic realization) is enough to guarantee complete integrability.
- However, if the coalgebra symmetry provides the complete integrability for $H^{(N)}$, the integrals of the motion can always be obtained in arbitrary dimension N in an explicit form. Hence the coalgebra symmetry arises as a unified approach to integrability since we obtain families of ND systems that share a very large common set of integrals of the motion. Furthermore, in this case $H^{(N)}$ will probably be superintegrable (to some extent) [7].
- In the case of q-deformations of Poisson coalgebras, the Hamiltonians obtained through the coalgebra approach are integrable deformations of the $z = 0$ $(q = 1)$ cases in such a manner that different quantum algebras give rise to different integrable deformations. Moreover, this fact can provide some relevant information concerning the geometric/physical interpretation of the deformation parameter.
- This construction holds for noncommutative coalgebras aswell. Thus, quantum mechanical systems can also be constructed (although ordering problems have to be fixed).

Before entering into explicit examples of the coalgebra construction, two important and general aspects have to be more deeply analysed in order to provide a global overview on the subject: they are the role of the symplectic realizations (which has been thoroughly studied in [8]) and the superintegrability features of the coalgebra symmetry (which were presented for the first time in [7]).

2.1. *Symplectic realizations and complete integrability*

Let (A, Δ) a Poisson coalgebra with r Casimir functions $\mathcal{C}_j$ $(j = 1, \dots, r)$. A symplectic leaf of A (which is always even-dimensional) will be denoted by $A_{(k_1,k_2,\dots,k_r)}$, where the leaf is characterized by a given set of constant values $(k_1, k_2, \dots, k_r)$ for the Casimirs.

An *s-dimensional symplectic realization* D for $A_{(k_1,k_2,\dots,k_r)}$ is given (locally) in terms of s pairs (q_i, p_i) of canonical Darboux variables

$$D : x \to x(q_1, p_1, q_2, p_2, \dots, q_s, p_s) \tag{8}$$

where x is any point on $A_{(k_1,k_2,\dots,k_r)}$. We remark that, in principle, different symplectic leaves $A_{(k_{1,i},k_{2,i},\dots,k_{r,i})}$ can be chosen for each copy i of A within $A \otimes A \otimes \dots^{N)} \otimes A$.

In particular, if we consider symplectic realizations with the same s for all the N sites in the tensor product $A \otimes A \otimes \dots^{N)} \otimes A$, the Hamiltonian (3)

$$H^{(N)} = (D \otimes D \dots^{N)} \otimes D)(\Delta^{(N)}(\mathcal{H})) \tag{9}$$

turns out to be a function of $N \cdot s$ pairs of canonical variables, i.e., it defines a system with $N \cdot s$ degrees of freedom. Then for each nonlinear Casimir $\mathcal{C}_j$, we get at most $(N - 1)$ integrals coming from its m-th coproducts (4) written under the m-th tensor product of the symplectic realization D (8):

$$C_j^{(m)} = (D \otimes D \dots^{m)} \otimes D)(\Delta^{(m)}(\mathcal{C}_j)) \tag{10}$$

where $m = 2, \dots, N$; notice that $C_j^{(1)} = D(\mathcal{C}_j) = k_j$. Note that under symplectic realizations m-th coproducts of linear Casimirs always give just numerical constants.

Thus, if we have R nonlinear Casimirs we find a maximum possible number of integrals in involution given by

$$(N-1)\cdot R.$$

In order to get complete integrability, we should have that

$$N\cdot s-1\leq (N-1)\cdot R$$

thus we need that the chosen symplectic realization D of A fulfils (for any N)

$$s\leq R-\frac{R-1}{N}. \tag{11}$$

Consequently, the *necessary condition for complete integrability* connects the dimension of the symplectic realization and the number of nonlinear Casimir functions through the conditions [8]:

- $s=1$ for coalgebras with $R=1$.
- $s<R$ for coalgebras with $R>1$.

Let us now consider a particular type of symplectic realizations whose dimension is fixed by the dimension l of the Poisson coalgebra and its number r of Casimir functions. We shall call 'generic' to the symplectic realization with maximal dimension s_m given by

$$s_m=\frac{l-r}{2}. \tag{12}$$

Hence, the *integrability condition* for the generic symplectic realization is

$$s_m=\frac{l-r}{2}\leq R-\frac{R-1}{N}$$

which leads to the final expression

$$l\leq (2\,R+r)-\frac{2}{N}(R-1). \tag{13}$$

Therefore, complete integrability for the generic symplectic realization can be achieved if [8]:

- $l\leq 2+r$ for coalgebras with $R=1$.
- $l<2\,R+r$ for coalgebras with $R>1$.

As a byproduct of the above results, if we consider *simple Lie algebras*, with rank $r=R$, we find that for coalgebras with $R=1$, their dimension must be $l\leq 3$, and for coalgebras with $R>1$, then we get the condition $l<3R$. This, in turn, shows that only simple Lie algebras of rank 1 can provide complete integrable systems in the generic symplectic realization, whilst all higher-rank ones do not fulfill the condition.

Another family of interesting cases is given by Lie coalgebras in which the generic symplectic realization has $s_m=1$, since the coalgebra symmetry always satisfies the integrability condition provided that $R\geq 1$. This possibility covers many *non-simple Lie algebras* fulfilling $l-r=2$.

Table 2. Coalgebra symmetry and QMS: the $(2N-3)$ integrals coming from a Casimir $\mathcal{C}$.

'Left' set of $(N-1)$ integrals $C^{(m)}$ in involution	Tensor product space
$C^{(2)} \equiv \Delta^{(2)}(\mathcal{C})$	$1 \otimes 2$
$C^{(3)} \equiv \Delta^{(3)}(\mathcal{C})$	$1 \otimes 2 \otimes 3$
$\vdots$	$\vdots$
$C^{(m)} \equiv \Delta^{(m)}(\mathcal{C})$	$1 \otimes 2 \otimes \cdots \otimes m$
$\vdots$	$\vdots$
$C^{(N)} \equiv \Delta^{(N)}(\mathcal{C})$	$1 \otimes 2 \otimes \cdots \otimes (N-1) \otimes N$
'Right' set of $(N-1)$ integrals $C^{(m)}$ in involution	Tensor product space
$C_{(2)} \equiv \Delta_R^{(2)}(\mathcal{C})$	$(N-1) \otimes N$
$C_{(3)} \equiv \Delta_R^{(3)}(\mathcal{C})$	$(N-2) \otimes (N-1) \otimes N$
$\vdots$	$\vdots$
$C_{(m)} \equiv \Delta_R^{(m)}(\mathcal{C})$	$(N-m+1) \otimes (N-m+2) \otimes \cdots \otimes N$
$\vdots$	$\vdots$
$C_{(N)} = C^{(N)} \equiv \Delta_R^{(N)}(\mathcal{C})$	$1 \otimes 2 \otimes \cdots \otimes (N-1) \otimes N$

2.2. Coalgebra superintegrability from coassociativity

Instead of (2), another recursion relation for the m-th coproduct map can be defined as [7]:

$$\Delta_R^{(m)} := (\Delta^{(2)} \otimes \mathrm{id} \otimes \ldots^{m-2)} \otimes \mathrm{id}\,) \circ \Delta_R^{(m-1)}. \tag{14}$$

Due to the coassociativity property of the coproduct, this new expression will provide exactly the same expressions for the N-th coproduct of any generator of A:

$$\Delta^{(N)}(X_i) \equiv \Delta_R^{(N)}(X_i). \tag{15}$$

However, if we label from 1 to N the sites of the chain $A \otimes A \otimes \ldots^{N)} \otimes A$, lower dimensional mth-coproducts (with $2 < m < N$) will be 'different' in the sense that the 'left' coproducts $\Delta^{(m)}$ (2) will contain objects living on the tensor product space $1 \otimes 2 \otimes \cdots \otimes m$, whilst the 'right' coproducts $\Delta_R^{(m)}$ will be defined on the sites $(N-m+1) \otimes (N-m+2) \otimes \cdots \otimes N$.

Therefore, the coalgebra symmetry of a given Hamiltonian gives rise to two 'pyramidal' sets of $r \cdot N$ integrals of the motion in involution that Poisson-commute with $H^{(N)}$ [7], under the corresponding symplectic realization D (8). Namely, both sets are just the 'left' integrals above considered $C_j^{(m)}$ (4) together with the 'right' ones given by

$$C_{j,(m)} := \Delta_R^{(m)}(\mathcal{C}_j(X_1, \ldots, X_l)) = \mathcal{C}_j(\Delta_R^{(m)}(X_1), \ldots, \Delta_R^{(m)}(X_l)) \tag{16}$$

such that $C_j^{(N)} \equiv C_{j,(N)}$, while both $C_j^{(1)}$ and $C_{j,(1)}$ are constants.

The very same arguments discussed in section 2.1 on the connection between integrability and the dimensionality of the chosen symplectic realizations can be applied to the *additional* set of 'right' integrals $C_{j,(m)}$. In this way, a completely integrable Hamiltonian $H^{(N)}$ with coalgebra symmetry will be, in principle, *superintegrable*.

For the sake of symplicity let us assume that we are dealing with a Poisson coalgebra with $R = 1$, that is, with N degrees of freedom and with a single nonlinear Casimir. Therefore if we take a symplectic realization with $s = 1$ the necessary condition for integrability (11) is fulfilled. In this case, the coalgebra approach gives rise to $(2N-3)$ functionally independent integrals, which are displayed in table 2, and with $\Delta^{(N)}(\mathcal{C}) \equiv \Delta_R^{(N)}(\mathcal{C})$ being a common 'left-right' integral. Then each set of 'left' and 'right' integrals is then formed by $(N-1)$ functions in involution. Therefore there is *only one* missing integral in order to ensure the *maximal superintegrability* of the system, which means that $H^{(N)}$ is always, at least, QMS. Such a 'remaining' integral, in case it does exist, will not be provided by the coalgebra symmetry and it will have to be found by alternative methods. Hence, we can conclude that the coalgebra symmetry for this class of Hamiltonian systems with N degrees of freedom implies the following integrability hierarchy:

- $N = 2$. The Hamiltonian is only *integrable* with a single constant of motion $C^{(2)} = C_{(2)}$.
- $N = 3$. The Hamiltonian is *minimally* (or weakly) superintegrable with three integrals given by $\{C^{(2)}, C^{(3)} = C_{(3)}, C_{(2)}\}$.
- $N > 3$. The Hamiltonian is *QMS* with $(2N-3)$ integrals $\{C^{(m)}, C^{(N)} = C_{(N)}, C_{(m)}\}$ for $m = 2, \ldots, N-1$.

3. Three Poisson coalgebras

We stress that the coalgebra approach to complete integrability is completely general and constructive for any Poisson coalgebra endowed with a *suitable* symplectic realization. In order to illustrate the above ideas, we will mainly consider in this contribution the following three Poisson coalgebras: (i) $\mathfrak{sl}(2,\mathbb{R})$, (ii) its non-standard q-deformation $\mathfrak{sl}_z(2,\mathbb{R})$ and (iii) the (two-photon) algebra h_6. With them a bunch of physically interesting ND Hamiltonians can be obtained, and some of them will be explicitly presented in sections 4, 5 and 6, respectively.

3.1. The $\mathfrak{sl}(2,\mathbb{R})$ Lie–Poisson coalgebra

This coalgebra is defined by the following Lie–Poisson brackets and comultiplication map:

$$\{J_3, J_+\} = 2J_+ \qquad \{J_3, J_-\} = -2J_- \qquad \{J_-, J_+\} = 4J_3 \tag{17}$$

$$\Delta(1) = 1 \otimes 1 \qquad \Delta(J_i) = J_i \otimes 1 + 1 \otimes J_i \qquad i = +, -, 3. \tag{18}$$

The Casimir function for $\mathfrak{sl}(2,\mathbb{R})$ reads

$$\mathcal{C} = J_- J_+ - J_3^2. \tag{19}$$

A one-particle symplectic realization of this coalgebra is given by

$$D(J_-) = q_1^2 \qquad D(J_+) = p_1^2 + \frac{b_1}{q_1^2} \qquad D(J_3) = q_1 p_1, \tag{20}$$

where $\{q_1, p_1\} = 1$. Note that, under this realization, $C^{(1)} = D(\mathcal{C}) = b_1$. Hence, according to the notation and results presented in section 2.1 we are dealing with a Poisson coalgebra with dimension $l = 3$ and with a single nonlinear Casimir. Thus $r = R = 1$ and this implies that (20) is the generic symplectic realization with $s \equiv s_m = (l-r)/2 = 1$. The corresponding N-particle symplectic realization of $\mathfrak{sl}(2,\mathbb{R})$, living on $\mathfrak{sl}(2,\mathbb{R}) \otimes \ldots^{N)} \otimes \mathfrak{sl}(2,\mathbb{R})$, is obtained by applying

$$J_i^{(N)} = (D \otimes D \otimes \ldots^{N)} \otimes D)(\Delta^{(N)}(J_i))$$

which gives [3, 12]

$$J_-^{(N)} = \sum_{i=1}^{N} q_i^2 \equiv \mathbf{q}^2 \qquad J_3^{(N)} = \sum_{i=1}^{N} q_i p_i \equiv \mathbf{q} \cdot \mathbf{p}$$

$$J_+^{(N)} = \sum_{i=1}^{N} \left(p_i^2 + \frac{b_i}{q_i^2} \right) \equiv \mathbf{p}^2 + \sum_{i=1}^{N} \frac{b_i}{q_i^2}, \tag{21}$$

where b_i are N arbitrary real parameters. This means that the N-particle generators (21) fulfil the commutation rules (17) with respect to the ND canonical Poisson bracket.

Next, since $s_m = 1$ and $R = 1$ we are in the case of a Poisson coalgebra endowed with the $(2N-3)$ integrals displayed in table 2; these turn out to be [7, 12]:

$$C^{(m)} = \sum_{1\le i<j}^{m} I_{ij} + \sum_{i=1}^{m} b_i \qquad C_{(m)} = \sum_{N-m+1\le i<j}^{N} I_{ij} + \sum_{i=N-m+1}^{N} b_i \tag{22}$$

where $m = 2, \ldots, N$ and

$$I_{ij} = (q_i p_j - q_j p_i)^2 + \left(b_i \frac{q_j^2}{q_i^2} + b_j \frac{q_i^2}{q_j^2} \right) \tag{23}$$

are the b_i-generalization of the square of the 'angular momentum' generators $J_{ij} = q_i p_j - q_j p_i$ which span an $\mathfrak{so}(N)$ Lie–Poisson algebra. As a consequence of the coalgebra symmetry, the generators (21) Poisson commute with these $(2N-3)$ functions. Therefore, any arbitrary function $\mathcal{H}$ defined as

$$H^{(N)} = \mathcal{H}\left(J_-^{(N)}, J_+^{(N)}, J_3^{(N)}\right) = \mathcal{H}\left(\mathbf{q}^2, \mathbf{p}^2 + \sum_{i=1}^{N} \frac{b_i}{q_i^2}, \mathbf{q}\cdot\mathbf{p}\right) \tag{24}$$

gives rise to an ND QMS Hamiltonian system which is always endowed, at least, with the $(2N-3)$ integrals (22).

3.2. q-Poisson coalgebras: the $\mathfrak{sl}_z(2,\mathbb{R})$ *case*

The Poisson analogues of quantum algebras and groups [23, 24] are also (deformed) coalgebras (A_z, Δ_z) $(q = \mathrm{e}^z)$, which means that any function of the generators of a given 'quantum' Poisson algebra (with deformed Casimir elements $\mathcal{C}_{z,j}$) will provide a deformation of the Hamiltonian generated by the undeformed coalgebra. Such a q-coalgebraic deformation will preserve, by construction, the (super)integrability properties of the system defined on the undeformed Lie–Poisson coalgebra. Therefore, q-deformations can be understood in this context as the algebraic machinery suitable for generating integrable deformations of Hamiltonian systems.

In particular, let us focus on the non-standard $\mathfrak{sl}_z(2,\mathbb{R})$ Poisson coalgebra defined by the following (deformed) Poisson brackets and coproduct map (see [3, 26]):

$$\{J_3, J_+\} = 2J_+ \cosh zJ_- \qquad \{J_3, J_-\} = -2\frac{\sinh zJ_-}{z} \qquad \{J_-, J_+\} = 4J_3 \tag{25}$$

$$\begin{aligned} \Delta_z(1) &= 1\otimes 1 \qquad \Delta_z(J_-) = J_-\otimes 1 + 1\otimes J_- \\ \Delta_z(J_i) &= J_i \otimes \mathrm{e}^{zJ_-} + \mathrm{e}^{-zJ_-}\otimes J_i \qquad (i = +, 3). \end{aligned} \tag{26}$$

The Casimir function for $\mathfrak{sl}_z(2,\mathbb{R})$ reads

$$\mathcal{C}_z = \frac{\sinh zJ_-}{z} J_+ - J_3^2. \tag{27}$$

A one-particle (deformed) symplectic realization of $\mathfrak{sl}_z(2,\mathbb{R})$ is:

$$D_z(J_-) = q_1^2 \qquad D_z(J_+) = \frac{\sinh zq_1^2}{zq_1^2} p_1^2 + \frac{zb_1}{\sinh zq_1^2} \qquad D_z(J_3) = \frac{\sinh zq_1^2}{zq_1^2} q_1 p_1 \tag{28}$$

such that $C_z^{(1)} = D_z(\mathcal{C}_z) = b_1$. Hence we are dealing again with a coalgebra with $l = 3$, $r = R = 1$ and $s_m = 1$. The Nth-coproduct of (26) through the one-particle representation (28) gives rise to an N-particle symplectic realization on $\mathfrak{sl}_z(2,\mathbb{R}) \otimes \ldots^{N)} \otimes \mathfrak{sl}_z(2,\mathbb{R})$ through

$$J_i^{(N)} = (D_z \otimes D_z \otimes \ldots^{N)} \otimes D_z)(\Delta_z^{(N)}(J_i)),$$

namely

$$J_-^{(N)} = \sum_{i=1}^{N} q_i^2 \equiv \mathbf{q}^2 \qquad J_3^{(N)} = \sum_{i=1}^{N} \frac{\sinh zq_i^2}{zq_i^2} q_i p_i \exp\left\{-z\sum_{k=1}^{i-1} q_k^2 + z\sum_{l=i+1}^{N} q_l^2\right\} \equiv (\mathbf{q}\cdot\mathbf{p})_z$$
$$J_+^{(N)} = \sum_{i=1}^{N}\left(\frac{\sinh zq_i^2}{zq_i^2} p_i^2 + \frac{zb_i}{\sinh zq_i^2}\right)\exp\left\{-z\sum_{k=1}^{i-1} q_k^2 + z\sum_{l=i+1}^{N} q_l^2\right\} \equiv \tilde{\mathbf{p}}_z^2 \tag{29}$$

where the b_i's are again N arbitrary real parameters that label the representation on each 'lattice' site.

In this case the $(2N-3)$ functions written in table 2 are explicitly given by [3, 7]:

$$C_z^{(m)} = \sum_{1\le i<j}^{m} I_{ij}^z \exp\left\{-2z\sum_{k=1}^{i-1} q_k^2 - zq_i^2 + zq_j^2 + 2z\sum_{l=j+1}^{m} q_l^2\right\}$$
$$+\sum_{i=1}^{m} b_i \exp\left\{-2z\sum_{k=1}^{i-1} q_k^2 + 2z\sum_{l=i+1}^{m} q_l^2\right\}$$
$$C_{z,(m)} = \sum_{N-m+1\le i<j}^{N} I_{ij}^z \exp\left\{-2z\sum_{k=N-m+1}^{i-1} q_k^2 - zq_i^2 + zq_j^2 + 2z\sum_{l=j+1}^{N} q_l^2\right\}$$
$$+\sum_{i=N-m+1}^{N} b_i \exp\left\{-2z\sum_{k=N-m+1}^{i-1} q_k^2 + 2z\sum_{l=i+1}^{N} q_l^2\right\} \tag{30}$$

where $m = 2, \ldots, N$ and

$$I_{ij}^z = \frac{\sinh zq_i^2}{zq_i^2}\frac{\sinh zq_j^2}{zq_j^2}(q_i p_j - q_j p_i)^2 + \left(b_i \frac{\sinh zq_j^2}{\sinh zq_i^2} + b_j \frac{\sinh zq_i^2}{\sinh zq_j^2}\right). \tag{31}$$

Consequently, any smooth function $\mathcal{H}_z$ defined on the N-particle symplectic realization (29) of the generators of $\mathfrak{sl}_z(2,\mathbb{R})$ in the form

$$H_z^{(N)} = \mathcal{H}_z\left(J_-^{(N)}, J_+^{(N)}, J_3^{(N)}\right) = \mathcal{H}_z\left(\mathbf{q}^2, \tilde{\mathbf{p}}_z^2, (\mathbf{q}\cdot\mathbf{p})_z\right) \tag{32}$$

defines a QMS Hamiltonian system. We stress that all the choices for $\mathcal{H}_z$ share the 'universal' set of $(2N-3)$ constants of motion given by (30).

Clearly, the non-deformed limit $z \to 0$ of all the above expressions gives rise to the ones corresponding to the $\mathfrak{sl}(2,\mathbb{R})$-coalgebra presented in section 3.1, which shows that quantum algebras may provide (super)integrable generalizations of non-deformed Hamiltonian systems.

We also remark that this quantum deformation has been interpreted in [3] as an algebraic way to introduce long-range interactions in the underlying underformed systems due to the exponentials of the type $\exp(z\sum_j q_j^2)$ coming from the deformed symplectic realization (29) (compare with (21) and (24)).

3.3. The two-photon Lie–Poisson coalgebra

The third coalgebra that we explicitly consider also satisfies the *necessary condition* (11) for complete integrability. Nevertheless, we shall use this example to show that the latter is *not* a *sufficient condition*, thus making evident the essential role played by the symplectic realization of the chosen coalgebra.

The two-photon coalgebra (h_6, Δ) is spanned by six generators $\{K, A_+, A_-, B_+, B_-, M\}$ with Lie–Poisson brackets given by [27]

$$\begin{array}{lll} \{K, A_+\} = A_+ & \{K, A_-\} = -A_- & \{A_-, A_+\} = M \\ \{K, B_+\} = 2B_+ & \{K, B_-\} = -2B_- & \{B_-, B_+\} = 4K + 2M \\ \{A_+, B_-\} = -2A_- & \{A_+, B_+\} = 0 & \{M, \cdot\} = 0 \\ \{A_-, B_+\} = 2A_+ & \{A_-, B_-\} = 0 & \end{array} \tag{33}$$

together with the usual nondeformed coproduct map given by (7). This is just the Poisson analogue of the (non-simple) h_6 Lie algebra which has been widely applied in quantum optics [28, 29]. The h_6 algebra is isomorphic to the $(1+1)$D Schrödinger algebra [30] and the following embedding of the Heisenberg–Weyl h_3, harmonic oscillator h_4 and two-photon coalgebras can be easily identified:

$$(h_3, \Delta) \subset (h_4, \Delta) \subset (h_6, \Delta) \qquad h_3 = \langle A_-, A_+, M\rangle \qquad h_4 = \langle K, A_-, A_+, M\rangle.$$

The h_6-coalgebra is endowed with two Casimir operators [31], namely

$$\mathcal{C}_1 = M \qquad \mathcal{C}_2 = (MB_+ - A_+^2)(MB_- - A_-^2) - (MK - A_-A_+ + M^2/2)^2. \tag{34}$$

Let us now introduce the one-particle symplectic realization of h_6 given by [16, 27]

$$\begin{array}{lll} D(A_+) = \lambda_1 p_1 & D(A_-) = \lambda_1 q_1 & D(K) = q_1 p_1 - \frac{1}{2}\lambda_1^2 \\ D(B_+) = p_1^2 & D(B_-) = q_1^2 & D(M) = \lambda_1^2 \end{array} \tag{35}$$

where λ_1 is a non-vanishing constant that labels the realization such that $C_1^{(1)} = D(\mathcal{C}_1) = \lambda_1^2$ and $C_2^{(1)} = D(\mathcal{C}_2) = 0$. Notice that if $\lambda_1 = 0$ the h_6-coalgebra reduces to the $\mathfrak{sl}(2,\mathbb{R}) = \langle B_-, B_+, K\rangle$ one (20) with $b_1 = 0$. In fact, for any $\lambda_1 \neq 0$, $\mathfrak{gl}(2,\mathbb{R}) = \langle B_-, B_+, K, M\rangle$ is also a sub-coalgebra of h_6.

Therefore, by taking into account section 2.1 we are now dealing with a coalgebra with $l = 6$, $r = 2$, $R = 1$ ($\mathcal{C}_1 = M$ is linear and leads to trivial integrals) and the chosen symplectic realization has $s = 1$. This means that the necessary condition for integrability (11) is fulfilled since $s = 1 \equiv R = 1$, although note that the symplectic realization is not a generic one (12) since $s_m = 1 \neq \frac{l-r}{2} = 2$.

By making use of the coproduct map it is immediate to show that the N-particle symplectic realization on $h_6 \otimes \ldots^{N)} \otimes h_6$ reads

$$\begin{array}{lll} A_+^{(N)} = \sum_{i=1}^N \lambda_i p_i \equiv \boldsymbol{\lambda}\cdot\mathbf{p} & A_-^{(N)} = \sum_{i=1}^N \lambda_i q_i \equiv \boldsymbol{\lambda}\cdot\mathbf{q} & M^{(N)} = \sum_{i=1}^N \lambda_i^2 \equiv \boldsymbol{\lambda}^2 \\ B_+^{(N)} = \sum_{i=1}^N p_i^2 \equiv \mathbf{p}^2 & B_-^{(N)} = \sum_{i=1}^N q_i^2 \equiv \mathbf{q}^2 & K = \sum_{i=1}^N \left(q_i p_i - \frac{\lambda_i^2}{2}\right) \equiv \mathbf{q}\cdot\mathbf{p} - \frac{\boldsymbol{\lambda}^2}{2} \end{array} \tag{36}$$

where $\boldsymbol{\lambda} = (\lambda_1, \ldots, \lambda_N)$. From this viewpoint it is clear that the h_6-coalgebra may give rise to a generalization of the integrable systems with $\mathfrak{sl}(2,\mathbb{R})$-coalgebra symmetry, provided that *all* $b_i = 0$ but, as we shall show in what follows, one has to pay, in principle, the price of loosing *complete integrability*. We stress that the h_6 Poisson algebra was formerly considered in [32] in the framework of some 3D integrable systems, but only the additional coalgebra structure here presented makes possible to consider h_6 as a useful symmetry in ND systems.

We can now compute the $(2N-3)$ integrals of motion displayed in table 2 and coming from the nonlinear Casimir

$$\mathcal{C} = \mathcal{C}_2/\mathcal{C}_1 = MB_+B_- - B_+A_-^2 - B_-A_+^2 - M(K+M/2)^2 + 2A_-A_+(K+M/2). \tag{37}$$

A straightforward computation shows that $C^{(2)} = C_{(2)} = 0$ and the remaining ones read

$$\begin{aligned} C^{(m)} &= \sum_{1\leq i<j<k}^{m} \left(\lambda_i(p_jq_k - p_kq_j) + \lambda_j(p_kq_i - p_iq_k) + \lambda_k(p_iq_j - p_jq_i)\right)^2 \\ C_{(m)} &= \sum_{N-m+1\leq i<j<k}^{N} \left(\lambda_i(p_jq_k - p_kq_j) + \lambda_j(p_kq_i - p_iq_k) + \lambda_k(p_iq_j - p_jq_i)\right)^2 \end{aligned} \tag{38}$$

where $m = 3, \ldots, N$. This means that any smooth function

$$\begin{aligned} H^{(N)} &= \mathcal{H}\left(K^{(N)}, B_-^{(N)}, B_+^{(N)}, A_-^{(N)}, A_+^{(N)}, M^{(N)}\right) \\ &= \mathcal{H}\left(\mathbf{q}\cdot\mathbf{p} - \frac{1}{2}\boldsymbol{\lambda}^2, \mathbf{q}^2, \mathbf{p}^2, \boldsymbol{\lambda}\cdot\mathbf{q}, \boldsymbol{\lambda}\cdot\mathbf{p}, \boldsymbol{\lambda}^2\right) \end{aligned} \tag{39}$$

Poisson commutes with the $2N-5$ functions (38) but now the sets $\{C^{(3)}, \ldots, C^{(N)}\}$ and $\{C_{(3)}, \ldots, C_{(N)}\}$ are only formed by $N-2$ integrals in involution so that, in general, $H^{(N)}$ does *not* provide a completely integrable system. Since there is only one constant left, say $\mathcal{I}$, to obtain complete integrability we have called these systems *quasi-integrable* ones [16]. Nevertheless, this initial failure of the coalgebra approach can be circumvented by making use of the rich subalgebra structure of h_6 and, for some particular choices of $H^{(N)}$, such a remaining integral can be obtained as follows (see [16] for a complete discussion on the subject).

Firstly, in order to ensure the existence of $\mathcal{I}$ for any dimension N, we shall assume that this additional integral is also h_6-coalgebra invariant, which means that it can be written as a function

$$\mathcal{I} = \mathcal{I}(K, B_+, B_-, A_+, A_-, M) \tag{40}$$

where the h_6 generators are written in their N-particle symplectic realization (36). In this way, if $\mathcal{I}$ is functionally independent with respect to both the h_6 Casimir (37) and the Hamiltonian $\mathcal{H}$, the coalgebra symmetry guarantees –by construction– the involutivity of $\mathcal{I}$ with respect to the $(N-2)$ 'left' integrals $C^{(m)}$ $(m = 3, \ldots, N)$ and its functional independence with respect to them. And the very same result holds for the $(N-3)$ 'right' integrals $C_{(m)}$ where $m = 3, \ldots, N-1$ (we recall that $C_{(N)} = C^{(N)}$). Therefore, if $\mathcal{I}$ can be finally found then the corresponding Hamiltonian is not only integrable but furthermore superintegrable with $2N-4$ functionally independent constants of motion (one less than a QMS system).

In particular, we can consider two different situations in which the existence of $\mathcal{I}$ is guaranteed by construction:

• *Subalgebra integrability*. If the Hamiltonian $\mathcal{H}$ is defined within a subalgebra of h_6 that has a nonlinear Casimir invariant, the N-particle realization of the Casimir of the subalgebra provides automatically the integral $\mathcal{I}$.

• *Generator integrability*. Now, let us choose a given generator X of h_6. If we look for all the generators X_j $(j = 1, \ldots, n)$ commuting with X and we also look for all the subalgebras

g_k $(k = 1, \ldots, t)$ containing X as generator, then the Hamiltonian constructed through any function of the type

$$\mathcal{H}_X = \mathcal{H}_X \left(\mathcal{C}_{g_1}, \ldots, \mathcal{C}_{g_t}, X, X_1, \ldots, X_n\right), \tag{41}$$

where $\mathcal{C}_{g_k}$ is the Casimir function of the subalgebra g_k, is such that

$$\{\mathcal{H}_X, X\} = 0. \tag{42}$$

Moreover, the N-th particle symplectic realization of both X and $\mathcal{H}_X$ will Poisson-commute with the two sets of integrals $C^{(m)}$ and $C_{(m)}$ (38). Under such hypotheses, $\mathcal{H}_X$ is completely integrable since the N-th particle symplectic realization of the generator X is just the additional constant of motion $\mathcal{I}$. Since we have five relevant generators $\{K, B_+, B_-, A_+, A_-\}$ this procedure will give rise to five families of completely integrable systems that have been studied in detail in [16]. As we shall see in section 6, h_6-coalgebra systems include natural Hamiltonians as well as static electromagnetic fields and curved geodesic flows.

But even in the case that a given Hamiltonian does not fit within the two previous cases, the search for the remaining integral $\mathcal{I}$ –in case it does exist– can also be performed by using direct methods that can be computerized. In fact, the additional integral $\mathcal{I}$ can be searched among h_6 functions with cubic or higher dependence on the momenta (note that all the integrals that we have presented so far are, at most, quadratic in the momenta). Indeed, some particular solutions to this problem have recently been found and lead to new completely integrable ND Hamiltonians (see [16, 33]).

Finally, it could also happen that for a certain $\mathcal{H}$ the additional integral $\mathcal{I}$ does exist, but it cannot be written as a function (40) of the h_6 generators (*i.e.*, $\mathcal{I}$ is not coalgebra-invariant). This implies that the explicit form for $\mathcal{I}$ has to be found for each dimension, and –hopefully– a generic ND expression can be found inductively. Nevertheless, due to the physical interest of h_6 invariant systems (see section 6) this possiblity is worth to be explored with the aid of symbolic computation packages.

4. QMS Hamiltonians with $\mathfrak{sl}(2,\mathbb{R})$-coalgebra symmetry

In the sequel we will show the potentialities of the coalgebra framework by summarizing some of the $\mathfrak{sl}(2,\mathbb{R})$-coalgebra symmetric Hamiltonians of the form (24) that have been so far described in the literature. Some of them were already known but many others have been constructed for the first time by making use of the constructive approach detailed in section 3.1. Since we are always dealing with ND systems and for the sake of simplicity of the notation, from now on we shall drop the index '(N)' in both the ND symplectic realization and the Hamiltonians.

4.1. Evans systems

The following generalization of the motion of a classical particle on an ND Euclidean space $\mathbb{E}^N$ under a spherically symmetric potential is $\mathfrak{sl}(2,\mathbb{R})$-coalgebra invariant [3]

$$H = \frac{1}{2}J_+ + \mathcal{F}(J_-) = \frac{1}{2}\mathbf{p}^2 + \mathcal{F}\left(\mathbf{q}^2\right) + \sum_{i=1}^{N} \frac{b_i}{2q_i^2} \tag{43}$$

where $\mathcal{F}$ is a smooth function and the terms depending on the b_i's constants are just additional 'centrifugal barriers' coming from a non-zero symplectic realization of $\mathfrak{sl}(2,\mathbb{R})$. In fact, the $\mathfrak{sl}(2,\mathbb{R})$-coalgebra provides a common set of (right and left) integrals of the motion (22) for any choice of $\mathcal{F}$ [12]. Note that the ND Smorodinsky–Winternitz Hamiltonian [34] is recovered when $\mathcal{F}(\mathbf{q}^2) = \omega^2\mathbf{q}^2/2$, and the Kepler–Coulomb potential corresponds to $\mathcal{F}(\mathbf{q}^2) = -k/|\mathbf{q}|$, where ω and k are real constants (and $|\mathbf{q}| = \sqrt{\mathbf{q}^2}$). Therefore, both outstanding MS systems do have coalgebra symmetry.

4.2. Superintegrable electromagnetic field Hamiltonians

Certain velocity-dependent potentials on $\mathbb{E}^N$ giving rise to superintegrable electromagnetic field Hamiltonians [35, 36] can also be obtained in this framework. The most general example of this class is given by [12]:

$$\begin{aligned} H &= \frac{1}{2} J_+ - eJ_3 \mathcal{G}(J_-) + e\mathcal{F}(J_-) \\ &= \frac{1}{2}\,\mathbf{p}^2 - e(\mathbf{q}\cdot\mathbf{p})\mathcal{G}(\mathbf{q}^2) + e\mathcal{F}(\mathbf{q}^2) + \sum_{i=1}^{N} \frac{b_i}{2q_i^2} \end{aligned} \tag{44}$$

where e is the electric charge, while $\mathcal{G}$ and $\mathcal{F}$ are smooth functions.

When $N = 3$, the (time-independent) scalar ψ and vector $\mathbf{A}$ potentials read

$$\psi(\mathbf{q}) = \mathcal{F}(\mathbf{q}^2) - \frac{e}{2}\,\mathbf{q}^2\mathcal{G}^2(\mathbf{q}^2) + \sum_{i=1}^{3} \frac{b_i}{2eq_i^2} \qquad \mathbf{A}(\mathbf{q}) = \mathbf{q}\,\mathcal{G}(\mathbf{q}^2).$$

Then the electric $\mathbf{E} = -\nabla\psi$ and magnetic $\mathbf{H} = \nabla \times \mathbf{A}$ fields turn out to be

$$\mathbf{E} = \left(e\mathcal{G}^2 + 2e\mathbf{q}^2\mathcal{G}\mathcal{G}' - 2\mathcal{F}'\right)\mathbf{q} + \frac{1}{e}\left(\frac{b_1}{q_1^3}, \frac{b_2}{q_2^3}, \frac{b_3}{q_3^3}\right) \qquad \mathbf{H} = 0$$

where $\mathcal{G}'$ and $\mathcal{F}'$ are the derivatives with respect to the variable $\mathbf{q}^2$. This kind of construction can also be applied to obtain certain ND Fokker–Planck Hamiltonians (see [37] and references therein).

4.3. Free motion on Riemannian spaces of constant curvature

The kinetic energy $\mathcal{T}$ of a particle on the ND Euclidean space $\mathbb{E}^N$ is just given by the generator J_+ in the symplectic realization (21) with *all* $b_i = 0$:

$$H \equiv \mathcal{T} = \frac{1}{2} J_+ = \frac{1}{2}\,\mathbf{p}^2. \tag{45}$$

Surprisingly enough, the kinetic energy on ND Riemannian spaces with constant sectional curvature κ can also be expressed in Hamiltonian form as a function of the ND symplectic realization of the $\mathfrak{sl}(2,\mathbb{R})$ generators (21). In fact, this can be done in two different ways [12]:

$$\begin{aligned} H^{\mathrm{P}} &\equiv \mathcal{T}^{\mathrm{P}} = \frac{1}{2}(1+\kappa J_-)^2 J_+ = \frac{1}{2}\left(1+\kappa\mathbf{q}^2\right)^2 \mathbf{p}^2 \\ H^{\mathrm{B}} &\equiv \mathcal{T}^{\mathrm{B}} = \frac{1}{2}(1+\kappa J_-)\left(J_+ + \kappa J_3^2\right) = \frac{1}{2}(1+\kappa\mathbf{q}^2)\left(\mathbf{p}^2 + \kappa(\mathbf{q}\cdot\mathbf{p})^2\right). \end{aligned} \tag{46}$$

In these expressions the function $\mathcal{T}^{\mathrm{P}}$ is just the kinetic energy for a free particle on the spherical $\mathbb{S}^N$ ($\kappa > 0$) and hyperbolic $\mathbb{H}^N$ ($\kappa < 0$) spaces when Poincaré coordinates $\mathbf{q}$ and their canonical momenta $\mathbf{p}$ (coming from a stereographic projection in $\mathbb{R}^{N+1}$ [38]) are used. On the other hand $\mathcal{T}^{\mathrm{B}}$ corresponds to Beltrami coordinates and momenta (central projection). The canonical transformation between both sets of phase space variables can be found in [39].

We can immediately conclude that, by construction, both Hamiltonians are QMS ones since they are $\mathfrak{sl}(2,\mathbb{R})$-coalgebra symmetric. In other words, we can say that the ND Riemannian spaces with constant curvature are $\mathfrak{sl}(2,\mathbb{R})$-coalgebra spaces.

4.4. Superintegrable potentials on spaces with constant curvature
Without breaking the $\mathfrak{sl}(2,\mathbb{R})$-coalgebra symmetry, QMS potentials on constant curvature spaces can now be constructed by adding some suitable functions depending on J_- to (46) and also by considering arbitrary centrifugal terms that come from symplectic realizations of the J_+ generator with generic b_i's. With all these ingredients, the full $\mathfrak{sl}(2,\mathbb{R})$ Hamiltonian will be of the form

$$H = \mathcal{T}(J_-, J_+, J_3) + \mathcal{U}(J_-). \tag{47}$$

The QMS systems obtained in this way are just the curved counterpart of the Euclidean systems, and by making use of the curvature parameter κ we can simultaneously describe the spherical $\mathbb{S}^N$ ($\kappa > 0$), hyperbolic $\mathbb{H}^N$ ($\kappa < 0$) and Euclidean $\mathbb{E}^N$ ($\kappa = 0$) cases.

We remark that special choices for $\mathcal{U}$ lead to the many interesting QMS systems on constant curvature spaces, that can always be expressed in both Poincaré and Beltrami phase space coordinates. In what follows we write three types of examples of this kind [12] which share the *same* set of constants of the motion (22) although the geometric meaning of the canonical coordinates and momenta can be different for each example.

4.4.1. Curved Evans systems The ND constant curvature generalization of the generic 3D Euclidean system with radial symmetry [40] is given by

$$\begin{aligned} H^{\mathrm{P}} &= \mathcal{T}^{\mathrm{P}} + \mathcal{U}\left(\frac{4J_-}{(1-\kappa J_-)^2}\right) = \frac{1}{2}\left(1+\kappa\mathbf{q}^2\right)^2\mathbf{p}^2 + \mathcal{U}\left(\frac{4\mathbf{q}^2}{(1-\kappa\mathbf{q}^2)^2}\right) + \left(1+\kappa\mathbf{q}^2\right)^2\sum_{i=1}^{N}\frac{b_i}{2q_i^2} \\ H^{\mathrm{B}} &= \mathcal{T}^{\mathrm{B}} + \mathcal{U}(J_-) = \frac{1}{2}(1+\kappa\mathbf{q}^2)\left(\mathbf{p}^2+\kappa(\mathbf{q}\cdot\mathbf{p})^2\right) + \mathcal{U}\left(\mathbf{q}^2\right) + (1+\kappa\mathbf{q}^2)\sum_{i=1}^{N}\frac{b_i}{2q_i^2} \end{aligned} \tag{48}$$

where $\mathcal{U}$ is the smooth function that gives the central potential and its specific dependence in terms of J_- corresponds to the square of the radial distance in each coordinate system [12]. Hence, the Hamiltonians (48) are the curved generalization of the (flat) Evans systems on $\mathbb{E}^N$ (43), which are recovered under the limit (contraction) $\kappa \to 0$.

4.4.2. The curved Smorodinsky–Winternitz system This well-known Hamiltonian [41, 42, 43, 44, 45, 46] is just the Higgs oscillator [47, 48] (that comes from the function $\mathcal{U}$ in (48)) plus the corresponding N centrifugal terms:

$$\begin{aligned} H^{\mathrm{P}} &= \mathcal{T}^{\mathrm{P}} + \frac{\omega^2 J_-}{2(1-\kappa J_-)^2} = \frac{1}{2}\left(1+\kappa\mathbf{q}^2\right)^2\mathbf{p}^2 + \frac{\omega^2\mathbf{q}^2}{2(1-\kappa\mathbf{q}^2)^2} + \left(1+\kappa\mathbf{q}^2\right)^2\sum_{i=1}^{N}\frac{b_i}{2q_i^2} \\ H^{\mathrm{B}} &= \mathcal{T}^{\mathrm{B}} + \frac{1}{2}\omega^2 J_- = \frac{1}{2}(1+\kappa\mathbf{q}^2)\left(\mathbf{p}^2+\kappa(\mathbf{q}\cdot\mathbf{p})^2\right) + \frac{1}{2}\omega^2\mathbf{q}^2 + (1+\kappa\mathbf{q}^2)\sum_{i=1}^{N}\frac{b_i}{2q_i^2}. \end{aligned} \tag{49}$$

This is a MS Hamiltonian whose integrals of the motion are quadratic in the momenta. The only integral of the motion that does not come from the coalgebra symmetry has been explicitly written, in this variables, in [12]. Clearly the limit $\kappa \to 0$ of both expressions (49) gives rise to the superposition of the ND isotropic harmonic oscillator with N centrifugal terms in Cartesian coordinates:

$$H = \frac{1}{2}J_+ + \frac{1}{2}\omega^2 J_- = \frac{1}{2}\mathbf{p}^2 + \frac{1}{2}\omega^2\mathbf{q}^2 + \sum_{i=1}^{N}\frac{b_i}{2q_i^2}.$$

4.4.3. The curved generalized Kepler–Coulomb system The Kepler–Coulomb potential on constant curvature spaces [42, 43, 44, 49, 50, 51, 52, 53, 54] with coupling constant k and N centrifugal terms is given by

$$H^{\rm P} = \mathcal{T}^{\rm P} - k\left(\frac{J_-}{(1-\kappa J_-)^2}\right)^{-1/2} = \frac{1}{2}(1+\kappa\mathbf{q}^2)^2\,\mathbf{p}^2 - k\,\frac{(1-\kappa\mathbf{q}^2)}{|\mathbf{q}|} + (1+\kappa\mathbf{q}^2)^2\sum_{i=1}^{N}\frac{b_i}{2q_i^2}$$

$$H^{\rm B} = \mathcal{T}^{\rm B} - kJ_-^{-1/2} = \frac{1}{2}(1+\kappa\mathbf{q}^2)\left(\mathbf{p}^2+\kappa(\mathbf{q}\cdot\mathbf{p})^2\right) - \frac{k}{|\mathbf{q}|} + (1+\kappa\mathbf{q}^2)\sum_{i=1}^{N}\frac{b_i}{2q_i^2}. \tag{50}$$

This has recently been shown to be a MS system [39, 55]. When, at least, one of the centrifugal terms vanishes ($b_i = 0$) the remaining constant of the motion is quadratic in the momenta [12], whilst when *all* the constants $b_i \neq 0$, the additional integral is quartic in the momenta and has been recently presented in [39]. We recall that the MS nature of the Euclidean case

$$H = \frac{1}{2}J_+ - kJ_-^{-1/2} = \frac{1}{2}\mathbf{p}^2 - \frac{k}{|\mathbf{q}|} + \sum_{i=1}^{N}\frac{b_i}{2q_i^2}$$

was proven in [55], and it can be recovered from our formalism when $\kappa = 0$.

4.5. Free motion on spherically symmetric spaces of nonconstant curvature
With the previous examples in mind, it is easy to realize that any metric of the type

$$\mathrm{d}s^2 = f(|\mathbf{q}|)^2\mathrm{d}\mathbf{q}^2 \tag{51}$$

leads to the free Hamiltonian given by

$$H \equiv \mathcal{T} = \frac{J_+}{2f(\sqrt{J_-})} = \frac{\mathbf{p}^2}{2f(|\mathbf{q}|)^2} \tag{52}$$

provided that all $b_i = 0$. Thus this system determines the geodesic motion on the ND spherically symmetric space (51) with conformal factor $f(|\mathbf{q}|)$; this is generically a Riemannian space of nonconstant curvature. In particular, its scalar curvature turns out to be

$$R = -(N-1)\,\frac{2f''(|\mathbf{q}|) + 2(N-1)|\mathbf{q}|^{-1}f'(|\mathbf{q}|) + (N-2)f'(|\mathbf{q}|)^2}{f(|\mathbf{q}|)^2} \tag{53}$$

where $f'(|\mathbf{q}|) = \mathrm{d}f/\mathrm{d}|\mathbf{q}|$ and $f''(|\mathbf{q}|) = \mathrm{d}^2f/\mathrm{d}|\mathbf{q}|^2$.

Therefore, since the Hamiltonian (52) corresponds to an $\mathfrak{sl}(2,\mathbb{R})$-coalgebra space then its geodesic flows define a QMS system for any choice of $f(|\mathbf{q}|)$.

Clearly, the geodesic flows on the Riemannian spaces of constant curvature described in section 4.3 are particular examples of the family (52) corresponding to set $f(|\mathbf{q}|) = (1+\kappa\mathbf{q}^2)^{-1}$ in Poincaré coordinates. Let us now describe other physically relevant examples with coalgebra symmetry.

4.5.1. Darboux spaces Four interesting examples of this class of spherically symmetric spaces with nonconstant curvature are the ND generalizations of the four 2D Darboux spaces introduced in [13]. We recall that Darboux surfaces are the only 2D spaces with nonconstant curvature for which there exists two functionally independent quadratic Killing tensors, i.e., the geodesic motion on these 2D spaces is (quadratically) MS. There are only four spaces of this

Table 3. Examples of spherically symmetric spaces and their corresponding QMS free Hamiltonian with $\mathfrak{sl}(2,\mathbb{R})$-coalgebra symmetry.

Space	Metric ds^2	Hamiltonian $\mathcal{T} = \mathcal{T}(J_-, J_+)$
Darboux I	$\dfrac{\ln \lvert\mathbf{q}\rvert \, d\mathbf{q}^2}{\mathbf{q}^2}$	$\dfrac{J_- J_+}{2 \ln \sqrt{J_-}}$
Darboux II	$\dfrac{1+\ln^2 \lvert\mathbf{q}\rvert}{\mathbf{q}^2 \ln^2 \lvert\mathbf{q}\rvert} \, d\mathbf{q}^2$	$\dfrac{J_- \ln^2 \sqrt{J_-} \, J_+}{2(1+\ln^2 \sqrt{J_-})}$
Darboux IIIa	$\dfrac{1+\lvert\mathbf{q}\rvert}{\mathbf{q}^4} \, d\mathbf{q}^2$	$\dfrac{J_-^2 J_+}{2(1+\sqrt{J_-})}$
Darboux IIIb	$(k+\mathbf{q}^2) \, d\mathbf{q}^2$	$\dfrac{J_+}{2(k+J_-)}$
Darboux IV	$\dfrac{a+\cos(\ln \lvert\mathbf{q}\rvert)}{\mathbf{q}^2 \sin^2(\ln \lvert\mathbf{q}\rvert)} \, d\mathbf{q}^2$	$\dfrac{J_- \sin^2(\ln \sqrt{J_-}) J_+}{2(a+\cos(\ln \sqrt{J_-}))}$
Multifold Kepler	$\dfrac{(a+b\lvert\mathbf{q}\rvert^{\frac{1}{\nu}}) \, d\mathbf{q}^2}{\lvert\mathbf{q}\rvert^{2-\frac{1}{\nu}}}$	$\dfrac{J_-^{1-\frac{1}{2\nu}} J_+}{2(a+bJ_-^{\frac{1}{2\nu}})}$
Taub-NUT	$\dfrac{(4m+\lvert\mathbf{q}\rvert) \, d\mathbf{q}^2}{\lvert\mathbf{q}\rvert}$	$\dfrac{\sqrt{J_-} J_+}{2(4m+\sqrt{J_-})}$

type, that were characterized by Koenigs in a note included in the famous Darboux treatise [56]. It turns out that, by using appropriate charts, the four 2D Darboux metrics [57, 58, 59, 60] can be rewritten in the form (51). Therefore, ND spherically symmetric generalizations of such spaces can be obtained by expressing their kinetic energy Hamlitonians with $\mathfrak{sl}(2,\mathbb{R})$-coalgebra symmetry. The four ND Darboux metrics [13] are explicitly written in table 3 (where a and k are real constants).

Once again, we have that the $\mathfrak{sl}(2,\mathbb{R})$-coalgebra symmetry ensures that the geodesic flows on all these spaces are QMS Hamiltonian systems. In fact, one more independent integral of the motion is expected to exist for all of them, since all these spaces are presumably MS in any dimension. Such an additional integral has already been found for the Darboux metric of type III [14].

4.5.2. Iwai–Katayama spaces These are the ND counterpart of the 3D spaces underlying the so called 'multifold Kepler' systems introduced in [61, 62], which depend on two real constants, a and b, as well on a *rational* parameter ν as shown in table 3. The physical interest of these systems relies on the fact that they are generalizations of the Taub-NUT metric which is recovered as the particular case with $\nu = 1$, $a = 4m$ and $b = 1$. We stress that in the 3D case the multifold Kepler Hamiltonians have been shown to be MS [63], but the additional integral of motion is not, in general, quadratic in the momenta.

4.5.3. Potentials We stress that centrifugal terms and central potentials can directly be added to the free system (52) by considering again an ND symplectic realization with arbitrary b_i's plus a function $\mathcal{U}(\sqrt{J_-})$ as

$$H = \frac{J_+}{2f(\sqrt{J_-})} + \mathcal{U}(\sqrt{J_-}) = \frac{\mathbf{p}^2}{2f(\lvert\mathbf{q}\rvert)^2} + \mathcal{U}(\lvert\mathbf{q}\rvert) + \frac{1}{2f(\lvert\mathbf{q}\rvert)^2} \sum_{i=1}^{N} \frac{b_i}{q_i^2}. \tag{54}$$

Several examples, such as oscillator and Kepler–Coulomb potentials on spherically symmetric spaces, can be found in [15].

5. QMS Hamiltonians with $\mathfrak{sl}_z(2,\mathbb{R})$-coalgebra symmetry

A generalization of the construction presented in the previous section can be obtained by making use of the non-standard quantum deformation of $\mathfrak{sl}(2,\mathbb{R})$ described in section 3.2 as the dynamical symmetry for the Hamiltonians corresponding to geodesic motion. In this case, the spaces so obtained are, in general, of nonconstant curvature, and the latter depends on the deformation parameter z.

5.1. Free motion

In general, we can consider an infinite family of QMS geodesic flows with $\mathfrak{sl}_z(2,\mathbb{R})$-coalgebra symmetry (sharing the set of integrals (30)) through the family of Hamiltonians

$$H_z = \frac{1}{2}J_+\, g(zJ_-) = \frac{1}{2}\, g(z\mathbf{q}^2) \sum_{i=1}^{N} \frac{\sinh zq_i^2}{zq_i^2}\, p_i^2 \exp\left\{ -z\sum_{k=1}^{i-1} q_k^2 + z\sum_{l=i+1}^{N} q_l^2 \right\} \tag{55}$$

where g is a smooth function such that $\lim_{z\to 0} g(zJ_-) = 1$, so that $\lim_{z\to 0} H_z = \frac{1}{2}\mathbf{p}^2$. This, in turn, means that H_z is a QMS deformation of the free Euclidean motion which defines a geodesic flow on an ND $\mathfrak{sl}_z(2,\mathbb{R})$-coalgebra space with metric given by:

$$\mathrm{d}s^2 = \frac{1}{g(z\mathbf{q}^2)} \sum_{i=1}^{N} \frac{2zq_i^2}{\sinh zq_i^2}\, \mathrm{d}q_i^2 \exp\left\{ z\sum_{k=1}^{i-1} q_k^2 - z\sum_{l=i+1}^{N} q_l^2 \right\}. \tag{56}$$

For $N = 2$, the Gaussian curvature of the space turns out to be [9]

$$K(x) = z\left(g'(x)\cosh x + \left(g''(x) - g(x) - g'^2(x)/g(x) \right)\sinh x \right) \tag{57}$$

where $x \equiv zJ_- = z(q_1^2 + q_2^2) \equiv z\mathbf{q}^2$, $g' = \frac{\mathrm{d}g(x)}{\mathrm{d}x}$ and $g'' = \frac{\mathrm{d}^2 f(x)}{\mathrm{d}x^2}$. For $N = 3$, the scalar curvature reads [64]

$$R(x) = z\left(6g'(x)\cosh x + \left(4g''(x) - 5g(x) - 5g'^2(x)/g(x) \right)\sinh x \right) \tag{58}$$

where $x \equiv zJ_- = z(q_1^2 + q_2^2 + q_3^2) \equiv z\mathbf{q}^2$.

Now we briefly comment on some specific $\mathfrak{sl}_z(2,\mathbb{R})$-coalgebra spaces that arise for simple choices of the function g (see [9, 11, 64]). For each of them we write the Gaussian (57) and scalar (58) curvatures corresponding to $N = 2, 3$.

- The simplest choice is to set $g(zJ_-) = 1$. The curvatures reduce to

  $$K = -z\sinh\left(z\mathbf{q}^2\right) \qquad R = -5z\sinh\left(z\mathbf{q}^2\right)$$

  that is, they are nonconstant and *negative* so the space is of hyperbolic type.
- Another possibility, which generalizes the above one, is to take $g(zJ_-) = \exp(azJ_-)$ where a is a real constant. The curvatures are given by

  $$K = z\mathrm{e}^{az\mathbf{q}^2}\left(a\cosh\left(z\mathbf{q}^2\right) - \sinh\left(z\mathbf{q}^2\right)\right)$$
  $$R = z\mathrm{e}^{az\mathbf{q}^2}\left(6a\cosh\left(z\mathbf{q}^2\right) - (5+a^2)\sinh\left(z\mathbf{q}^2\right)\right).$$

  It is worthy to remark that the very special cases with $a = \pm 1$ gives $K = \pm z$ and $R = \pm 6z$, so this choice for g includes the three classical Riemannian spaces of constant curvature (see section 4.3), for which the role of the sectional curvature κ is now played by the real deformation parameter z.

- As a third example we consider $g(zJ_-) = \cosh(zJ_-)^b$ where b is a real constant. This yields

$$K = z\left(\cosh(z\mathbf{q}^2)\right)^{b-2}\sinh(z\mathbf{q}^2)\left(b + (b-1)\cosh^2(z\mathbf{q}^2)\right)$$
$$R = -z\left(\cosh(z\mathbf{q}^2)\right)^{b-2}\sinh(z\mathbf{q}^2)\left((5-10b)\cosh^2(z\mathbf{q}^2) + b(4+b)\sinh^2(z\mathbf{q}^2)\right).$$

When $b = 1$, the expressions are rather simplified as $K = z\tanh(z\mathbf{q}^2)$ and $R = 5K$.

We stress that the geodesic motion of a classical particle on any of these spaces will have as integrals of the motion the functions (30).

5.2. Potentials

We can also introduce more general ND QMS Hamiltonians based on $\mathfrak{sl}_z(2,\mathbb{R})$ (29) by considering symplectic realizations with arbitrary b_i's (thus giving rise to 'deformed centrifugal terms') and by adding functions depending on J_-. In particular, we have already studied the Hamiltonians for the explicit 2D and 3D construction (see [10] and [64], respectively). In this case the ND Hamiltonian reads

$$\begin{aligned} H_z &= \frac{1}{2}J_+\, g(zJ_-) + \mathcal{U}(zJ_-) \\ &= \frac{1}{2}\, g(z\mathbf{q}^2)\sum_{i=1}^{N}\left(\frac{\sinh zq_i^2}{zq_i^2}\, p_i^2 + \frac{zb_i}{\sinh zq_i^2}\right)\exp\left\{-z\sum_{k=1}^{i-1} q_k^2 + z\sum_{l=i+1}^{N} q_l^2\right\} + \mathcal{U}(z\mathbf{q}^2) \end{aligned} \tag{59}$$

where the smooth functions g and $\mathcal{U}$ are such that

$$\lim_{z\to 0}\mathcal{U}(zJ_-) = \mathcal{V}(J_-) \qquad \lim_{z\to 0} g(zJ_-) = 1.$$

This condition means that the non-deformed/flat limit for these Hamiltonians is

$$\lim_{z\to 0} H_z = \frac{1}{2}\mathbf{p}^2 + \mathcal{V}(\mathbf{q}^2) + \sum_{i=1}^{N}\frac{b_i}{2q_i^2},$$

which are just the Evans systems (43). So in this way we could define QMS analogues of the Smorodinsky–Winternitz and Kepler potentials [64] on the $\mathfrak{sl}_z(2,\mathbb{R})$-coalgebra spaces.

In any case, a common feature of interacting systems with $\mathfrak{sl}_z(2,\mathbb{R})$-coalgebra symmetry is the long-range nature of the 'interaction terms' depending on the coordinates and appearing in (59).

6. Quasi-integrable Hamiltonians with h_6-coalgebra symmetry

We present three families of Hamiltonians with the underlying h_6-coalgebra symmetry described in section 3.3. We recall that all of them are endowed with the $2N-5$ integrals (38) but, in principle, they are only quasi-integrable since only $N-2$ integrals are in involution. Nevertheless, for some specific choices of the function H the subalgebra structure of h_6 has allowed us to obtain the remaining integral by following the procedure already described in section 3.3. We omit here the explicit expressions for the families of integrable cases, that are fully described in [16].

6.1. Natural systems

The Hamiltonian

$$H = \frac{1}{2}B_+ + \mathcal{F}(A_-, B_-) \tag{60}$$

Journal of Physics: Conference Series **175** (2009) 012004 doi:10.1088/1742-6596/175/1/012004

where $\mathcal{F}$ is a function playing the role of a potential, gives rise to the following quasi-integrable system on $\mathbb{E}^N$:

$$H = \frac{1}{2}\mathbf{p}^2 + \mathcal{F}\left(\boldsymbol{\lambda}\cdot\mathbf{q}, \mathbf{q}^2\right). \tag{61}$$

Notice that central potentials arise if $\mathcal{F}$ does not depend on A_-; in the case with generic $\mathcal{F}(A_-, B_-)$, spherical symmetry is broken and QMS is, in principle, reduced to quasi-integrability.

6.2. Electromagnetic Hamiltonians

The most general h_6-Hamiltonian including linear terms in the momenta is given by

$$H = \frac{1}{2}B_+ + K\,\mathcal{F}(A_-, B_-) + A_+\,\mathcal{G}(A_-, B_-) + \mathcal{R}(A_-, B_-) \tag{62}$$

where $\mathcal{F}$, $\mathcal{G}$ and $\mathcal{R}$ are smooth functions. This means that

$$H = \frac{1}{2}\mathbf{p}^2 + \left(\mathbf{q}\cdot\mathbf{p} - \frac{\boldsymbol{\lambda}^2}{2}\right)\mathcal{F}\left(\boldsymbol{\lambda}\cdot\mathbf{q}, \mathbf{q}^2\right) + (\boldsymbol{\lambda}\cdot\mathbf{p})\,\mathcal{G}\left(\boldsymbol{\lambda}\cdot\mathbf{q}, \mathbf{q}^2\right) + \mathcal{R}\left(\boldsymbol{\lambda}\cdot\mathbf{q}, \mathbf{q}^2\right). \tag{63}$$

In 3D, this Hamiltonian describes the motion of a particle on $\mathbb{E}^3$ under the action of a static electromagnetic field with vector and scalar potentials given by

$$\begin{aligned} A_i &= -\frac{q_i}{e}\mathcal{F}(A_-, B_-) - \frac{\lambda_i}{e}\mathcal{G}(A_-, B_-) \qquad i = 1, 2, 3 \\ \psi &= \frac{1}{e}\mathcal{R}(A_-, B_-) - \frac{1}{2e}M\,\mathcal{F}(A_-, B_-) \\ &\quad -\frac{1}{2e}\left[B_-\mathcal{F}(A_-, B_-)^2 + 2A_-\mathcal{F}(A_-, B_-)\mathcal{G}(A_-, B_-) + M\mathcal{G}(A_-, B_-)^2\right] \end{aligned} \tag{64}$$

where e is the electric charge. The h_6 quasi-integrability implies that, for any choice of the functions $\mathcal{F}$, $\mathcal{G}$ and $\mathcal{R}$, the Hamiltonian (63) commutes with the function

$$C^{(3)} = \left(\lambda_1(p_2q_3 - p_3q_2) + \lambda_2(p_3q_1 - p_1q_3) + \lambda_3(p_1q_2 - p_2q_1)\right)^2.$$

Note that these systems can be thought of as a generalization of the electromagnetic Hamiltonians presented in section 4.2 and coming from the $\mathfrak{sl}(2,\mathbb{R})$-coalgebra symmetry (provided that all $b_i = 0$).

6.3. Geodesic flow Hamiltonians

A third family of relevant systems is given by ND Hamiltonians of the type

$$H = \sum_{i,j=1}^{N} g^{ij}(q_1, \ldots, q_N)\, p_i\, p_j$$

that are obtained by considering

$$\begin{aligned} H &= B_+\mathcal{F}(A_-, B_-) + A_+^2\mathcal{G}(A_-, B_-) \\ &\quad + \left(K + \frac{M}{2}\right)^2\mathcal{R}(A_-, B_-) + \left(K + \frac{M}{2}\right)A_+\mathcal{S}(A_-, B_-) \end{aligned} \tag{65}$$

since for any choice of the functions $\mathcal{F}$, $\mathcal{G}$, $\mathcal{R}$ and $\mathcal{S}$ we obtain a Hamiltonian which is a quadratic homogeneous function in the momenta; namely,

$$H = \mathbf{p}^2\mathcal{F}\left(\boldsymbol{\lambda}\cdot\mathbf{q},\mathbf{q}^2\right) + (\boldsymbol{\lambda}\cdot\mathbf{p})^2\,\mathcal{G}\left(\boldsymbol{\lambda}\cdot\mathbf{q},\mathbf{q}^2\right) + (\mathbf{q}\cdot\mathbf{p})^2\,\mathcal{R}\left(\boldsymbol{\lambda}\cdot\mathbf{q},\mathbf{q}^2\right) + (\mathbf{q}\cdot\mathbf{p})\,(\boldsymbol{\lambda}\cdot\mathbf{p})\,\mathcal{S}\left(\boldsymbol{\lambda}\cdot\mathbf{q},\mathbf{q}^2\right). \tag{66}$$

The specific form of the metric g^{ij} is so determined by $\mathcal{F}$, $\mathcal{G}$, $\mathcal{R}$ and $\mathcal{S}$ which, in general, give rise to an ND space of *nonconstant* curvature. In any case, the set of constants of motion (38) is universal and does not depend on the specific choice of the functions in the Hamiltonian.

Moreover, additional potentials on these h_6-coalgebra spaces can be naturally considered by adding functions such as, e.g., $\mathcal{U}(A_-, B_-)$ to the free Hamiltonian (65). In this way the natural Euclidean systems (60) can be generalized to the curved spaces defined through (65) without breaking the quasi-integrability of the free Hamiltonian.

7. Integrable Hamiltonians with other coalgebra symmetries

For the sake of completeness, we briefly present some integrable Hamiltonians which are based on different coalgebras to the above considered. We recall that a systematic study of Lie–Poisson coalgebras with dimensions 3, 4 and 5 has been performed in [8], and some more examples can be found in [4, 5, 6].

7.1. The Calogero–Gaudin system

The Calogero–Gaudin (CG) Hamiltonian [65, 66]

$$H = \sum_{1\le i<j}^{N} 2\,p_i\,p_j\,(1-\cos(q_i-q_j)) \tag{67}$$

was proven in [1, 2] to have an underlying $\mathfrak{so}(2,1)$-coalgebra symmetry and its constants of the motion were identified with the coproducts of the $\mathfrak{so}(2,1)$ Casimir under a certain symplectic realization. The generalized CG system

$$H = \left(\sum_{i=1}^{N} p_i\right)^2 + \frac{1}{2}\sum_{i,j=1}^{N} p_i\,p_j\big((\kappa_i+\kappa_j)\cos(q_i-q_j) - (\kappa_i-\kappa_j)\cos(q_i+q_j)\big) \tag{68}$$

where κ_i are free parameters, was also proven to be completely integrable in [2].

We stress that, in general, the modification of the chosen symplectic realization drastically changes the 'shape' of the Hamiltonian through an associated canonical transformation. For instance, by using the Gelfan'd–Dyson symplectic map the very same CG system reads [5]

$$H = \sum_{1\le i<j}^{N}\left\{-p_i\,p_j\,(q_i-q_j)^2 - b\,(p_i-p_j)\,(q_i-q_j)\right\} + \frac{b^2}{4}N^2 \tag{69}$$

where b is a constant. On the other hand, the $N=2$ rational Calogero–Moser Hamiltonian [67] has also been proven to have $\mathfrak{sl}(2,\mathbb{R})$-coalgebra symmetry [5].

7.2. An integrable deformation of the CG system from $\mathfrak{so}_z(2,1)$

As we have pointed out previously, given a Hamiltonian with certain coalgebra symmetry any quantum deformation of the underlying coalgebra provides an integrable deformation of the initial Hamiltonian. This was explicitly shown for the first time by constructing the following deformed CG system [1]:

$$H_z = \sum_{1\le i<j}^{N} 2\,\pi_i\,\pi_j\,(1-\cos(q_i-q_j)) \tag{70}$$

where the non-local deformations of the momenta are

$$\pi_k = 2\,\frac{\sinh(\frac{z}{2}p_k)}{z}\prod_{i=1}^{k-1}\mathrm{e}^{-\frac{z}{2}p_i}\prod_{j=k+1}^{N}\mathrm{e}^{\frac{z}{2}p_j}.$$

The corresponding constants of the motion come from the (deformed) coproduct of the (deformed) Casimir of $\mathfrak{so}_z(2,1)$ (the standard deformation of $\mathfrak{so}(2,1)$) and, as expected, in the limit $z\to 0$ we recover the 'classical' CG system given in (67). We remark that the quantum mechanical version of this deformed CG system has been explicitly solved [17, 18] and a 'twisted' version of the Gaudin magnet has already been introduced [68] through a quantum deformation of the $\mathfrak{gl}(2,\mathbb{R})$-coalgebra.

7.3. Systems defined on the harmonic oscillator coalgebra
The Hamiltonian [69]

$$H = (\lambda+\mu)\sum_{i=1}^{N} p_i + 2\,\mu\sum_{i<j}^{N}\sqrt{p_i\,p_j}\,\cosh(q_i-q_j) \tag{71}$$

where λ and μ are real constants, can also be proven to be coalgebra invariant under the harmonic oscillator coalgebra (h_4,Δ). The quantum version of this Hamiltonian has also been solved [70] and the system turns out to be equivalent to a chain of coupled oscillators [70, 71].

7.4. Ruijsenaars–Schneider-like systems from a quantum Poincaré coalgebra
Another interesting example of coalgebra-invariant system is the following analogue [2] of the Ruijsenaars–Schneider model [72]:

$$H_z = \sum_{i=1}^{N}\cosh\theta_i\,\exp\left(-\frac{z}{2}\left(\sum_{j=1}^{i-1} q_j\right)+\frac{z}{2}\left(\sum_{k=i+1}^{N} q_k\right)\right) \tag{72}$$

where (q_i,θ_i) are canonically conjugate variables such that $\{q_i,\theta_j\}=\delta_{ij}$. This completely integrable Hamiltonian was obtained by using the Poisson analogue of the quantum deformation of the (1+1)D Poincaré algebra introduced in [73].

8. Generalizations of coalgebra symmetry

So far we have shown how the coalgebra formalism provides an algebraic approximation to the integrability properties of Hamiltonian systems defined on a chain of N copies, $A\otimes A\otimes\cdots\otimes A$, of a given coalgebra A. Amongst the possible generalizations of this approach, two of them have already been analysed. The first one deals with integrable systems defined on the N-chain $V\otimes A\otimes A\otimes\cdots\otimes A$ in which V is an A-comodule algebra [4, 74]. The second extension of the formalism deals with the so-called *loop coproducts* [75], that generalize in terms of loop algebras all the algebraic machinery presented in this paper. As a concluding section of this contribution, we briefly sketch both approaches.

8.1. Comodule algebra symmetry

We recall that a (right) *coaction* of a coalgebra (A, Δ) on a vector space V is a linear map $\phi : V \to V \otimes A$ such that the following diagram is commutative:

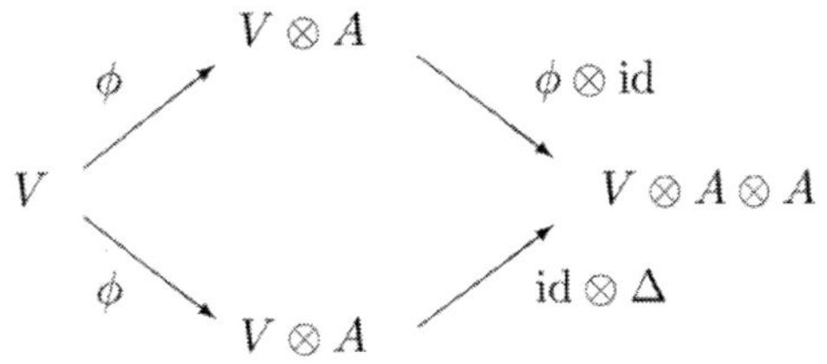

If V is an algebra, we shall say that V is an A-comodule algebra if the coaction ϕ is a homomorphism with respect to the product on the algebra V:

$$\phi(ab) = \phi(a)\,\phi(b) \qquad \forall a, b \in V.$$

Moreover, if V is a Poisson algebra and

$$\phi(\{a, b\}) = \{\phi(a), \phi(b)\} \qquad \forall a, b \in V$$

we will say that V is a Poisson A-comodule algebra.

It is straightforward to realize that this comodule structure can be used to mimic the coalgebra construction (see [4, 74]) in order to define *completely integrable systems* on

$$V \otimes A \otimes A \otimes \cdots \otimes A.$$

However, in this case the superintegrability is lost since 'right' integrals do not exist (the left-right symmetry of the chain of algebras has been broken).

Clearly, A itself is an A-comodule algebra, with the coproduct Δ playing the role of the coaction ϕ, so that the comodule symmetry is indeed a generalization of the coalgebra one. In [74] some examples of classical and quantum integrable systems with comodule symmetry have been presented. We illustrate here one of them. As we said in section 3.3, the two photon algebra $h_6 = \langle K, A_+, A_-, B_+, B_-, M\rangle$ with Poisson brackets (33) contains the $\mathfrak{gl}(2,\mathbb{R}) = \langle B_-, B_+, K, M\rangle$ subalgebra. Moreover, it is easy to realize that $\mathfrak{gl}(2,\mathbb{R})$ is a (deformed) h_6-comodule algebra with the coaction ϕ defined by:

$$\begin{aligned}
&\phi(M) = 1 \otimes M + M \otimes 1 \\
&\phi(K) = 1 \otimes K + K \otimes \frac{1}{1-\sigma A_-} + M \otimes \frac{\sigma A_-}{1-\sigma A_-} \\
&\phi(B_+) = 1 \otimes B_+ + B_+ \otimes \frac{1}{(1-\sigma A_-)^2} - 2\sigma K \otimes \frac{A_+}{1-\sigma A_-} - \sigma^2 K^2 \otimes \frac{M}{(1-\sigma A_-)^2} \\
&\phi(B_-) = 1 \otimes A_- + A_- \otimes 1 - \sigma A_- \otimes A_-
\end{aligned} \tag{73}$$

where σ is a deformation parameter. If we consider the following symplectic realization of h_6

$$\begin{aligned}
D(B_+) = q_1^2 \qquad D(B_-) = p_1^2 \qquad D(K) = -p_1 q_1 + \frac{\lambda_1^2}{2} \\
D(M) = -\lambda_1^2 \quad D(A_+) = -\lambda_1 q_1 \quad D(A_-) = -\lambda_1 p_1
\end{aligned}$$

and we take as the Hamiltonian on $\mathfrak{gl}(2,\mathbb{R})$ the function

$$H = \frac{1}{2}(B_+ + B_-)$$

the symplectic realization D will give us the 1D harmonic oscillator:

$$H^{(1)} := D(H) = \frac{p_1^2}{2} + \frac{q_1^2}{2}.$$

Now, by using the coaction map (73) we can define a Hamiltonian function on $\mathfrak{gl}(2,\mathbb{R}) \otimes h_6$:

$$\phi(H) = \frac{1}{2}(\phi(B_+) + \phi(B_-)).$$

This Hamiltonian can be expressed in terms of canonical coordinates by taking the symplectic realization $D \otimes D$ of (73). It reads:

$$\begin{aligned} H_\sigma^{(2)} = (D\otimes D)(\phi(H)) &= \frac{1}{2}(p_1^2 + p_2^2) + \frac{q_2^2}{2} + \frac{q_1^2}{2(1+\sigma\,\lambda_2\,p_2)^2} \\ &+\sigma\,\lambda_2\left(p_1^2\,p_2 + \frac{q_2(\lambda_1^2 - 2q_1\,p_1)}{2(1+\sigma\,\lambda_2\,p_2)}\right) + \sigma^2\,\lambda_2^2\left(\frac{1}{2}p_1^2\,p_2^2 + \frac{(\lambda_1^2 - 2q_1\,p_1)^2}{8(1+\sigma\,\lambda_2\,p_2)^2}\right). \end{aligned}$$

This Hamiltonian is just an integrable deformation of the 2D isotropic oscillator, since $\lim_{\sigma\to 0} H_\sigma^{(2)} = \frac{1}{2}(p_1^2+p_2^2) + \frac{1}{2}(q_1^2+q_2^2)$. The integral of motion in involution with $H_\sigma^{(2)}$ is just obtained as the coaction of the $\mathfrak{gl}(2,\mathbb{R})$ Casimir:

$$\begin{aligned} C_\sigma^{(2)} &= (D\otimes D)(\phi(C)) \qquad (74)\\ &= -\frac{\{2(p_2q_1 - p_1q_2) + \sigma p_1(2p_1q_1 - 4p_2q_2 - \lambda_1^2) - \sigma^2\lambda_2^2 p_1p_2(-2p_1q_1 + 2p_2q_2 + \lambda_1^2)\}^2}{16\,(1+\sigma\lambda_2p_2)^2}. \end{aligned}$$

As expected, the limit $\sigma \to 0$ of (74) is just $-(p_2q_1 - p_1q_2)^2/4$. Obviously, further iterations of the coaction map would provide the corresponding integrable deformation of the isotropic oscillator in an arbitrary dimension. Nevertheless, the explicit form of such a Hamiltonian is quite involved due to the form of the coaction mapping ϕ.

8.2. Loop coproducts

In [75] a further generalization of the coalgebra approach is proposed by weakening the Poisson homomorphism property:

$$\Delta^{(m)}(\{a,b\}) = \left\{\Delta^{(m)}(a), \Delta^{(m)}(b)\right\}.$$

Here we recall how the generalization works for the standard coalgebra structure (7) associated with a Lie–Poisson algebra.

Let A be a Poisson algebra with generators $\{X^\alpha\}$, $\alpha = 1,\ldots,l = \dim(A)$, with Poisson brackets given by

$$\left\{X^\alpha, X^\beta\right\} = F^{\alpha\beta}(\vec{X}) \qquad \vec{X} = (X^1,\ldots,X^l)$$

and with r Casimirs functions $\mathcal{C}_j$, $j = 1,\ldots,r$. Let us assume that one succeeds in finding a set of m maps depending on a parameter λ:

$$\Delta_\lambda^{(k)} : A \to A\otimes A\otimes \ldots^{N)} \otimes A \qquad k = 1,\ldots,m \qquad (75)$$

such that for the generators of A the following relations hold:

$$\{\Delta_\lambda^{(i)}(X^\alpha), \Delta_\mu^{(k)}(X^\beta)\} = f^{ik}(\lambda,\mu) F^{\alpha\beta}\left(\Delta_\lambda^{(i)}(\vec{X})\right) \qquad k > i \tag{76}$$

$$\{\Delta_\lambda^{(k)}(X^\alpha), \Delta_\mu^{(k)}(X^\beta)\} = f^{k}(\lambda,\mu) F^{\alpha\beta}\left(\Delta_\lambda^{(k)}(\vec{X})\right) + g^{k}(\lambda,\mu) F^{\alpha\beta}\left(\Delta_\mu^{(k)}(\vec{X})\right) \tag{77}$$

where f^{ik}, f^k, g^k are some functions only depending on the parameters λ and μ. If the maps $\Delta_\lambda^{(k)}$ are defined on any smooth function h of the generators of A as:

$$\Delta_\lambda^{(k)}\left(h(X^1,\ldots,X^l)\right) = h\left(\Delta_\lambda^{(k)}(X^1),\ldots,\Delta_\lambda^{(k)}(X^l)\right),$$

then it can be proven that the following relations are satisfied [75]:

$$\{\Delta_\lambda^{(i)}(\mathcal{C}_j), \Delta_\mu^{(k)}(X^\beta)\} = 0 \qquad k > i \tag{78}$$

$$\{\Delta_\lambda^{(i)}(\mathcal{C}_j), \Delta_\mu^{(k)}(\mathcal{C}_n)\} = 0. \tag{79}$$

This result can be interpreted as a generalization of the coalgebra approach. If A is a Lie–Poisson algebra with commutation rules

$$\{X^\alpha, X^\beta\} = C_\gamma^{\alpha\beta} X^\gamma$$

and equipped with the standard coproduct (7), then

$$\{\Delta^{(i)}(X^\alpha), \Delta^{(k)}(X^\beta)\} = C_\gamma^{\alpha\beta}\Delta^{(i)}(X^\gamma) \qquad k \geq i,$$

i.e., this is of the form (76) and (77) with

$$f^{ik} = f^k = 1 \qquad g^k = 0 \qquad F^{\alpha\beta}\left(\Delta^{(i)}(\vec{X})\right) = C_\gamma^{\alpha\beta}\Delta^{(i)}(X^\gamma).$$

Equations (78) and (79), in turn, generalize equations (5) and (6). Consequently, the maps (75) can be properly called 'loop coproducts'.

In [75] it is shown that the loop coproducts given by

$$\Delta_\lambda^{(k)}(X^\alpha) = \frac{\Delta^{(k-1)}(X^\alpha)}{\lambda} + \frac{X_k^\alpha}{\lambda - \epsilon} \qquad k = 2,\ldots,N \tag{80}$$

with ϵ an arbitrary constant parameter, satisfy equations (76) and (77). From equation (79) it follows that the residues of the functions $\Delta_\lambda^{(k)}(\mathcal{C}_j)$ $(k = 2,\ldots,N,\ j = 1,\ldots,r)$ are in involution. Furthermore this family of functions in involution contains the standard coproduct of the Casimirs $\Delta^{(k)}(\mathcal{C}_j)$.

Finally, it can be shown that if the Lie–Poisson algebra A is a simple one and

$$\{I_1,\ldots,I_s\} \qquad s = \frac{l-r}{2}$$

(to be compared with (12)) define an integrable system on A, then the residues of the functions $\Delta_\lambda^{(k)}(\mathcal{C}_j)$ $(k = 2,\ldots,N,\ j = 1,\ldots,r)$ together with the functions $\Delta^{(N)}(I_1),\ldots,\Delta^{(N)}(I_s)$ define an integrable system on $A \otimes A \otimes \ldots^{N)} \otimes A$. This last result solves for this particular case the integrability problem mentioned in section 2.1 for the case of simple Lie–Poisson algebras.

Acknowledgments
This work was partially supported by the Spanish Ministerio de Ciencia e Innovación under grant MTM2007-67389 (with EU-FEDER support), by the Junta de Castilla y León (Project GR224) and by INFN–CICyT.

References

[1] Ballesteros A, Corsetti M and Ragnisco O 1996 *Czech. J.Phys.* **46** 1153
[2] Ballesteros A and Ragnisco O 1998 *J. Phys. A: Math. Gen.* **31** 3791
[3] Ballesteros A and Herranz F J 1999 *J. Phys. A: Math. Gen.* **32** 8851
[4] Grabowski J, Marmo G and Michor P W 1999 *Mod. Phys. Lett. A* **14** 2109
[5] Ballesteros A and Ragnisco O 2002 *J. Math. Phys.* **43** 954
[6] Ballesteros A and Ragnisco O 2003 *J. Phys. A: Math. Gen.* **36** 10505
[7] Ballesteros A, Herranz F J, Musso F and Ragnisco O 2004 in *Superintegrability in Classical and Quantum Systems (CRM Proc. and Lecture Notes vol. 37)* ed P Tempesta *et al* (Providence, RI: AMS) p 1 arXiv:math-ph/0412067
[8] Ballesteros A and Blasco A 2008 *J. Phys. A: Math. Gen.* **41** 304028
[9] Ballesteros A, Herranz F J and Ragnisco O 2005 *Phys. Lett. B* **610** 107
[10] Ballesteros A, Herranz F J and Ragnisco O 2005 *J. Phys. A: Math. Gen.* **38** 7129
[11] Ragnisco O, Ballesteros A, Herranz F J and Musso F 2006 *SIGMA* **3** 026
[12] Ballesteros A and Herranz F J 2007 *J. Phys. A: Math. Theor.* **40** F51
[13] Ballesteros A, Enciso A, Herranz F J and Ragnisco O 2007 *Phys. Lett. B* **652** 376
[14] Ballesteros A, Enciso A, Herranz F J and Ragnisco O 2008 *Physica D* **237** 505
[15] Ballesteros A, Enciso A, Herranz F J and Ragnisco O 2009 *Ann. Phys.* **324** 1219
[16] Ballesteros A, Blasco A and Herranz F J 2009 *J. Phys. A: Math. Gen.* to appear arXiv:0901.1483
[17] Musso F and Ragnisco O 2000 *J. Math. Phys.* **41** 7386
[18] Musso F and Ragnisco O 2001 *J. Phys. A: Math. Gen.* **34** 2625
[19] Calogero F and Van Diejen J F 1996 *J. Math, Phys.* **37** 4243
[20] Breiderhoff J, Musso F and Ragnisco O 2005 *J. Math. Phys.* **46** 032109
[21] Musso F, Petrera M, Satta G and Ragnisco O 2005 *Nucl. Phys. B* **716** 543
[22] Arnold V I, Kozlov V V and Neishtadt A I 1997 *Mathematical aspects of classical and celestial mechanics* (Berlin: Springer)
[23] Drinfel'd V G 1987 in *Proc. Int. Congress of Math. (Berkeley 1986)* ed A V Gleason (Providence, RI: American Mathematical Society) p 798
[24] Chari V and Pressley A 1994 *A Guide to Quantum Groups* (Cambridge: Cambridge University Press)
[25] Tjin T 1992 *Int. J. Mod. Phys. A* **7** 6175
[26] Ohn C 1992 *Lett. Math. Phys.* **25** 85
[27] Ballesteros A and Herranz F J 2001 *J. Nonlin. Math. Phys.* **8** Suppl. 18
[28] Zhang W M, Feng D H and Gilmore R 1990 *Rev. Mod. Phys.* **62** 867
[29] Brif C 1996 *Ann. Phys.* **251** 180
[30] Ballesteros A, Herranz F J and Parashar P 1997 *J. Phys. A: Math. Gen.* **30** 8587
[31] Burdet G, Patera J, Perrin M and Winternitz P 1978 *J. Math. Phys.* **19** 1758
[32] Wojciechowski S 1983 *Phys. Lett. A* **96** 389
[33] Ballesteros A and Blasco A 2009 in preparation.
[34] Fris J, Mandrosov V, Smorodinsky Y A, Uhlir M and Winternitz P 1965 *Phys. Lett.* **16** 354
[35] Dorizzi B, Grammaticos B, Ramani A and Winternitz P 1985 *J. Math. Phys.* **26** 3070
[36] McSween E and Winternitz P 2000 *J. Math. Phys.* **41** 2957
[37] Hietarinta J 1985 *J. Math. Phys.* **26** 1970
[38] Doubrovine B, Novikov S and Fomenko A 1982 *Géométrie Contemporaine, Méthodes et Applications* First Part (Moscow: MIR)
[39] Ballesteros A and Herranz F J 2009 *J. Phys. A: Math. Gen.* to appear arXiv:0903.2337
[40] Evans N W 1990 *Phys. Rev. A* **41** 5666
[41] Kalnins E G, Miller W Jr and Pogosyan G S 1997 *J. Math. Phys.* **38** 5416
[42] Rañada M F and Santander M 1999 *J. Math. Phys.* **40** 5026
[43] Kalnins E G, Miller W Jr and Pogosyan G S 2000 *J. Phys. A: Math. Gen.* **33** 6791
[44] Kalnins E G, Kress J M, Pogosyan G S and Miller W Jr 2001 *J. Phys. A: Math. Gen.* **34** 4705
[45] Ballesteros A, Herranz F J, Santander M and Sanz-Gil T 2003 *J. Phys. A: Math. Gen.* **36** L93
[46] Herranz F J, Ballesteros A, Santander M, Sanz-Gil T 2004 *Superintegrability in Classical and Quantum*

Systems (CRM Proc. and Lecture Notes vol. 37) ed P Tempesta *et al* (Providence, RI: AMS) p 75 `arXiv:math-ph/0501035`
[47] Higgs P W 1979 *J. Phys. A: Math. Gen.* **12** 309
[48] Leemon I 1979 *J. Phys. A: Math. Gen.* **12** 489
[49] Schrödinger E 1940 *Proc. R. Ir. Acad. A* **46** 9
[50] Kalnins E G, Miller W Jr and Pogosyan G S 2000 *J. Math. Phys.* **41** 2629
[51] Nersessian A and Pogosyan G S 2001 *Phys. Rev. A* **63** 020103
[52] Rodríguez M A and Winternitz P 2002 *J. Math. Phys.* **43** 1309
[53] Kalnins E G, Williams G C, Miller W Jr and Pogosyan G S 2002 *J. Phys. A: Math. Gen.* **35** 4755
[54] Herranz F J and Ballesteros A 2006 *SIGMA* **2** 010
[55] Verrier P E and Evans N W 2008 *J. Math. Phys.* **49** 022902
[56] Koenigs G 1972 in *Leçons sur la théorie géneralé des surfaces* vol 4 ed G Darboux (New York: Chelsea) p 368
[57] Kalnins E G, Kress J M and Winternitz P 2002 *J. Math. Phys.* **43** 970
[58] Kalnins E G, Kress J M, Miller W Jr and Winternitz P 2003 *J. Math. Phys.* **44** 5811
[59] Grosche C, Pogosyan G S and Sissakian A N 2007 *Phys. Part. Nuclei* **38** 299
[60] Grosche C, Pogosyan G S and Sissakian A N 2007 *Phys. Part. Nuclei* **38** 525
[61] Iwai T and Katayama N 1994 *J. Phys. A: Math. Gen.* **27** 3179
[62] Iwai T and Katayama N 1995 *J. Math. Phys.* **36** 1790
[63] Ballesteros A, Enciso A, Herranz F J and Ragnisco O 2009 *Comm. Math. Phys.* doi: 10.1007/s00220-009-0793-5 `arXiv:0810.0999`
[64] Ballesteros A and Herranz F J 2008 `arXiv: 0812.4124`
[65] Calogero F 1995 *Phys. Lett. A* **201** 306
[66] Gaudin M 1983 *La Fonction d'Onde de Bethe* (Paris: Masson)
[67] Calogero F 1971 *J. Math. Phys.* **12** 419
[68] Ballesteros A 1999 *Int. J. Mod. Phys. B* **13** 2903
[69] Calogero F 1995 *J. Math. Phys.* **36** 9
[70] Calogero F and Van Diejen J F 1995 *Phys. Lett. A* **205** 143
[71] Karimipour V 1998 *J. Math. Phys.* **39** 913
[72] Ruijsenaars S N M and Schneider H 1986 *Ann. Phys.* **170** 370
[73] Vaksman L and Korogodskii L I 1989 *Sov. Math. Dokl.* **39** 173
[74] Ballesteros A, Musso F and Ragnisco O 2002 *J. Phys. A: Math. Gen.* **35** 8197
[75] Musso F, Ballesteros A, and Ragnisco O in preparation

Black hole entropy in loop quantum gravity: the role of internal symmetries

J. Fernando Barbero G.

Instituto de Estructura de la Materia, CSIC, Serrano 123, 28006 Madrid, Spain

E-mail: `fbarbero@iem.cfmac.csic.es`

Abstract. I will discuss here the role of the internal symmetry group in the computations of black hole entropy in loop quantum gravity according to the standard prescription given by Domagala and Lewandowski [1]. In particular I will show how it is possible to take into account the possible choice of either $SO(3)$ or $SU(2)$ as the internal symmetry groups of general relativity in Loop Quantum Gravity and how this choice changes the combinatorial problem of counting the black hole degrees of freedom.

The problem of quantizing general relativity is one of the most difficult and important ones in high energy theoretical physics. This is so because some of the deepest questions in physics are expected to be answered by such a theory. Among them one could mention the resolution of space-time singularities –including the Big Bang and those appearing inside black holes. To this date there is no complete satisfactory theory of quantum gravity, though it can be argued that some proposals –led by string theory– are rather advanced and provide satisfactory answers to many issues. This contribution concentrates on a different candidate theory, Loop Quantum Gravity or LQG in short, that has also provided some tantalizing hints about the behavior of gravity in its quantum regime.

The problem of understanding the microphysics behind black hole thermodynamics is a good testing ground for candidate quantum gravity theories. In this respect both string theories and LQG can claim a reasonable success because both of them can explain the Bekenstein-Hawking law. In the first case this is done for extremal or almost extremal black holes whereas the computations in LQG can deal with the more physical Schwarzschild or Kerr spacetimes.

The Bekenstein-Hawking area law is arguably the most firmly rooted result obtained to date concerning the semiclassical behavior of black holes. However this is not the only prediction that has been put forward for the behavior of black hole entropy. Another interesting idea is the quantization of entropy and area suggested in the late seventies by Bekenstein and Mukhanov [2]. In their proposal the area spectrum of black holes is an equally space one (only quantized areas are possible) and, consequently, the entropy is also effectively quantized.

A remarkable achievement of LQG has been the construction of well defined operators in a suitable Hilbert space that represent geometrical quantities such as areas and volumes (see the nice review [3] and references therein). The quantization of the areas is of particular interest for us in the study of black hole entropy because they are predicted to be quantized as suggested by Bekenstein and Mukhanov. There are however some problems with this picture. The most obvious one is due to the fact that the area spectrum in LQG is not equally spaced.

A surprising and unexpected way to reconcile these two alternates views (the equally spaced

Journal of Physics: Conference Series **175** (2009) 012005 doi:10.1088/1742-6596/175/1/012005

spectrum of Bekenstein-Mukhanov and the LQG one) was suggested by the authors of [4]. In this paper they show that the actual computations for the entropy of small black holes performed by a direct brute force counting of the relevant microstates display an intriguing behavior: when the black hole entropy is plotted as a function of area the expected linear growth is modulated by an oscillating term that gives rise to a staircase structure for the entropy that becomes effectively quantized and jumps at equally spaced values for the area. This means that, at least for microscopic black holes, the area and the entropy are essentially quantized in the way suggested by Bekenstein and Mukhanov. This is a highly non-trivial result because of the non-evenness of the spectrum of the area operator in LQG. Subsequently a rather detailed description of this phenomenon has been obtained by using some techniques borrowed from number theory and combinatorics [5]. I will briefly discuss them here and, in particular, concentrate on how a proposal [6] to study and understand the effects of changing the internal gauge group of LQG from $SU(2)$ to $SO(3)$ can be incorporated in our framework.

The computation of the entropy of a black hole modeled as an isolated horizon was reduced to a concrete counting problem in [7]. This was subsequently simplified in [1] and reduced to a much simpler combinatorial problem[1]. This will be my starting point here so I will use the following prescription to obtain the black hole entropy [1].

Definition 1 *The entropy $S_{\rm bh}(a)$ of a quantum horizon of the classical area a, according to Quantum Geometry and the Ashtekar-Baez-Corichi-Krasnov framework [7], is*

$$S(a) = \log \mathfrak{n}(a)\,,$$

where $\mathfrak{n}(a)$ is 1 plus the number of all the finite, arbitrarily long, sequences $\vec{m} = (m_1, \ldots, m_n)$ of non-zero half integers, such that the following equality and inequality are satisfied:

$$\sum_{i=1}^{n} m_i = 0, \quad \sum_{i=1}^{n} \sqrt{|m_i|(|m_i|+1)} \leq \frac{a}{8\pi\gamma\ell_P^2}. \tag{1}$$

Here γ is the Barbero-Immirzi parameter of Quantum Geometry and ℓ_P the Planck length. The extra term 1 above comes from the trivial sequence.

The counting that must be performed to obtain the black hole entropy according to the previous prescription can be carried out in different ways. In fact, a proposal [8] that appeared simultaneously with [1] gives a solution that leads to the Bekenstein-Hawking area law. However, as pointed out in [9], this result is not useful to see the effective entropy quantization described in [4] (it does not easily lend itself to numerical computations). Also a full understanding of the asymptotic behavior of the entropy for large areas is difficult (though maybe not impossible) to get.

An alternative way of studying the entropy was proposed in [5, 10]. The main idea was to start from a precise characterization of the area spectrum in LQG, in particular the degeneracies of the different eigenvalues, and then use this to obtain a concrete algorithm to compute the black hole entropy. In fact it is not only possible to build this algorithm but also to "implement" it in the form of a *black hole generating function* [10] that can be used to check some of the results obtained to date on this issue and also as the starting point of an asymptotic analysis for large black hole areas.

In the Ashtekar-Baez-Corichi-Krasnov approach [7] to black hole entropy one works in a spacetime with an inner boundary that models the black hole horizon (technically it will become an isolated horizon [13]). The gravitational degrees of freedom can be classified either as bulk

[1] Although there are some subtle differences between both approaches for small black holes, it is expected that the values for the entropy obtained with either definition are the same, at least in the macroscopic regime.

degrees of freedom or horizon degrees of freedom. The latter are described by a $U(1)$ Chern-Simons theory on the sphere $\mathbb{S}^2$ whose level $\kappa \in \mathbb{N}$ is related to the classical area of the horizon a_0 according to $a_0 = 4\pi\gamma\ell_P^2\kappa$. In this sense the entropy definition appearing above is restricted only to such values of the area. However, it is possible to argue that the horizon area should be given by an eigenvalue in the spectrum of the area operator in LQG[2]. My point of view here will be then to extend the definition above to arbitrary values of the area a, actually, by taking units such that $4\pi\gamma\ell_P^2 = 1$ the inequality appearing in (1) will become

$$\sum_{i=1}^{n} m_i = 0, \quad \sum_{i=1}^{n} \sqrt{|m_i|(|m_i|+1)} \leq \frac{a}{2},$$

and hence the difference between the two different points of view boils down to choosing a as an integer or a real number.

The counting of the allowed sequences mentioned in the previous definition can be conveniently performed in four steps

(i) Start by fixing a value for the area a and obtain all the possible choices for the half integers $|m_i|$ compatible with the area, i.e. satisfying

$$\sum_{i=1}^{n} \sqrt{|m_i|(|m_i|+1)} = \frac{a}{2}.$$

Obviously this number will be zero if a does not belong to the spectrum of the area operator. At this point we do not worry about the ordering of the labels but, rather, we just try to find out how many times the different values of each $|m| \in \mathbb{N}/2$ actually appear. In this step we count then the number of different finite multisets of $|m_i|$ compatible with the area a.

(ii) Count the different ways in which the previous multisets can be reordered.

(iii) Count all the different ways to introduce signs in the sequences of positive half-integers obtained in the previous step in such a way that the projection constraint $\sum_{i=1}^{n} m_i = 0$ is satisfied.

(iv) Repeat the same procedure for all the area eigenvalues smaller than a and add the number of sequences obtained for each value of the area.

The first step can be thought of as a characterization of the part of the spectrum of the area operator relevant in the computation of black hole entropy. This is so because the area eigenvalues for the isolated horizon that represents the black hole have the form

$$a(\vec{j}) := 8\pi\gamma\ell_P^2 \sum_{i=1}^{n} \sqrt{j_i(j_i+1)}$$

where the j_i "spin" labels are associated to the edges that pierce the horizon of the graph labeling the state vector in the LQG Hilbert space. For convenience I introduce here the integer variables $k_i := 2|m_i|$ and write

$$\sum_{i=1}^{n} \sqrt{|m_i|(|m_i|+1)} = \frac{a}{2} \Rightarrow \sum_{i=1}^{n} \sqrt{(k_i+1)^2 - 1} = a \Rightarrow$$

$$\sum_{i=1}^{k_{max}(a)} n_k \sqrt{(k+1)^2 - 1} = a \qquad (2)$$

[2] It is possible to justify, also, that in the case of isolated horizons one can work with an area operator different from the standard one in LQG [14].

where the n_k in the last sum tell us the number of times that the label $k \in \mathbb{N}$ appears. Also I denote as $k_{max}(a)$ the maximum value of the positive integer k compatible with the area a. At this point we want to find out all the sets of pairs $\{(k, n_k) : k \in \mathbb{N},\ n_k \in \mathbb{N}\}$ such that the previous equation is satisfied[3]. It is obvious that we can always write $\sqrt{(k+1)^2-1}$ as the product of an integer times the square root of a square-free positive number[4] p_I by using its prime factor decomposition. This means that the area eigenvalue a must be an integer linear combination of square roots of squarefree numbers of the form

$$a = \sum_{I=1}^{r} q_I \sqrt{p_I}.$$

or else the condition given by equation (2) cannot be satisfied. This leads then to the following equation

$$\sum_{k=1}^{k_{max}(a)} n_k \sqrt{(k+1)^2-1} = \sum_{I=1}^{r} q_I \sqrt{p_I}. \tag{3}$$

This can be solved by first considering each of the square-free numbers p_I separately and computing the possible values of k such that $\sqrt{(k+1)^2-1}$ is an integer multiple of $\sqrt{p_I}$. This amounts to solving the equations

$$\sqrt{(k_I+1)^2-1} = y_I \sqrt{p_I}$$

in the two non-negative, integer unknowns k_I and y_I (here the label I refers to the square-free number p_I). These are equivalent to

$$x_I^2 - p_I y_I^2 = 1 \tag{4}$$

where $x_I := k_I + 1$. Equation (4) is the well known Pell equation [15]. Its general solution is known and can be found in the following way. Start by obtaining the so called *fundamental solution* [15] (x_1^I, y_1^I) corresponding to the smallest positive value for x. Once this is known the remaining solutions are given by the sequence $\{(x_\alpha^I, y_\alpha^I) : \alpha \in \mathbb{N}\}$

$$x_\alpha^I = \frac{1}{2}\left[(x_1^I + y_1^I\sqrt{p_I})^\alpha + (x_1^I - y_1^I\sqrt{p_I})^\alpha\right]$$
$$y_\alpha^I = \frac{1}{2\sqrt{p_I}}\left[(x_1^I + y_1^I\sqrt{p_I})^\alpha - (x_1^I - y_1^I\sqrt{p_I})^\alpha\right].$$

From these we get the solutions (k_α^I, y_α^I) to the original equation. For instance for $p_I = 2$ the previous sequence starts as $(2,2), (16,12), (98,70), (576,408), \ldots$ It is interesting to note at this point that its terms grow exponentially fast.

Once the values of the k_I are known we have to obtain the corresponding n_{k_I}. This is easily accomplished by solving a system of uncoupled, linear, diophantine equations (with non-negative, integer, unknowns). Indeed, by writing

$$\sum_{k=1}^{k_{max}} n_k \sqrt{(k+1)^2-1} = \sum_{I=1}^{r}\sum_{\alpha=1}^{\infty} n_{k_\alpha^I} y_\alpha^I \sqrt{p_I} = \sum_{i=I}^{r} q_I \sqrt{p_I},$$

[3] Notice that by requiring that $n_k \neq 0$ we restrict ourselves to find only the values of k that actually do appear.
[4] We will enumerate these numbers as $p_1 = 2$, $p_2 = 3$, $p_3 = 5$, and so on.

and using the fact that the square roots of the square-free numbers are linearly independent over the rationals (and hence also over the integers) the previous equation is equivalent to the following system of linear, uncoupled, diophantine equations

$$\sum_{\alpha=1}^{\infty} y_\alpha^I n_{k_\alpha^I} = q_I, \quad I = 1, \ldots, r.$$

Several comments are in order now. The first is that for a fixed value of the area a (necessarily an integer linear combination with a finite number of coefficients q_I corresponding to the square-free integers p_I) only a finite number of labels k_α^I come into play in these equations. Second, it may happen that some of these equations admit no solutions, in that case a does not belong to the spectrum of the area operator. Finally when they can be solved the solution tell us exactly what the allowed values for the spin labels

In the entropy definition introduced above the m_i labels are half-integers. This is a consequence of the fact that the internal symmetry group in the Ashtekar formulation of general relativity is $SU(2)$. As long as one does not want to couple fermionic matter[5] to gravity the Ashtekar formulation can also be obtained by using $SO(3)$ as the internal symmetry group. When considering quantization basically all the results corresponding to the standard $SU(2)$ formulation can be translated. The main difference lies in the fact that the edges in the spin network states are labeled only by integers. As far as the problem that I am considering here is concerned this has two main consequences. On one hand the spectrum of the area operator is modified, on the other we can only use integer values for the $|m_i|$ (and consequently only even values for the k_i).

Up to this point we have found all the possible choices of $|m|$ compatible with a given value of the area a, together with their multiplicities, that can be represented in the following schematic form[6] as the multiset

$$(\underbrace{\frac{1}{2}, \ldots, \frac{1}{2}}_{n_1}, \underbrace{1, \ldots, 1}_{n_2}, \ldots, \underbrace{\frac{k_{max}(a)}{2}, \ldots, \frac{k_{max}(a)}{2}}_{n_{k_{max}}}).$$

The number of different sequences obtained from each of these multisets by reordering its elements is immediately given by the multinomial coefficient

$$\frac{(\sum_{k=1}^{k_{max}(a)} n_k)!}{\prod_{k=1}^{k_{max}(a)} n_k!}.$$

The degeneracy originating in this type of relabeling is responsible for many of the features of the black hole degeneracy spectrum and, in particular, the staircase structure of the black hole entropy in LQG discovered in [4]. This completes the second step mentioned above.

Once we have identified all the possible sequences of positive half-integers satisfying the area condition in (1) step three consists of finding out how many of them satisfy the so-called projection constraint

$$\sum_{i=1}^{n} m_i = 0 \tag{5}$$

for each of the sequences of positive half-integers $\{|m_1|, \ldots, |m_n|\}$ obtained in the previous two steps. In fact we will consider the slightly more general problem of solving

$$\sum_{i=1}^{n} m_i = p, \quad p \in \mathbb{Z}/2 \Leftrightarrow \sum_{i=1}^{n} k_i = 2p, \quad k_i := 2m_i.$$

[5] At least according to the most simple choices for its dynamics.

[6] If some n_k is zero then there are no terms with the number $k/2$.

There are several approaches to solve this problem that can be found in the literature. Some of them are interesting because they suggest a connection of the present problem with conformal field theories [11] and problems arising in quantum computing [12]. Here I will give a simple solution based on the use of generating functions[7] Let us consider then that we have a multiset of positive half integers $k_1/2, k_2/2, \ldots, k_n/2$ ($k_i \in \mathbb{N}$) with multiplicities given by $n_i \in \mathbb{N}$ and take the following function of the variable z (a Laurent polynomial)

$$\prod_{i=1}^{n}(z^{k_i} + z^{-k_i})^{n_i}. \tag{6}$$

By expanding it in powers of z it is easy to see that the coefficient of the power z^{2p} is, *precisely*, the number of different ways to distribute signs among the elements of the multiset of k_i in such a way that

$$\sum_i k_i = 2p.$$

In particular if we look for the constant term in (6) we get the number of solutions for the proyection constraint that can be built from a given multiset. A convenient way to extract this type of information is by using Cauchy's theorem. This allows us to extract these coefficients (for a given p) by computing the contour integral

$$\frac{1}{2\pi i}\oint_\gamma \frac{\mathrm{d}z}{z^{p+1}}\prod_{i=1}^{n}(z^{k_i} + z^{-k_i})^{n_i} \tag{7}$$

where γ is an index 1 curve surrounding the origin. By choosing for it a unit circumference the previous integral can be written in the following useful alternative form

$$\frac{2^{N-1}}{\pi}\int_{-\pi}^{\pi}\mathrm{d}\theta\cos p\theta\prod_{i=1}^{n}\cos^{n_i}k_i\theta \tag{8}$$

where N denotes the number of elements in the multiset (considering multiplicities).

Up to this point we have given a procedure to compute what we call the *black hole degeneracy spectrum*. This is defined, for every value of the area spectrum, as the number of sequences of non-zero half-integers satisfying the two conditions

$$\sum_{i=1}^{n} m_i = 0, \quad \sum_{i=1}^{n}\sqrt{|m_i|(|m_i|+1)} = \frac{a}{2}.$$

The addition of the degeneracies corresponding to all the possible values of the area spectrum smaller or equal than the one for which the entropy is computed is the last remaining step to carry our program to completion. This can be done in principle by repeating the previous procedure for each of the relevant area eigenvalues and adding up the results. For an equally spaced spectrum this task can be easily accomplished by using generating functions (as has been done, for example, in [16] for a simplified, equally spaced, area spectrum). However, for the general case that we are considering here, one has to resort to more complicated methods based on the use of functional equations or a combination of generating functions and integral transforms [8, 9]. Here we only quote the result for completeness.

$$\exp S_{SU(2)}(a) = \frac{1}{8\pi^2 i}\int_0^{4\pi}\int_{x_0-i\infty}^{x_0+i\infty} s^{-1}\Big(1 - 2\sum_{k=1}^{\infty} e^{-s\sqrt{k(k+2)}}\cos\frac{\omega k}{2}\Big)^{-1} e^{as}\,\mathrm{d}s\,\mathrm{d}\omega\,, \quad a \geq 0$$

[7] Due precisely to this fact this solution is especially suited to obtain the black hole generating functions of [10].

where x_0 is a real number larger than the real part of all the singularities of the integrand in the previous expression. From the last formula it is especially easy to find the expression corresponding to the choice of $SO(3)$ [6] as the internal symmetry group: it suffices to restrict the values of the label k appearing in the sum in the denominator of the integrand to even values.

$$\exp S_{SO(3)}(a) = \frac{1}{8\pi^2 i}\int_0^{4\pi}\int_{x_0-i\infty}^{x_0+i\infty} s^{-1}\Big(1-2\sum_{k=1}^{\infty} e^{-s\sqrt{k(k+1)}}\cos\omega k\Big)^{-1} e^{as}\,\mathrm{d}s\,\mathrm{d}\omega\ , \quad a\geq 0$$

This result can also be derived either by the functional methods originally used by Meissner to study the entropy or by the use of generating functions of the type developed in [10]. Though the value of the Immirzi parameter changes [6], the staircase structure of the entropy survives (with a periodicity different from the one corresponding to the $SU(2)$ case [17]).

Finally the black hole generating function corresponding to the choice of either $SU(2)$ or $SO(3)$ as the internal group is

$$G(z, x_1, x_2, \ldots) = \left(1-\sum_{I=1}^{\infty}\sum_{m=1}^{\infty}(z^{k_m^I}+z^{-k_m^I})x_I^{y_m^I}\right)^{-1}, \qquad (9)$$

where the variables $x_1, x_2\ldots$ are associated to the different square-free numbers p_I, z is an additional variable, and the (k_m^I, y_m^I) are the solutions to the Pell equation. In the case of working with $SU(2)$ all the solutions should be taken, whereas for $SO(3)$ only even values of k_m^I (for each p_I) have to be considered. The coefficient of the term $z^0 x_1^{q_1}\cdots x_r^{q_r}$ tells us the value of the exponential of the entropy $e^{S(a)}$ for an area a given by $a=\sum_I q_I\sqrt{p_I}$.

In conclusion the results and techniques developed to study in detail black hole entropy in loop quantum gravity can be easily adapted to work with internal gauge groups different from the standard $SU(2)$ choice. This allows to study the kind of microstructure found in the black hole spectrum for this alternative choices, in particular it is straightforward to write down an integral expression for the entropy in the form of an inverse Laplace transform and also to derive the corresponding generating functions along the lines of [10].

Acknowledgments

The author wants to thank Rutwig Campoamor for his kind invitation to participate in the Higher Symmetries in Physics Workshop. This work was supported by the Spanish MICINN Project FIS2008-03221 and the Consolider-Ingenio 2010 Program CPAN (CSD2007-00042).

References

[1] M. Domagala and J. Lewandowski, Class. Quant. Grav. **21**, 5233 (2004).
[2] J. D. Bekenstein, Lett. Nuovo Cimento **11**, 467 (1974). V. F. Mukhanov, JETP Letters **44**, 63 (1986). J. D. Bekenstein. *Black Holes: Classical Properties, Thermodynamics and Heuristic Quantization*, in Cosmology and Gravitation, M. Novello, ed. (Atlantisciences, France 2000), pp. 1-85. arXiv: gr-qc/9808028
[3] A. Ashtekar and J. Lewandowski, *Background Independent Quantum Gravity: A Status Report*, Class. Quant. Grav. **21**, R53 Class. Quant. Grav. **21**, R53 (2004).
[4] A. Corichi, J. Diaz-Polo, and E. Fernandez-Borja, Phys. Rev. Lett. **98**, 181301 (2007). A. Corichi, J. Diaz-Polo, and E. Fernandez-Borja, Class. Quant. Grav. **24**, 243 (2007).
[5] I. Agulló, J. F. Barbero G., E. Fernandez-Borja, J. Díaz-Polo, and E. J. S. Villaseñor. Phys. Rev. Lett. **100**, 211301 (2008).
[6] C-H. Chou, Y. Ling., C. Soo, and H-L. Yu. Phys. Lett. **B637**, 12 (2006).
[7] A. Ashtekar, J. Baez, A. Corichi, and K. Krasnov, Phys. Rev. Lett. **80**, 904 (1998). A. Ashtekar, J. Baez, and K. Krasnov, Adv. Theor. Math. Phys. **4**, 1 (2000).
[8] K. Meissner, Class. Quant. Grav. **21**, 5245 (2004).

[9] J. F. Barbero G. and E. J. S. Villaseñor. Class. Quant. Grav. **26**, 035017 (2009).
[10] J. F. Barbero G. and E. J. S. Villaseñor. Phys. Rev **D77**, 121502(R) (2008).
[11] I. Agullo, J. Diaz-Polo, and E. Fernandez-Borja. Phys. Rev **D77**, 104024 (2009).
[12] H. De Raedt, K. Michielsen, K. De Raedt, and S. Miyashita, Phys. Lett. **A290**, 227 (2001).
[13] A. Ashtekar, A. Corichi, and K. Krasnov. Adv. Theo. Math. Phys. **3**, 418.
[14] J. F. Barbero G. J. Lewandowski and E. J. S. Villaseñor. In preparation.
[15] D. M. Burton. *Elementary Number Theory*. McGraw-Hill, New York (2002).
[16] H. Sahlmann, Class. Quantum Grav.**25**, 055004 (2008).
[17] I. Agullo, J. Diaz-Polo, and E. Fernandez-Borja. In preparation.

Journal of Physics: Conference Series **175** (2009) 012006 doi:10.1088/1742-6596/175/1/012006

Vafa's Approach: F-Theory and GUTs

Luis J. Boya†

† Dpto. Física Teórica
Facultad de Ciencias
Universidad de Zaragoza
E-50009 Zaragoza, Spain

E-mail: luisjo@unizar.es

Abstract. After a short mention of String- and M- and F-Theory, we introduce the new methods of Vafa (2008), which aim to understand some features of the Standard Model (without gravity), by following a "bottom-up" approach in F-Theory. We recall briefly the role of the 7-Branes, which encode the gauge groups and can generate chiral matter in their intersections. Prospects for a realistic Grand Unified Theory (GUT) are also pointed out.

1. Review of String Theory

The *aim* of String Theory is to deduce the Standard Model (SM) of particles and forces from scratch, including gravitational interactions; the fundamental objects are *strings*, not point particles. Starting with an inglorious birth in 1968, String Theory was proposed first as a possible theory for strong interactions; but by 1972 we had Quantum Chromodynamics (QCD) for hadrons, and String Theory was proposed then as a theory of everything (essentially because it included gravitation), (Yoneda; Scherk and Schwarz; 1973/75). A bright moment arose in 1984, when Green and Schwarz proved that the absence of anomalies selected dimensions and gauge groups, to the extent that there were few viable so-called SuperString theories. By early 1985, the situation was: there were only FIVE SuperString theories, all living in 10 dimensions, all included gravitation, all were supersymmetric on the worldsheet (2 dimensions) *and* in the target space (10 dimensions):

(1) Type IIA: [closed, nonchiral, $\mathcal{N} = 2$ Susy, no gauge group].
(2) Type IIB: [closed, chiral, $\mathcal{N} = 2$ Susy, no gauge group].
(3) Heterotic Exceptional H-E: [closed, $\mathcal{N} = 1$ Susy, $E_8 \times E_8$ gauge group].
(4) Heterotic Orthogonal H-O: [closed, $\mathcal{N} = 1$ Susy, $O(32)$ gauge group].
(5) Type I: [open, $\mathcal{N} = 1$, $O(32)$ gauge group].

Here $\mathcal{N}$ is the number of Supercharges. Notice the exotics: besides working only (!) in $10 = (9, 1)$ dimensions, the gauge groups were also very peculiar (the largest exceptional Lie group E_8 by duplicate, or $SO(32)$, which has the same dimension (496) and rank(16)).

The favourite *girl* was the Heterotic Exceptional SuperString, as E_8^2 could give rise, in principle, to E_6 (with another hidden E_8 sector) and other Grand Unified Theory Groups, down to the Standard Model (SM) gauge group $SU(3) \times SU(2) \times U(1) \approx$ "3, 2, 1". The method was

compactification (dimensional reduction) from 10 to our mundane 4 dimensions, by a compact complex manifold of type Calabi-Yau (CY_3), as to provide only $\mathcal{N} = 1$ Supersymmetry in four dimensions:
10-dimensional Heterotic-Exceptional Theory [$\mathcal{N} = 1 \Leftrightarrow 16$ elementary Supercharges, group E_8^2] descending to 4-dim (Minkowski space), $\mathcal{N} = 1$; 4 elementary Supercharges, gauge group "3, 2, 1". This might be achieved by means of a Calabi-Yau space with holonomy $SU(3)$, allowing a surviving spinor (as $SU(3) \subset SU(4) = Spin(6)$, $6 = 10 - 4$), so to mantain Susy $\mathcal{N} = 1$ down to 4 dimensions to allow for chiral couplings.

No model was realistic, but Candelas and coll. in Austin, TX, discovered a new phenomenon, *Mirror Symmetry*, of enormous interest in pure mathematics [4]. However, this heterotic model will play a role in our story.

Best reference for this "classical" period of SuperStrings is the book [3].

2. M- and F- Theories

In 1995 Paul Townsend, Ed Witten and Joe Polchinski established "M-Theory", linking the 10-dimensional five SuperStrings among themselves and also with Supergravity in 11-dimensions and allowing for extended objects other than strings:

1) Supergravity in 11 dimensions was known since 1978 (Cremmer, Julia and Scherk), and its particle content

$$h \text{ graviton}, \ 44, \quad \Psi \text{ gravitino}, 128, \quad C \text{ (3-form}, 84) \tag{1}$$

became the particle content of the low-energy limit of the IIA theory by descending to ten dimensions: $h \to (h + A + \Phi)$, or $44 \to 35 + 8 + 1$; $C \to (B + C)$, etc. Besides, the membrane "$M2$" seen in M-theory became the fundamental string in 10-dimensions, after wraping a circle S^1. *Viceversa*, and that was really the big breakthrough, Witten showed explicitly (March, 1995) how the strong limit of the IIA-theory in 10 dimensions generated a 11-th one, becoming M-Theory by definition.

2) M-theory compactified in a segment *could* be equivalent to the H-E theory (Horawa and Witten, 1996).

3) Besides, it was already known that by T-duality (descending to 9 dimensions) the IIA and IIB theories were equivalent (there is no chirality in odd dimensions), and so were the two heterotic strings, H-E and H-O.

4) Also, Polchinsky and Witten proved that the strong limit of the H-O theory corresponds to the weak limit of the Type I (open) string: recall they share the same gauge group, O(32).

This unique and ill-defined theory, encompassing the five superstring theories, and whose low-energy particle limit was coincidental with the particle content of 11-dim SuperGravity, and containing the $M2$ brane and other extended objects, was christened M-Theory, and it was, for a (short) time, the "Gran Esperanza Blanca". It included, besides strings, other extended objects, among them the D-p-Branes, radiating $(p + 1)$-forms, which played an important role as terminal endings of open strings. The best reference here is the preprint book [5]

Now, in mid-2009, M-theory has not lived up the expectations: The Standard Model is not recovered, neither some of its conspicuous features explained, as number of generations, chiral matter, Grand Unification, the very different mass scales, etc.; neither has been any clarification of cosmological issues, inspite of several attemps.

Before we delve in the extension to 12-dim theories (originally from C. Vafa, 1996), we need to show the particle content of the diverse superstring theories.

The primordial theory is a $\mathcal{N}=1$ Yang-Mills gauge theory in 10 dimensions; no counting the gauge group, the symmetry is $8_v = 8_s$, namely the irreducible representation vector 8_v of $O(8)$ and the two spinor representations (*irreps*) 8_s, $s = L$ or $s = R$ of $Spin(8)$ are equivalent under triality. From this situation we construct *four* supersymmetric theories:

1) IIA Theory: $(8_v - 8_L) \times (8_v - 8_R) = \text{NS sector} + \text{RR sector} - \text{Fermi sector} = 35\,(\text{h, graviton}) + 28(\text{B, } 2-\text{form}) + 1(\Phi,\ \text{dilaton}) + 56(\text{C}, 3-\text{form}) + 8(\text{A}, 1-\text{form}) - \text{fermions}$

2) IIB Theory: $(8_v - 8_L) \times (8_v - 8_L) = (h + B + \Phi + D^+ + B' + \Phi' - \text{fermions})$

3) Type I and heterotic: $(8_v - 8_s) \wedge (8_v - 8_s) = (h + B + \Phi, - \text{fermions})$

The Neveu-Schwarz (NS) sector is the Bose $\times$ Bose product, the Ramond (RR) sector is Fermi $\times$ Fermi product; the fermions come from the Bose $\times$ Fermi product. The symmetrizing symbol ($\wedge$) means: symmetrize the bose product and antisymmetrize the fermi product.

11-dim SuperGravity has the particle content as the IIA theory, but seen from eleven dimensions:

4) 11d Supergravity: 44 (h, graviton) + 84 (C, 3-form) -128 (Ψ, gravitino).

As particles couple to paths, strings couple to surfaces (world-sheet), and in general even dimensional p-*branes* couple to odd (p+1)-forms. In particular, in IIA theory, we have even-dimensional Branes, in IIB odd ones. The 7-Brane of IIB comes from dualizing the axion (or RR scalar, a 0-form), and it will play an important role.

The 3-form C in 11-dim SuGra couples to a 2-Brane (called $M2$-Brane), which becomes the fundamental string of the IIA theory after the descent $11d \to 10d$.

Notice there is *no* direct route to IIB from M-theory: it is only via compactification to 9 dimensions that IIA and IIB are duals of each other. This (and other reasons) prompted C. Vafa in 1996 to introduce F-Theory:

There is a "fundamental" theory in 12 dimensions, becoming M-theory with circle (S^1) compactation, but also it becomes IIB theory with torus ($T^2 = S^1 \times S^1$) compactation:

$$T^2 \to \text{"F"} - \text{Theory} \to \text{IIB} \quad \text{(elliptic fibration)} \tag{2}$$

The two scalars (a, Φ) in IIB combine to parameterize the Torus T^2, $\chi = a + i\exp(\Phi)$, as they vary along the 10-dim IIB theory. There was much reluctance to accept F-Theory at face value: for one thing, to keep the number of supersymmetries, F-Theory has to include two time dimensions. For another, the whole particle content is unclear. The most conservative approach is to consider just the IIB theory, with the 2-Torus to vary over the surface of the 10-dim theory, allowing even the "fibers" to shrink at some points: this is the difference between fiber bundle and fibration, which is the case here. Notice the Torus T^2 is the only oriented surface with

Journal of Physics: Conference Series **175** (2009) 012006 doi:10.1088/1742-6596/175/1/012006

Euler number = 0, hence the only one that can accept a nonzero vector field, that is, it admits a metric with $(1, 1)$ signature, as F-Theory would imply perhaps two time dimensions.

There are several additional reasons to believe in F-theory [1]:

1) There is a Chern-Simons "boundary" term $(dC) \cdot (dC) \cdot C$ in the Lagrangian of 11d SuGra, which is best understood as coming from a conventional term $(dC)^3$ in 12 dimensions.

2) There is $(2, 2)$ Brane extant in the Brane scan, which lives naturally in $12 = (10, 2)$ dimensions, and generates the $M2$ Brane in eleven and the string in ten dimensions.

3) The algebra of 32 elementary supercharges of 11d SuGra still operates in 12 dimensions, as $2^{12/2-1} = 32$, and the $(10 - 2 = 8)$ type is real, so there is no need to enlarge the supersymmetries of the system.

4) We remarked above that $|8_v - 8_s|^2$ gives the particle content of 11-dim Sugra. Well, it turns out that another squaring $(44(h) - 128(\Psi) + 84(C))^2$ [6] gives the content of a putative 27-plet of particles of F-Theory. Details can be found in [1].

5) Compactification from 10d to 4d via CY_3 is complemented with compactation from 11 to 4 via manifolds with exceptional holonomy G_2, but also (and finally!) compactation from 12 to 4 can occur either with CY_4 manifolds, or $Spin(7)$ exceptional holonomy manifolds [7]. It is remarkable the (hidden) relation to octonions: $Spin(7)$ leaves the S^7-sphere of unit octonions invariant, G_2 leaves the S^6-sphere of unit imaginary octonions fixed, and $SU(3)$ leaves the equator of the later (S^5) also invariant.

Best reference for this "modern" period of strings are the Lectures [2].

3. The New Theory of Vafa (2008)

The philosphy of the new approach of Cumrum Vafa is based on two main ideas:

1) *Bottom-up* method: do not get trapped in the labyrinth of infinite vacua landscapes ($\approx 10^{500}$!!) which occur in the latest versions of string theories, and try instead to retrieve some of the features of the Standard Model (SM) model.

2) Please decouple *gravitation*: totally insignificant for microscopic physics; even the possible GUT scale, 10^{16} GeV, is far from the scale of Planck, 10^{19} GeV. This philosophy is totally contrary to the old wisdom, in which the automatic presence of gravitation was one of the atractiveness of string models (see our Sect.1). One of the present reasons to neglect gravity with respect to the other forces is this: people are getting more and more convinced that the extreme UV limit of gravitation would require new physics (for example, non-commutative geometry), in order to cope with exotic phenomena, like black holes and that like.

Vafa addresses directly the Grand Unification (GUT) issue: how the gauge groups of the standard model $SU(3) \times SU(2) \times U(1)$ would fit in a higher gauge group. He observes the "3, 2, 1" group is close to the E_3 group $(SU(3) \times SU(2))$, which appears naturally in the E_8 to $SU(2) = A_1 = E_1$ chain. He also addresses another important feature of the SM: the origin of chiral matter, a conspicuous aspect of nature, implying that matter multiplets are not invariant under parity; we even observe maximal parity violation in the beta decay sector of the electroweak theory.

There are three basic assumptions in Vafa's theory:

1) Low-scale 4-dim $\mathcal{N} = 1$ *Susy* is present at energies around the TeV (hence testable soon at the forthcoming accelerator LHC). Is this evidence for the Minimal Supersymmetric Standard Model (MSSM)?

2) At high energies, the matter content of the MSSM unifies into a GUT (Grand Unified Theory) scheme: either $SU(5)$, the original Georgi-Glashow invention (1974), or the offsprings $SO(10)$ (Fritsch and Minkowski) or E_6 (Gürsey, Ramond) [16]. Partial evidence for this stems from high-energy coupling constant unification, pointing out to $E_{GUT} \approx 10^{16}\ GeV$, well below the $10^{19}\ GeV$ of Plancks scale. This leads to the third assumption:

3) There is a well-separated limit

$$M_{GUT}/M_{Pl} \to 0 \tag{3}$$

which is the numerical reason for the decoupling of gravity (Cfr. remarks above).

The starting point to concrete calculations in this new Vafa's theory is the following: Take F-Theory, in the above sense, as a variable fibration of the 10-dimensional IIB theory by a 2-Torus. The main ingredient is *strong duality*, implying for example equivalence with the heterotic string suitably compactified. For example

$$F - Theory/K3 \quad \approx Het - Excep/T^2 \tag{4}$$

We recall $K3$ is the (topologically) unique manifold with $SU(2)$ holonomy, i.e. CY_2. How on earth the gauge group E_8^2 of the heterotic string is hidden in this compactation of F-theory? The answer is that it is hidden in the $K3$ manifold: the Hirzebruch signature of $K3$ is $(19, 3)$, and this 16 is related to the two E_8 groups!

How F-Theory (seen as a IIB Theory in 10 dimensions with a variable complex coupling χ) generates the physics can be seen in the following general scheme:

Gravitation (that we neglect) is extant in ten dimensions. Reduction to 8 makes it appear the all-important 7-Brane, with the time as the 8-th direction. This brane can be seen as dual to the RR scalar, the axion a.

In the main body of the 7-Brane the gauge groups emerge as singularities of the fibration: the membrane wraps around some cycles of the compactifying manifold; when the cycles collapse, the brane becomes massless, and hence supports gauge groups, which are also massless: there is a well-known relation between ADE singularities and the emergence of the ADE Lie groups (Arnold). Here A are the unitary groups, D the even orthogonal, E the exceptional E-groups.

Now intersection of 7-Branes gives rise, in 6 dimensions, to the presence of chiral matter (we do not enter into the details). Finally, descent to realistic, 4-dimensions makes it appear the Yukawa couplings among the scalars and fermi matter.

Here are some references for this new Vafa's theory: [8, 9, 10, 11].

4. Descent by even-dimensional manifolds

This is a synoptic scheme of the several phases on the descent from the hypothetical F-Theory living on 12 dimensions to the realistic, 4-dimensional spacetime; the process is better understood in steps of *two* dimensions each:

1) In twelve dimensions we have the realm of F-Theory. Here gravitation is supposed to dominate, but indeed the particle content is obscure. Best proposal seems to be a supersymmetric 27-plet of particles associated to the symmetric space $E_6/Spin^c(10)$ [1], which is the complexification of $F_4/Spin(9)$, the determinant of the trio of particles in 11-dim SuGra [12]. For the future.

2) Ten dimensions. The realm of the original IIB theory. Descent by an elliptic fibration, that is, F-Th/Ell $\approx IIB$, where "Ell" is a Torus (an elliptic curve): this is close to conventional string theory. The theory is self-dual: there is a discrete $SL_2(Z)$ symmetry, with the half-4-form and the graviton fixed, and with varying complex parameter $z = a + i\ exp(\Phi)$ undergoing an homographic transformation.

3) Eight dimensions. The realm of the space of the D-7-Brane. It could be understood directly as a $K3$ compactation from 12 dimensions, with $K3$ itself an elliptic fibration. When the fibration degenerates, as a membrane wraps around one of the 22 two-cycles of $K3$, there appear ADE-type singularities, and the cycles collapse so there are massless excitations, signalling the generation of gauge groups: this is the correspondence between singularity types and ADE gauge groups. Exceptional groups are favoured, as well as spin representations. (For example, the E_7 group may appear by combining the $SU(8)$ group with its four-form representation: $8^2 - 1 + 70 = 133$. Also a dual version consists on a stack of collapsing D-7 Branes on top of each other, with the connecting strings shrinking: this duality has been shown by Sen [14].

4) Six dimensions. Here the string duality dominates, one of the possible forms being similar to the IIB-Heterotic duality mentioned above, $IIB/K3 \approx HE/T^4$. In particular, the E_8^2 gauge group in the HE theory emerges from the hidden homology of $K3$; see above. From the physical point of view, two $D7$-Branes intersect also in six dimensions, and this generates, by an obscure procedure, chiral matter with bi-fundamental representations, another genuine feature of the standard model SM.

5) In the final descent to realistic four dimensions, we witness the appearance of Yukawa couplings, completing the ingredients of the SM. Hopefully one should describe a $\mathcal{N}$=1 Supersymmetric theory, at a scale of the order of the TeV: this is a prediction for the forthcoming experiments with the LHC at CERN. So we expect $1 + 4$ Higgses and plenty of Susy partners... Finally, even Susy is broken, and we have the bare SM at energies of 100 GeV, with no trace of Susy.

6) There are still many open problems, besides the problematic character of some of the above steps; after excluding gravitation, the most sensible absent feature is perhaps the unexplained existence of the three generations of quarks and leptons; recently some progress is being made on this issue also, but by a totally different approach: finite nonabelian groups as remnants of another broken gauge symmetry [15].

It should be clear to the reader that this is a wishful thinking development, although each step exhibits a well-defined mathematical structure. The general scheme seems clear, with the main ingredients fairly well understood: generation of gauge groups, presence of chiral matter and Yukawa couplings, similar to hoped-for Higgs couplings. Details will keep theoreticians busy for (many?) years to come.

Acknowledgements

I thank the organizers of the Madrid meeting, especially R. Campoamor, for the very stimulating workshop. Discussions with J. Figueroa-O'Farrill (in Madrid) and with C. Luhn (in Zaragoza) are dutyfully acknowledged.

References

[1] Boya L J 2006 *Mod. Phys. Lett. A* **21** 287
[2] Vafa C 1997 *Lectures on Strings and Dualities*, arXiv 97 92 201
[3] Green M, Schwarz J and Witten Ed 1985 *Superstring Theory*, (Cambridge U.P., Cambridge)
[4] Candelas P et al. 1993 *Nucl. Phys.* **B 407** 115
[5] Duff M. (ed.) 2000 *The World in Eleven Dimensions* (Inst. of Phys.,)
[6] Bars I 1995 *Phys. Rev.* **D 52** 3567
[7] Joyce D 2001 *Compact Manifolds with Special Holonomy.* (Oxford U.P., Cambridge)
[8] Vafa C et al. 2008, arXiv 08 02 3391 (hep-th).
[9] Vafa C et al. 2008, arXiv 08 06 0102 (hep-th).
[10] Vafa C et al. 2009, arXiv 09 04 3101 (hep-th).
[11] Wijnholt M 2008, arXiv 08 09 3878 (hep-th)
[12] Teparkson T and Ramond P 1999 *Phys. Rep.* **315** 137
[13] Northcott D G 1960 *Homological Algebra*, (Cambridge U.P.: Cambridge)
[14] Sen A 1999, arXiv hep-th 99 05 120.
[15] Luhn C et al. 2008 *Phys.Rev.* **D77** 056002
[16] Zee A (ed.) 1982 *Unity of Forces in the Universe* (World Scientific: Singapore)

Workshop on Higher Symmetries in Physics IOP Publishing
Journal of Physics: Conference Series **175** (2009) 012007 doi:10.1088/1742-6596/175/1/012007

Superintegrable Hamiltonian Systems: an algebraic approach

J. A. Calzada [1], J. Negro [2] and M. A. del Olmo [2]

[1] Departamento de Matemática Aplicada,
[2] Departamento de Física Teórica,
Universidad de Valladolid, E-47011, Valladolid, Spain

E-mail: juacal@eis.uva.es, jnegro@fta.uva.es, olmo@fta.uva.es

Abstract. Using an algebraic approach in terms of a complete set of shape-invariant intertwinig operators we study a family of quantum superintegrable hamiltonian systems. These intertwining operators close a certain Lie algebra in such a way that the eigenstates of these Hamiltonians belong to some of its unitary representations.

1. Introduction

Let us consider a Hamiltonian system living in a n-dimensional configuration space. The notion of integrability of this system is related with the existence of constants of motion allowing to solve or reduce the equations of motion by quadratures. In this way, a Hamiltonian system is said to be integrable (or completely integrable) if there are n constants of motion including the Hamiltonian H, which are independent and in involution. Incidentally, the maximal number of constants of motion, being the Hamiltonian one of them, functionally independents in involution is n for a Hamiltonian system whose (symplectic) phase space is $2n$-dimensional. But, if it has $0 < k \leq n-1$ additional constants of motion the system is called superintegrable and when there exist $2\,n{-}1$ invariants (well defined in phase-space) we say that the system is maximally superintegrable. In this case there exist more than one set of n motion constants, including the Hamiltonian itself, independent and in involution. The constants of motion are related with (partially hidden) symmetries of the corresponding physical system. Normally these properties of the superintegrable Hamiltonian systems (SHS) are exhibited by the possibility of separation of variables in more than one coordinate system for the Hamilton-Jacobi equation in the classical case and for the Schrödinger equation in the quantum case, and by the fact that the finite classical trajectories are closed (periodic) while the discrete energy levels are degenerate in the quantum case.

Obviously, in the real world the rule is chaotic systems and physical systems exhibiting complete integrability are exceptional. A list of SHS can be found in the classification works by Evans [1]. In the last years superintegrability in constant curvature configuration spaces has been studied in a systematic way [2, 3, 4]. However, we can also mention two former references, among a long list of contributions in relation with some generalizations of SHS to constant curvature configuration spaces. Lakshmanan and Eswaran [5] analyzed the isotropic oscillator on a 3-sphere and Higgs [6] studied versions of the Coulomb potential and of the harmonic oscillator living in the N-dimensional sphere and having $SO(N+1)$ and $SU(N)$ symmetry in classical and in quantum mechanics, respectively.

From an algebraic point of view, we can also mention the works by Daskaloyannis and collaborators [7, 8] where they present an approach to superintegrability using deformed algebras. Also Ballesteros and collaborators have obtained new completely integrable systems and recover others well-known from a coalgebra approach [9] (see [10] in this volumen for a review).

Another case of algebraic approach to integrable systems is the construction of a family of SHS from a group-theoretical method based on the (Marsden–Weinstein) symmetry reduction [11]. These systems were obtained from free systems in $\mathbb{C}^{p,q}$ with an original $u(p,q)$-symmetry [12]

$$H = \frac{c}{4} g^{\bar{\mu}\nu} \bar{p}_\mu p_\nu \quad \stackrel{\text{MW reduction}}{\longrightarrow} \quad H^r = \frac{c}{4} g^{\mu\nu} p_{s^\mu} p_{s^\nu} + V(s), \tag{1}$$

(the bar stands for the complex conjugate) with $V(s)$ a potential in terms of the real coordinates (s^μ). The constant curvature configuration spaces of these SHS are $SO(p,q)$-homogeneous spaces. Many aspects of these SHS have been studied using standard procedures [12]–[16] and their quantum versions are well known (for instance, the Morse potential [17] and the Pöschl–Teller potential [18]). In this work we present a review of a new framework based on intertwining operators (IO), a form of Darboux transformations [19], that will allow us to study them from an algebraic point of view. The interest is that the IO's obtained by this procedure close certain Lie algebras that show the symmetry properties, some times hidden, of the system under consideration and permit to describe these SHS in terms of representations of such "intertwining symmetry" algebras (or "dynamical algebras" [20]).

The intertwining operators (first order differential operators) A, connect different Hamiltonians, H, H', belonging to the same hierarchy

$$AH = H'A.$$

From this family of SHS a complete set of such IO's, in the sense that any of the Hamiltonians of the hierarchy can be expressed in terms of these operators, may be obtained. The IO's are associated to separable coordinate systems for the Hamiltonians. The algebraic analysis of the IO's associated to integrable Hamiltonians has been made, for instance, in Refs. [21, 22, 23]. The quantum physical systems factorized in terms of IO's have been till now one-dimensional systems. Recently, we introduced a natural extension to higher dimensions of the intertwining (Darboux) transformations of the Schrödinger equation for one-dimensional quantum systems [24], since the Hamiltonians of this family of SHS originate a coupled set of n–diferential equations, when a system of separable coordinates is used, which can be factorized one by one. We have used this approach to study two particular, but well representative cases, of this SHS family: a system [25] with "$u(3)$-symmetry" and other one with "$u(2,1)$-symmetry" [26] (see also for both cases [27]). As the reader will see later the generalization to systems with higher "$u(p,q)$-symmetry" is evident. This "$u(p,q)$-symmetry" is interpreted in the sense of (1), i.e., the original system before symmetry reduction has $u(p,q)$-symmetry. The case presented here is a $u(2,1)$-system, i.e., a quantum system with configuration space a $SO(2,1)$-homogeneous space (2D hyperboloid of two-sheets). We will be able to obtain a wide set of IO's closing a $u(2,1)$ dynamical Lie algebras. However, there is also a "hidden symmetry" since considering discrete symmetry operators we can enlarge $u(2,1)$ to $so(4,2)$.

Note that we have recovered the algebraic structure of the system that was involved in the Marsden-Weinstein reduction procedure

$$su(p,q) \stackrel{\text{symmetry reduction}}{\longrightarrow} so(p,q) \stackrel{\text{IO factorization}}{\longrightarrow} u(p,q) \stackrel{\text{discrete symmetries}}{\longrightarrow} so(2p,2q).$$

Journal of Physics: Conference Series **175** (2009) 012007 doi:10.1088/1742-6596/175/1/012007

This algebraic approach gives a simple explanation of the main features of these SHS. It allows us to characterize the discrete spectrum and the corresponding eigenfunctions of the system in terms of irreducible unitary representations (IUR) of the intertwining symmetry algebras.

We have shown the relation of eigenstates and eigenvalues with unitary representations of the $su(2)$, $su(2,1)$ and $so(4,2)$ Lie algebras. In particular, we have studied the degeneration problem as well as the number of bound states. Here, we remark that such a detailed study for a 'non-compact' superintegrable system had not been realized till now, up to our knowledge. We have restricted to IUR's, but a wider analysis can be done for hierarchies associated to representations with a not well defined unitary character.

The IO's can also be used to find second order integrals of motion for a Hamiltonian and their algebraic relations, which is the usual approach to (super)-integrable systems. As a general consideration we can say that it is easier to deal directly with the IO's, because they are more elementary and simpler, than with constants of motion.

2. Superintegrable classical systems

Let us define the free Hamiltonian

$$H = 4\, g^{\mu\bar{\nu}}\, p_\mu\, \bar{p}_\nu, \qquad \mu,\nu = 0,\ldots,n = p+q-1, \tag{2}$$

on the (complex) configuration space

$$\frac{SU(p,q)}{SU(p-1,q)\times U(1)}.$$

This space is an Hermitian hyperbolic space with metric $g_{\mu\nu}$ and coordinates $y^\mu \in \mathbb{C}$ [28, 14] verifying that

$$g_{\bar{\mu}\nu}\, \bar{y}^\mu\, y^\nu = 1.$$

A symmetry reduction procedure by means of a maximal abelian subalgebra (MASA) of $su(p,q)$ [29] originates a (non free) real Hamiltonian on a homogeneous $SO(p,q)$-space [12, 13]

$$H = \frac{1}{2} g^{\mu\nu}\, p_{s^\mu}\, p_{s^\nu} + V(s)$$

from the original one (2). Now $V(s)$ is a potential depending on the real coordinates s^μ which satisfy $g_{\mu\nu}\, s^\mu\, s^\nu = 1$.

The symmetry reduction procedure transforms the set of complex coordinates y^μ into a set of ignorable variables x^μ and the actual real coordinates s^μ. Let us consider a basis of a MASA of $u(p,q)$ constituted only by pure imaginary matrices Y_μ $(\mu = 0,\ldots,n)$. The relation between old y^μ and new coordinates x^μ, s^μ is given by

$$y^\mu = B(x)^\mu_{\ \nu}\, s^\nu\,, \qquad B(x) = \exp\left(x^\mu\, Y_\mu\right).$$

The ignorability of the x coordinates (i.e., the vector fields corresponding to the MASA are straightened out in these coordinates) is due to the fact that the x^μ are the parameters of the transformation associated to the MASA of $u(p,q)$ used in the reduction. The Jacobian matrix, J, of the coordinate transformation $((y,\bar{y}) \to (x,s))$ is

$$J = \frac{\partial(y\,,\bar{y})}{\partial(x\,,s)} = \begin{pmatrix} A & B \\ \bar{A} & \bar{B} \end{pmatrix},$$

where

$$A^\mu_{\ \nu} = \frac{\partial y^\mu}{\partial x^\nu} = (Y_\nu)^\mu_{\ \rho}\, y^\rho\,.$$

Then, the original Hamiltonian (2) reduces to the Hamiltonian

$$H = c\left(\frac{1}{2}g^{\mu\nu}p_\mu p_\nu + V(s)\right), \qquad V(s) = p_x^T\,(A^\dagger K\,A)^{-1}\,p_x\,,$$

which depends only in the new real coordinates s, with p_x the constant momenta associated to the ignorable coordinates x and K is the matrix defined by the metric g.

3. A superintegrable $u(2,1)$-Hamiltonian system

A classical superintegrable Hamiltonian associated, for instance, to $su(2,1)$ can be constructed using the symmetry reduction procedure presented before. A suitable basis of $su(2,1)$ in terms of 3×3 matrices $X_1,\ldots,X_8$ (with metric $K = \mathrm{diag}(-1,-1,+1)$) is

$$X_1 = \begin{pmatrix} i & 0 & 0\\ 0 & -i & 0\\ 0 & 0 & 0\end{pmatrix},\quad X_2 = \begin{pmatrix} 0 & 0 & 0\\ 0 & i & 0\\ 0 & 0 & -i\end{pmatrix},\quad X_3 = \begin{pmatrix} 0 & 1 & 0\\ -1 & 0 & 0\\ 0 & 0 & 0\end{pmatrix},\quad X_4 = \begin{pmatrix} 0 & i & 0\\ i & 0 & 0\\ 0 & 0 & 0\end{pmatrix},$$

$$X_5 = \begin{pmatrix} 0 & 0 & 1\\ 0 & 0 & 0\\ -1 & 0 & 0\end{pmatrix},\quad X_6 = \begin{pmatrix} 0 & 0 & i\\ 0 & 0 & 0\\ i & 0 & 0\end{pmatrix},\quad X_7 = \begin{pmatrix} 0 & 0 & 0\\ 0 & 0 & 1\\ 0 & -1 & 0\end{pmatrix},\quad X_8 = \begin{pmatrix} 0 & 0 & 0\\ 0 & 0 & i\\ 0 & i & 0\end{pmatrix}.$$

There are four MASAS for $su(2,1)$ [29]: the compact Cartan subalgebra, the noncompact Cartan subalgebra, the orthogonally decomposable subalgebra and the nilpotent subalgebra. In this example we will consider the symmetry reduction by the compact Cartan subalgebra, but other three SHS can be obtained with the other MASAs. The compact Cartan subalgebra can be generated by the matrices

$$\mathrm{diag}(i,-i,0) \qquad \mathrm{diag}(0,i,-i).$$

However, in order to simplify the computations we use the following matrices

$$Y_0 = \mathrm{diag}(i,0,0), \qquad Y_1 = \mathrm{diag}(0,i,0), \qquad Y_2 = \mathrm{diag}(0,0,i),$$

that determine a basis for the corresponding MASA in $u(2,1)$.

The old complex coordinates y are related to the real ones s by

$$y^\mu = e^{i\,x^\mu Y_\mu}\,s^\mu, \qquad \mu = 0,1,2,$$

and the Hamiltonian can be written as

$$H = \frac{1}{2}c\left(-p_0^2 - p_1^2 + p_2^2\right) + \frac{m_0^2}{s_0^2} + \frac{m_1^2}{s_1^2} - \frac{m_2^2}{s_2^2}\,, \tag{3}$$

being c a constant. It lives in a 2-dimensional two-sheet hyperboloid $-s_0^2 - s_1^2 + s_2^2 = 1$. Non-negative real numbers m_i can be chosen for the potential constants.

The system has three independent invariants of motion

$$\begin{aligned}
R_{01} &= (s_0\,p_1 - s_1\,p_0)^2 + \left(m_0\frac{s_1}{s_0} + m_1\frac{s_0}{s_1}\right)^2,\\
R_{02} &= (s_0\,p_2 + s_2\,p_0)^2 + \left(m_0\frac{s_2}{s_0} - m_2\frac{s_0}{s_2}\right)^2,\\
R_{12} &= (s_1\,p_2 + s_2\,p_1)^2 + \left(m_1\frac{s_2}{s_1} - m_2\frac{s_1}{s_2}\right)^2,
\end{aligned}$$

that can be written in terms of the basis of $su(2,1)$ (in the realization as function of s_μ and p_μ)

$$Q_1 \equiv R_{01} = X_3^2 + X_4^2, \qquad Q_2 \equiv R_{02} = X_5^2 + X_6^2, \qquad Q_3 \equiv R_{12} = X_7^2 + X_8^2.$$

The sum of these three invariants is the Hamiltonian (3) up to an additive constant

$$H = -Q_1 + Q_2 + Q_3 + \text{cnt}. \tag{4}$$

The quadratic Casimir of $su(2,1)$ can be also written in terms of the constants of motion and the second order operators in the enveloping algebra of the compact Cartan subalgebra of $su(2,1)$

$$\mathcal{C}_{su(2,1)} = 3Q_1 - 3Q_2 - 3Q_3 + 4X_1^2 + 2[X_1, X_2]_+ + 4X_2^2.$$

The Hamiltonian is in involution with all the three constants of motion (i.e. $[H, Q_i] = 0$, $i = 1,2,3$), but the Q_i's do not commute among themselves

$$\begin{aligned} [Q_1, Q_2] &= [Q_3, Q_1] = [Q_3, Q_2] \\ &= -[X_3, [X_5, X_7]_+]_+ - [X_3, [X_6, X_8]_+]_+ + [X_4, [X_5, X_8]_+]_+ - [X_4, [X_6, X_7]_+]_+. \end{aligned}$$

So, the system (3) is superintegrable.

The solutions of the motion problem for this system can be obtained solving the corresponding Hamilton–Jacobi (HJ) equation in an appropriate coordinate system, such that the HJ equation separates into a system of ordinary differential equations.

The 2-dimensional two-sheet hyperboloid can be parametrized, for instance, on pseudo-spherical coordinates (θ, ξ)

$$s_0 = \sinh\xi\,\cos\theta, \qquad s_1 = \sinh\xi\,\sin\theta, \qquad s_2 = \cosh\xi, \tag{5}$$

with $0 \le \theta < \pi/2$ and $0 \le \xi < \infty$. Now in these coordinates and specializing $c = -1$ the Hamiltonian takes the form

$$H = \frac{1}{2}\left(p_\xi^2 + \frac{p_\theta^2}{\sinh^2\xi}\right) + \frac{1}{\sinh^2\xi}\left(\frac{m_0^2}{\cos^2\theta} + \frac{m_1^2}{\sin^2\theta}\right) - \frac{m_2^2}{\cosh^2\xi}\,.$$

The potential (Fig. 1) is regular inside the domain of the variables and there is a saddle point for the values $\theta = \arctan\sqrt{m_1/m_0}$ and $\xi = \arg\tanh\sqrt{m_2(m_0+m_1)})$ if $m_0 + m_1 > m_2$.

The explicit form of the invariants of motion, Q_i, in terms of the coordinates (ξ, θ) is

$$\begin{aligned} Q_1 &= \frac{1}{2}p_\theta^2 + \frac{m_0^2}{\cos^2\theta} + \frac{m_1^2}{\sin^2\theta}, \\ Q_2 &= \coth^2\xi\left(\frac{1}{2}p_\theta^2\sin^2\theta + \frac{m_0^2}{\cos^2\theta}\right) + \cos^2\theta\left(\frac{1}{2}p_\xi^2 + \frac{m_2^2}{\coth^2\xi}\right) \\ &\quad + \frac{1}{2}p_\theta\,p_\xi\,\sin 2\theta\,\coth\xi, \\ Q_3 &= \coth^2\xi\left(\frac{1}{2}p_\theta^2\cos^2\theta + \frac{m_1^2}{\sin^2\theta}\right) + \sin^2\theta\left(\frac{1}{2}p_\xi^2 + \frac{m_2^2}{\coth^2\xi}\right) \\ &\quad - \frac{1}{2}p_\theta\,p_\xi\,\sin 2\theta\,\coth\xi. \end{aligned}$$

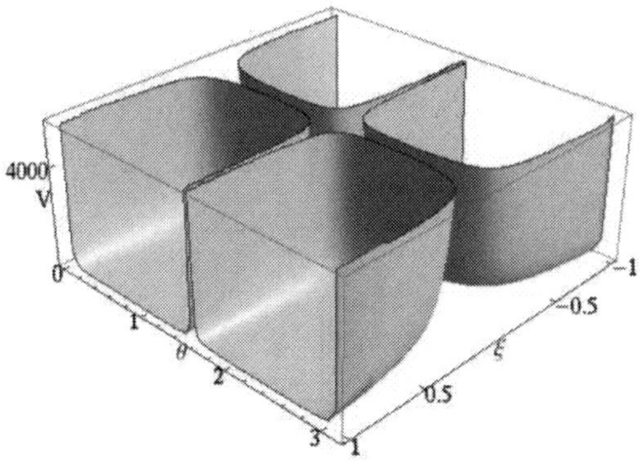

Figure 1. Graphical representation of the potential for $m_0 = 1, m_1 = 2, m_2 = 2.5$.

The Hamilton-Jacobi equation for this system is

$$\frac{1}{2}\left(\frac{\partial S}{\partial \xi}\right)^2 - \frac{m_2^2}{\cosh^2 \xi} + \frac{1}{\sinh^2 \xi}\left(\frac{1}{2}\left(\frac{\partial S}{\partial \theta}\right)^2 + \frac{m_0^2}{\cos^2 \theta} + \frac{m_1^2}{\sin^2 \theta}\right) = E,$$

which separates into two ordinary differential equations, if one takes into account that its solution can be written as $S(\theta, \xi) = S_1(\theta) + S_2(\xi) - Et$,

$$\begin{aligned} \frac{1}{2}\left(\frac{\partial S_1}{\partial \theta}\right)^2 + \frac{m_0^2}{\cos^2 \theta} + \frac{m_1^2}{\sin^2 \theta} &= \alpha_1, \\ \frac{1}{2}\left(\frac{\partial S_2}{\partial \xi}\right)^2 - \frac{m_2^2}{\cosh^2 \xi} + \frac{\alpha_1}{\sinh^2 \xi} &= \alpha_2, \end{aligned}$$

where $\alpha_2 = E$ and α_1 are the separation constants being α_1 always positive. Both equations are formally similar to those of the corresponding one-dimensional problem [15]. Their solutions can be found as particular cases of a more general approach [14]. The solutions depend on the value of the energy. So, we can distinguish three cases according with $E < 0, E = 0$ and $E > 0$ for the second Hamilton-Jacobi equation

$$\begin{aligned} E < 0 \qquad \cosh^2 \xi &= \frac{1}{2E}\left[-b_2 - \sqrt{b_2^2 + 4m_2^2 E} \sin\left(2\sqrt{2|E|}(\beta_2 + t)\right)\right], \\ E = 0 \qquad \cosh^2 \xi &= \frac{m_2^2}{m_2^2 - \alpha_1} + 2(m_2^2 - \alpha_1)(\beta_2 + t)^2, \\ E > 0 \qquad \cosh^2 \xi &= \frac{1}{2E}\left[-b_2 + \sqrt{b_2^2 + 4m_2 E} \sinh\left(2\sqrt{2E}(\beta_2 + t)\right)\right], \end{aligned}$$

where $b_2 = -E - \alpha_1 + m_2^2$ and $\beta_2 = \partial S/\partial \alpha_2 - t$. The remaining Hamilton-Jacobi equation has the following solution

$$\cos^2 \theta = \frac{1}{2\alpha_1}\left[b_1 - \sqrt{b_1^2 - 4\alpha_1 m_0^2} \, \sin\left(2\sqrt{2\alpha_1}\beta_1\right)\right],$$

with $b_1 = \alpha_1 + m_0^2 - m_1^2$ and $\beta_1 = \partial S/\partial \alpha_1$.

4. Superintegrable quantum systems

In order to obtain a quantum version of the classical system described on the previous section, we can proceed as follows. Taking into account the group theoretical method of construction of the Hamiltonian (3) and the constants of motions together with the inspection of all these objects we can relate the terms like $s_\mu p_\nu \pm s_\nu p_\mu$ with generators of "rotations" when $p_\mu \to \partial_\mu$ in the plane $X_\mu X_\nu$. Since the Hamiltonian is the sum of the three constants of motion up to constants (4) we can make the following Ansatz for a quantum Hamiltonian as a linear combination of the infinitesimal generators of "rotations" in the plane X_1X_2, X_0X_2 and X_0X_1 around the axis X_0, X_1 and X_2, respectively (i.e. J_0^2, J_1^2 and J_2^2). The signature of the metric of the configuration space imposes that the rotations be compact or noncompact. Hence, the generators will span $so(3)$ or $so(2,1)$ and it is enough to take a differential realization of them. In this way, the Casimir operator of $so(2,1)$ ($C = J_0^2+J_1^2-J_2^2$) will give the "kinetic" part of the Hamiltonian. The "potential" term will be obtain directly from the potential of (3) making the change $m_i^2 \to l_i^2 - 1/4$ in order to simplify the computations. Thus, we obtain the Hamiltonian

$$H_\ell = -J_0^2 - J_1^2 + J_2^2 + \frac{l_0^2-\frac14}{s_0^2} + \frac{l_1^2-\frac14}{s_1^2} - \frac{l_2^2-\frac14}{s_2^2},$$

defined in the 2-dimensional two-sheet hyperboloid $-s_0^2 - s_1^2 + s_2^2 = 1$, with $\ell = (l_0, l_1, l_2) \in \mathbb{R}^3$, $2\,m = 1$ and $\hbar = 1$. The differential operators

$$J_0 = s_1\partial_2 + s_2\partial_1, \qquad J_1 = s_2\partial_0 + s_0\partial_2, \qquad J_2 = s_0\partial_1 - s_1\partial_0,$$

determine a differential realization of $so(2,1)$ with Lie commutators

$$[J_0, J_1] = -J_2, \qquad [J_2, J_0] = J_1, \qquad [J_1, J_2] = J_0.$$

In the coordinates (5) the expressions of the J_i are

$$J_0 = \sin\theta\,\partial_\xi + \cos\theta\coth\xi\,\partial_\theta, \qquad J_1 = \cos\theta\,\partial_\xi - \sin\theta\coth\xi\,\partial_\theta, \qquad J_2 = \partial_\theta\,,$$

and constitute a set of anti-Hermitian operators acting on the space of square-integrable functions with invariant measure $d\mu(\theta,\xi) = \sinh\xi\,d\theta\,d\xi$.

Now the Hamiltonian H_ℓ (4) has the expression

$$H_\ell = -\partial_\xi^2 - \coth\xi\,\partial_\xi - \frac{l_2^2-\frac14}{\cosh^2\xi} + \frac{1}{\sinh^2\xi}\left[-\partial_\theta^2 + \frac{l_1^2-\frac14}{\sin^2\theta} + \frac{l_0^2-\frac14}{\cos^2\theta}\right].$$

We see that it can be separated in the variables ξ and θ, by choosing its eigenfunctions Φ_ℓ as $\Phi_\ell(\theta,\xi) = f(\theta)\,g(\xi)$. Thus, we get a pair of equations

$$H^\theta_{l_0,l_1} f(\theta) \equiv \left[-\partial_\theta^2 + \frac{l_1^2-\frac14}{\sin^2\theta} + \frac{l_0^2-\frac14}{\cos^2\theta}\right] f(\theta) = \alpha\, f(\theta), \tag{6}$$

$$\left[-\partial_\xi^2 - \coth\xi\,\partial_\xi - \frac{l_2^2-\frac14}{\cosh^2\xi} + \frac{\alpha}{\sinh^2\xi}\right] g(\xi) = E\, g(\xi),$$

with $\alpha > 0$ a separation constant.

5. Algebraic approach to superintegrable quantum systems

The algebraic analysis of this quantum system involves the construction of a complete set of intertwining operators that determine algebraic structures (Lie algebras) that are closely related with the symmetry of the system.

5.1. Intertwining symmetry algebras
We can consider the differential operator associated to the first equation (6) as the one-dimensional Hamiltonian $H^{\theta}_{l_0,l_1}$. It can be factorized as a product of first order operators $A^{\pm}$ and a constant λ

$$H^{\theta}_{(l_0,l_1)} = A^{+}_{(l_0,l_1)}\, A^{-}_{(l_0,l_1)} + \lambda_{(l_0,l_1)},$$

$$A^{\pm}_{l_0,l_1} = \pm\partial_\theta - (l_0+1/2)\tan\theta + (l_1+1/2)\cot\theta,$$

$$\lambda_{(l_0,l_1)} = (l_0+l_1+1)^2 .$$

A hierarchy of Hamiltonians

$$\dots, H^{\theta}_{l_0-1,l_1-1}, H^{\theta}_{l_0,l_1}, H^{\theta}_{l_0+1,l_1+1}, \dots, H^{\theta}_{l_0+n,l_1+n}, \dots,$$

satisfying the recurrence relations

$$A^{-}_{(l_0,l_1)}\, H^{\theta}_{(l_0,l_1)} = H^{\theta}_{(l_0+1,l_1+1)}\, A^{-}_{(l_0,l_1)}, \qquad A^{+}_{(l_0,l_1)}\, H^{\theta}_{(l_0+1,l_1+1)} = H^{\theta}_{(l_0,l_1)}\, A^{+}_{(l_0,l_1)} \tag{7}$$

is constructed using the fundamental relation between contiguous pairs of operators $A^{\pm}$

$$H^{\theta}_{(l_0,l_1)} = A^{+}_{(l_0,l_1)}\, A^{-}_{(l_0,l_1)} + \lambda_{(l_0,l_1)} = A^{-}_{(l_0+1,l_1+1)}\, A^{+}_{(l_0+1,l_1+1)} + \lambda_{(l_0+1,l_1+1)}.$$

Relations (7) show that $A^{\pm}_{(l_0,l_1)}$ act as shape invariant intertwining operators and that $A^{-}_{(l_0,l_1)}$ transforms eigenfunctions of $H^{\theta}_{(l_0,l_1)}$ on eigenfunctions of $H^{\theta}_{(l_0+1,l_1+1)}$, and viceversa for $A^{+}_{(l_0,l_1)}$, such that the original and the transformed eigenfunctions have the same eigenvalue.

Free-index operators $\hat{H}^{\theta}, \hat{A}^{\pm}, \hat{A}$ are defined from the set of index-depending operators $\{H^{\theta}_{(l_0+n,l_1+n)}, A^{\pm}_{(l_0+n,l_1+n)}\}_{n\in\mathbb{Z}}$ as follows:

$$\begin{aligned}
\hat{H}^{\theta} f_{(l_0,l_1)} &:= H^{\theta}_{(l_0,l_1)} f_{(l_0,l_1)},\\
\hat{A}^{-} f_{(l_0,l_1)} &:= \tfrac{1}{2} A^{-}_{(l_0,l_1)} f_{(l_0,l_1)},\\
\hat{A}^{+} f_{(l_0+1,l_1+1)} &:= \tfrac{1}{2} A^{+}_{(l_0,l_1)} f_{(l_0+1,l_1+1)},\\
\hat{A}\, f_{(l_0,l_1)} &:= -\tfrac{1}{2}(l_0+l_1)\, f_{(l_0,l_1)},
\end{aligned}$$

where $f_{(l_0+n,l_1+n)}$ are the eigenfunctions of $H^{\theta}_{(l_0+n,l_1+n)}$. These free-index operators close a $su(2)$-algebra

$$[\hat{A}, \hat{A}^{\pm}] = \pm A^{\pm}, \qquad [\hat{A}^{+}, \hat{A}^{-}] = 2\hat{A}, \tag{8}$$

and together with the operator $Df_{(l_0,l_1)} := (l_0-l_1) f_{(l_0,l_1)}$, that commutes with the other three ones, determine a $u(2)$-algebra.

The fundamental states of some of these Hamiltonians are related with the IUR's of $su(2)$. The Hamiltonian H^{θ} can be written in terms of the Casimir of $su(2)$, $\mathcal{C} = A^{+}A^{-} + A(A-1)$

$$H^{\theta} = 4(\mathcal{C} + 1/4).$$

An eigenstate $f^{0}_{(l_0+n,l_1+n)}$ of $H^{\theta}_{(l_0+n,l_1+n)}$ will be a fundamental (highest or lowest weight) vector if it verifies the equation

$$A^{-}\, f^{0}_{(l_0+n,l_1+n)} \equiv A^{-}_{(l_0+n,l_1+n)}\, f^{0}_{(l_0+n,l_1+n)} = 0.$$

Its solution is

$$f^0_{(l_0+n,l_1+n)}(\theta) = N\cos^{l_0+1/2+n}(\theta)\ \sin^{l_1+1/2+n}(\theta),$$

(being N a normalization constant) with eigenvalue

$$E^0_{(l_0+n,l_1+n)} = \lambda_{(l_0+n,l_1+n)} = (l_0+l_1+1+2\,n)^2.$$

This solution is regular and square-integrable if $l_0, l_1 \geq -1/2$. Moreover, it is also eigenfunction of the free-index operator A

$$A\, f^0_{(l_0+n,l_1+n)} \equiv A_{(l_0+n,l_1+n)}\, f^0_{(l_0+n,l_1+n)} = -\frac{1}{2}(l_0+l_1+2n)\, f^0_{(l_0+n,l_1+n)}. \tag{9}$$

From (9) we can made the following identification

$$f^0_{(l_0+n,l_1+n)} \simeq |j_n, -j_n\rangle, \tag{10}$$

with $j_n = \frac{1}{2}(l_0+l_1+2n)$, $n = 0,1,2,\ldots$ Hence, the representation, D^{j_n}, fixed by (10), will be a IUR of $su(2)$ of dimension $2j_n+1 = l_0+l_1+2n+1$ if $l_0+l_1 \in \mathbb{Z}^+$, $n \in \mathbb{Z}^+$. The other eigenstates belonging to D^{j_n} are obtained by recursion with A^+. Thus,

$$f^n_{(l_0,l_1)} = (A^+)^n\, f^0_{(l_0+n,l_1+n)} = A^+_{(l_0,l_1)}\, A^+_{(l_0+1,l_1+1)} \cdots A^+_{(l_0+n-1,l_1+n-1)}\, f^0_{(l_0+n,l_1+n)},$$

and

$$f^n_{(l_0,l_1)} \simeq |j_n, -j_n+n\rangle.$$

The explicit form of $f^n_{(l_0,l_1)}$ is

$$f^n_{(l_0,l_1)} = \sin^{l_1+1/2}(\theta)\ \cos^{l_0+1/2}(\phi_1)\ P_m^{(l_1,l_0)}[\cos(2\,\theta)],$$

being P_m a Jacobi polynomial, with eigenvalue

$$E^n_{(l_0,l_1)} = (l_0+l_1+1+2\,n)^2,\quad n \in \mathbb{Z}^+.$$

However, different fundamental states with values of l_0 and l_1, such that $j_0 = (l_0+l_1)/2$ is fixed, would lead to the same j-IUR of $su(2)$, and different $u(2)$-IUR's (since $Df_{(l_0,l_1)} = (l_0-l_1)f_{(l_0,l_1)}$) may correspond to states with the same energy. This fact suggests the existence of a larger algebra of operators such that all the eigenstates with the same energy belong to only one of its IUR's.

Another interesting fact is that although $A^{\pm}_{l_0,l_1}$ depend only on the θ-variable, they can act also as IO's of the complete Hamiltonians H_ℓ (4) and its global eigenfunctions Φ_ℓ, leaving the parameter l_2 unchanged

$$A^-_{\ell'}H_{\ell'} = H_\ell A^-_{\ell'}, \qquad A^+_{\ell'}H_\ell = H_{\ell'}A^+_{\ell'},$$

where $\ell = (l_0,l_1,l_2)$ and $\ell' = (l_0-1, l_1-1, l_2)$. In this sense, many of the above relations can be straightforwardly extended under this global point of view.

5.2. A larger symmetry algebra: $su(2,1)$

The construction of a larger algebra starts with the recognition of other coordinate sets that allow us to obtain new sets of IO's whose associated free-index operators close new $su(2)$ or $su(1,1)$ Lie algebras.

A second coordinate set that parametrizes the hyperboloid and separates the Hamiltonian is

$$s_0 = \cosh\psi \sinh\chi, \qquad s_1 = \sinh\psi, \qquad s_2 = \cosh\psi\cosh\chi,$$

with $-\infty < \psi < +\infty$ and $0 \le \chi < +\infty$. In these coordinates the $so(2,1)$-generators take the expressions

$$J_0 = -\tanh\psi\sinh\chi\,\partial_\chi + \cosh\chi\,\partial_\psi, \qquad J_1 = \partial_\chi, \qquad J_2 = \sinh\chi\,\partial_\psi - \tanh\psi\cosh\chi\,\partial_\chi,$$

and the Hamiltonian

$$H_\ell = -\partial_\psi^2 - \tanh\psi\,\partial_\psi + \frac{l_1^2-\frac14}{\sinh^2\psi} + \frac{1}{\cosh^2\psi}\left[-\partial_\chi^2 + \frac{l_0^2-\frac14}{\sinh^2\chi} - \frac{l_2^2-\frac14}{\cosh^2\chi}\right]$$

separates in the variables ψ and χ when the eigenfunctions H_ℓ are like $\Phi(\chi,\psi) = f(\chi)\,g(\psi)$. So,

$$H^\chi_{l_0,l_2} f(\chi) \equiv \left[-\partial_\chi^2 + \frac{l_0^2-\frac14}{\sinh^2\chi} - \frac{l_2^2-\frac14}{\cosh^2\chi}\right] f(\chi) = \alpha f(\chi), \tag{11}$$

$$\left[-\partial_\psi^2 - \tanh\psi\,\partial_\psi + \frac{l_1^2-\frac14}{\sinh^2\psi} + \frac{\alpha}{\cosh^2\psi}\right] g(\psi) = E g(\psi),$$

with α a separation constant.

The Hamiltonian $H^\chi_{l_0,l_2}$ (11) is factorized as a product of first order operators $B^\pm$

$$H^\chi_{l_0,l_2} = B^+_{l_0,l_2}B^-_{l_0,l_2} + \lambda_{l_0,l_2} = B^-_{l_0-1,l_2-1}B^+_{l_0-1,l_2-1} + \lambda_{l_0-1,l_2-1},$$

such that

$$B^\pm_{l_0,l_2} = \pm\partial_\chi + (l_2+1/2)\tanh\chi + (l_0+1/2)\coth\chi,$$
$$\lambda_{l_0,l_2} = -(1+l_0+l_2)^2.$$

The intertwining relations are now

$$B^-_{l_0-1,l_2-1}H^\chi_{l_0-1,l_2-1} = H^\chi_{l_0,l_2}B^-_{l_0-1,l_2-1}, \qquad B^+_{l_0-1,l_2-1}H^\chi_{l_0,l_2} = H^\chi_{l_0-1,l_2-1}B^+_{l_0-1,l_2-1}.$$

Hence, the operators $B^\pm$ connect eigenfunctions of $H^\chi_{l_0,l_2}$ as follows

$$B^-_{l_0-1,l_2-1} : f_{l_0-1,l_2-1} \to f_{l_0,l_2}, \qquad B^+_{l_0-1,l_2-1} : f_{l_0,l_2} \to f_{l_0-1,l_2-1}\,.$$

It is worthy noticing that $B^\pm_{l_0,l_2}$ can be also expressed in terms of the original coordinates (ξ,θ)

$$B^\pm_{l_0,l_2} = \pm(\cos\theta\,\partial_\xi - \sin\theta\coth\xi\,\partial_\theta) + (l_2+1/2)\tanh\xi\,\cos\theta + (l_0+1/2)\coth\xi\,\sec\theta.$$

In this case the free-index operators

$$\hat B^-\, f_{l_0,l_2} := \frac12\, B^-_{l_0,l_2}\, f_{l_0,l_2}, \quad \hat B^+\, f_{l_0,l_2} := \frac12\, B^+_{l_0,l_2}\, f_{l_0,l_2}, \quad \hat B\, f_{l_0,l_2} := -\frac12\,(l_0+l_2)\, f_{l_0,l_2},$$

close a $su(1,1)$ Lie algebra

$$[\hat{B}^+, \hat{B}^-] = -2\,\hat{B}, \qquad [\hat{B}, \hat{B}^\pm] = \pm\hat{B}^\pm. \tag{12}$$

Like in the previous case we will relate the (infinite dimensional) IUR's of $su(1,1)$ with the states of some Hamiltonians of the hierarchy. The appropriate IUR's of $su(1,1)$ are the discrete series having a fundamental state annihilated by the lowering operator

$$B^-\, f^0_{l_0,l_2} = 0.$$

Hence

$$f^0_{l_0,l_2}(\chi) = N(\cosh\chi)^{l_2+1/2}(\sinh\chi)^{l_0+1/2},$$

with N is a normalization constant. Imposing

$$l_0 \geq -1/2, \quad -k_1 \equiv l_0 + l_2 < -1\,.$$

we get a regular and square-integrable function $f^0_{l_0,l_2}(\chi)$. The lowest weight of the IUR is characterized by

$$j_1' = k_1/2 > 1/2.$$

because $\hat{B}\, f^0_{l_0,l_2} = -\frac{1}{2}\,(l_0 + l_2)\, f^0_{l_0,l_2}$.

On the other hand, the IO's $\hat{B}^\pm$ can also be considered as intertwining operators of the Hamiltonians H_ℓ linking their eigenfunctions Φ_ℓ, similarly to the $\hat{A}^\pm$, described before, but now with l_1 remaining unchanged.

A third set of coordinates,

$$s_0 = \sinh\phi, \quad s_1 = \cosh\phi\sinh\beta, \qquad s_2 = \cosh\phi\cosh\beta,$$

with $0 \leq \psi < +\infty$ and $-\infty < \beta < +\infty$, allows us to obtain a new $su(1,1)$ algebra following the same procedure like in the previous cases. The expression of the Hamiltonian is now

$$H_\ell = -\partial^2_\phi - \tanh\phi\,\partial_\phi + \frac{l_0^2 - \frac{1}{4}}{\sinh^2\phi} + \frac{1}{\cosh^2\phi}\left[-\partial^2_\beta + \frac{l_1^2 - \frac{1}{4}}{\sinh^2\beta} - \frac{l_2^2 - \frac{1}{4}}{\cosh^2\beta}\right].$$

It separates in the variables ϕ, β in terms of its eigenfunctions $\Phi(\beta,\phi) = f(\beta)\,g(\phi)$ as

$$H^\beta_{l_1,l_2} f(\beta) \equiv \left[-\partial^2_\beta + \frac{l_1^2 - \frac{1}{4}}{\sinh^2\beta} - \frac{l_2^2 - \frac{1}{4}}{\cosh^2\beta}\right] f(\beta) = \alpha f(\beta),$$

$$\left[-\partial^2_\phi - \tanh\phi\,\partial_\phi + \frac{l_0^2 - \frac{1}{4}}{\sinh^2\phi} + \frac{\alpha}{\cosh^2\phi}\right] g(\phi) = E\, g(\phi),$$

with the separation constant α.

The second order operator $H^\beta_{l_1,l_2}$ factorizes as

$$H^\beta_{l_1,l_2} = C^+_{l_1,l_2} C^-_{l_1,l_2} + \lambda_{l_1,l_2} = C^-_{l_1+1,l_2-1} C^+_{l_1+1,l_2-1} + \lambda_{l_1+1,l_2-1},$$

where

$$C^\pm_{l_1,l_2} = \pm\partial_\beta + (l_2 + 1/2)\tanh\beta + (-l_1 + 1/2)\coth\beta,$$

$$\lambda_{l_1,l_2} = -(1 - l_1 + l_2)^2.$$

The operators $C^{\pm}_{l_1,l_2}$ verify

$$C^{+}_{l_1+1,l_2-1}H^{\beta}_{l_1,l_2} = H^{\beta}_{l_1+1,l_2-1}C^{+}_{l_1+1,l_2-1}, \qquad C^{-}_{l_1+1,l_2-1}H^{\beta}_{l_1+1,l_2-1} = H^{\beta}_{l_1,l_2}C^{-}_{l_1+1,l_2-1}.$$

and they can also be rewritten in terms of the first set of coordinates (ξ,θ)

$$C^{\pm}_{l_1,l_2} = \pm(\sin\theta\,\partial_\xi + \cos\theta\coth\xi\,\partial_\theta) + (l_2+1/2)\tanh\xi\,\sin\theta + (-l_1+1/2)\coth\xi\,\csc\theta.$$

Free-index operators are defined by

$$\hat{C}^{-}\,f_{l_1,l_2} := \frac{1}{2}\,C^{-}_{l_1,l_2}\,f_{l_1,l_2}, \qquad \hat{C}^{+}\,f_{l_1,l_2} := \frac{1}{2}\,C^{+}_{l_1,l_2}\,f_{l_1,l_2}, \qquad \hat{C}\,f_{l_1,l_2} := -\frac{1}{2}\,(l_2-l_1)f_{l_1,l_2}.$$

They satisfy the commutation relations of the $su(1,1)$ algebra

$$[\hat{C}^{-},\hat{C}^{+}] = 2\,\hat{C}, \qquad [\hat{C},\hat{C}^{\pm}] = \pm\hat{C}^{\pm}. \tag{13}$$

The equation $\hat{C}^{-}\,f^{0}_{l_1,l_2} = 0$ gives the fundamental state for a representation of $su(1,1)$

$$f^{0}_{l_1,l_2}(\beta) = N(\cosh\beta)^{l_2+1/2}(\sinh\beta)^{-l_1+1/2},$$

with N a normalization constant. If besides

$$l_1 \le 1/2, \quad -k_2 \equiv l_2 - l_1 < -1$$

we obtain a (discrete series) IUR labeled by

$$j'_2 = k_2/2 > 1/2.$$

Also in this case the $C^{\pm}$ can be considered as IO's connecting global Hamiltonians H_ℓ and their eigenfunctions, having in mind that now the parameter l_0 is unaltered.

The set of IO's $\{\hat{A}^{\pm},\hat{A},\hat{B}^{\pm},\hat{B},\hat{C}^{\pm},\hat{C}\}$ obtained before generates a $su(2,1)$ Lie algebra with Lie commutators displayed in (8), (12) and (13) together with the crossed commutators that the intereted reader can compute himself using, for instance, the differential expression of these operators in the coordinates (ξ,θ). There is a constraint in the differential realization:

$$\hat{A} - \hat{B} + \hat{C} = 0.$$

The second order Casimir operator of $su(2,1)$ is

$$\mathcal{C} = \hat{A}^{+}\hat{A}^{-} - \hat{B}^{+}\hat{B}^{-} - \hat{C}^{+}\hat{C}^{-} + \frac{2}{3}\left(\hat{A}^{2}+\hat{B}^{2}+\hat{C}^{2}\right) - (\hat{A}+\hat{B}+\hat{C}).$$

The generator,

$$\mathcal{C}' = l_1 + l_2 - l_0,$$

commutes with the generators of $su(2,1)$ and we have

$$\langle\hat{A}^{\pm},\hat{A},\hat{B}^{\pm},\hat{B},\hat{C}^{\pm},\hat{C}\rangle \oplus \langle\mathcal{C}'\rangle \approx u(2,1).$$

The Hamiltonian (4) is a linear combination of of $\mathcal{C}$ and $\mathcal{C}'$ up an additive constant

$$\begin{aligned} H_\ell &= -4\,\mathcal{C} + \tfrac{1}{3}\,\mathcal{C}'^2 - \tfrac{15}{4} \\ &= -4\left(\hat{A}^{+}\hat{A}^{-} - \hat{B}^{+}\hat{B}^{-} - \hat{C}^{+}\hat{C}^{-} + \tfrac{2}{3}\left(\hat{A}^{2}+\hat{B}^{2}+\hat{C}^{2}\right) - (\hat{A}+\hat{B}+\hat{C})\right) + \tfrac{1}{3}\,\mathcal{C}'^2 - \tfrac{15}{4}. \\ &= -4\hat{A}^{+}\hat{A}^{-} + 4\hat{B}^{+}\hat{B}^{-} + 4\hat{C}^{+}\hat{C}^{-} + \text{constant}. \end{aligned}$$

The quadratic operators $\hat{A}^+\hat{A}^-$, $\hat{B}^+\hat{B}^-$ and $\hat{C}^+\hat{C}^-$ are constants of motion since they commute with the Hamiltonian, but they do not commute among themselves giving cubic expressions

$$[\hat{A}^+\hat{A}^-, \hat{B}^+\hat{B}^-] = -[\hat{A}^+\hat{A}^-, \hat{C}^+\hat{C}^-] = -[\hat{B}^+\hat{B}^-, \hat{C}^+\hat{C}^-] = \hat{A}^+\hat{C}^+\hat{B}^- - \hat{B}^+\hat{C}^-\hat{A}^-.$$

Obviously, we can implement a quadratic algebra by means of the generators

$$X_1 := \hat{A}^+\hat{A}^-, \quad X_2 := \hat{B}^+\hat{B}^-, \quad X_3 := \hat{C}^+\hat{C}^-, \quad Y_1 := \hat{A}^+\hat{C}^+\hat{B}^-, \quad Y_2 := \hat{B}^+\hat{C}^-\hat{A}^-.$$

The eigenfunctions of the Hamiltonians H_ℓ, that have the same energy, support an IUR of $su(2,1)$ characterized by a value of $\mathcal{C}$ and other of $\mathcal{C}'$. The fundamental state has to be simultaneously annihilated by $\hat{A}^-, \hat{C}^-$ and $\hat{B}^-$ ($\hat{A}^-_\ell\Phi^0_\ell = \hat{C}^-_\ell\Phi^0_\ell = \hat{B}^-_\ell\Phi^0_\ell = 0$). So,

$$\Phi^0_\ell(\xi,\theta) = N(\cos\theta)^{l_0+1/2}(\sin\theta)^{1/2}(\cosh\xi)^{l_2+1/2}(\sinh\xi)^{l_0+1},$$

where $\ell = (l_0, 0, l_2)$ and N is a normalization constant. The parameters l_0 and l_2 have to verify the following conditions

$$(l_0 + l_2) < -3/2, \qquad l_0 \geq -1/2, \qquad (l_0 + l_2) < -5/2.$$

This state Φ^0_ℓ also supports IUR's of the subalgebras $su(2)$ ($j = l_0/2$) and $su(1,1)$ (generated by $\hat{C}^\pm$ with $j'_2 = -l_2/2$). The energy of $\Phi^0_\ell(\xi,\theta)$ is

$$H_\ell\Phi^0_\ell = -(l_0 + l_2 + 3/2)(l_0 + l_2 + 5/2)\Phi^0_\ell \equiv E^0_\ell\Phi^0_\ell\,.$$

The other states of the $su(2,1)$ representation can be obtained applying the raising operators $\hat{A}^+, \hat{B}^+, \hat{C}^+$ over Φ^0_ℓ. Obviously, all of them share the same energy eigenvalue E^0_ℓ. All the states in the family of IUR's derived from fundamental states $\Phi^0_\ell(\xi,\theta)$, with the same value of $l_0 + l_2$, shall have the same energy. The energy of bound states is negative and the set of such bound states for each Hamiltonian H_ℓ is finite.

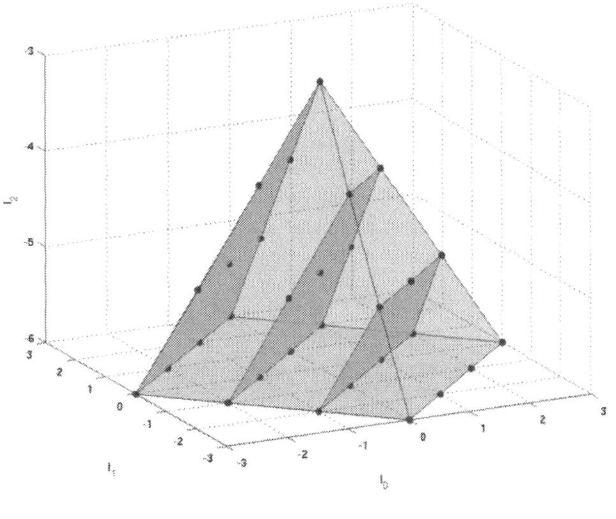

Figure 2. States of IUR's of $su(2,1)$ with the same energy and represented by points in the three dark planes associated to the ground state Φ^0_ℓ with $\ell = (0,0,-3)$, $\ell = (1,0,-4)$ and $\ell = (2,0,-5)$.

As we see in Fig. 2 the states of the IUR's of $su(2,1)$ are represented by points in parallel 2-dimensional planes placed inside a tetrahedral unbounded pyramid whose basis extends towards $-\infty$ along the axis l_2.

There are some points (in the parameter space (l_0, l_1, l_2)) which are degenerated because they correspond to an eigenspace with dimension bigger than 1. For instance, the states of the IUR whose fundamental state Φ^0_ℓ ($\ell = (0,0,-3)$) lie in a triangle and are nondegenerated. The IUR associated to the ground state with $\ell' = (1,0,-4)$ has states with the same energy, $E = -(-3+3/2)(-3+5/2)$, as the previous one, since in both cases $l_0 + l_2 = -3$. The states characterized by to $\ell'' = (0,0,-5)$, inside this IUR, are:

$$\Phi^2_{(0,0,-5)} = \hat{C}^+\hat{A}^+\Phi^0_{(0,0,-3)}, \qquad \tilde{\Phi}^2_{(0,0,-5)} = \hat{A}^+\hat{C}^+\Phi^0_{(0,0,-3)}\,.$$

Since $[A^+, C^+] = B^+$ both states are independent and span a 2-dimensional eigenspace of the Hamiltonian $H_{(0,0,-5)}$ with eigenvalue $E = -(-3+3/2)(-3+5/2)$. The ground state for $H_{(0,0,-5)}$ is the wavefunction $\Phi^0_{(0,0,-5)}$ with energy $E^0_{(0,0,-5)} = -(-5+3/2)(-5+5/2)$.

5.3. *The complete symmetry algebra* $so(4,2)$

It is trivial to see the Hamiltonian H_ℓ (4) is invariant under reflections in the space of parameters (l_0, l_1, l_2)

$$I_0 : (l_0,l_1,l_2) \to (-l_0,l_1,l_2), \quad I_1 : (l_0,l_1,l_2) \to (l_0,-l_1,l_2), \quad I_2 : (l_0,l_1,l_2) \to (l_0,l_1,-l_2),$$

however, the implications of this property are very interesting as we show now.

These reflections generate by conjugation other sets of intertwining operators from the ones already defined. Thus, for instance,

$$I_0 : \{\hat{A}^\pm, \hat{A}\} \longrightarrow \{\tilde{A}^\pm = I_0\hat{A}^\pm I_0, \tilde{A} = I_0\hat{A}I_0\},$$

where

$$\tilde{A}^\pm_{l_0,l_1} = \pm\partial_\theta - (-l_0+1/2)\tan\theta + (l_1+1/2)\cot\theta, \qquad \tilde{\lambda}_{l_0,l_1} = (1-l_0+l_1)^2.$$

They act on the eigenfunctions of the Hamiltonians (6)

$$\tilde{A}^-_{l_0,l_1} : f_{l_0,l_1} \to f_{l_0-1,l_1+1}, \qquad \tilde{A}^+_{l_0,l_1} : f_{l_0-1,l_1+1} \to f_{l_0,l_1}.$$

We can consider global operators $\tilde{A}^\pm$ as before. Then, $\tilde{A}^\pm$ together with $\tilde{A}$ ($\tilde{A}\, f_{l_0,l_1} := -\frac{1}{2}(-l_0+l_1)\, f_{l_0,l_1}$) close a new $\widetilde{su}(2)$.

In a similar procedure new sets of operators $\{\tilde{B}^\pm, \tilde{B}\}$ and $\{\tilde{C}^\pm, \tilde{C}\}$ closing $\widetilde{su}(1,1)$ algebras, can also be constructed under the action of these reflections The whole set of the operators obtained in this way $\{A^\pm, \tilde{A}^\pm, B^\pm, \tilde{B}^\pm, C^\pm, \tilde{C}^\pm\}$ together with the set of diagonal operators $\{L_0, L_1, L_2\}$, defined by

$$L_i\Psi_\ell = l_i\Psi_\ell,$$

span a Lie algebra of rank three: $o(4,2)$. All these generators link eigenstates of Hamiltonians H_ℓ with the same eigenvalue.

The fundamental state Ψ^0_ℓ for $so(4,2)$ will be characterized by

$$A^-_\ell\Psi^0_\ell = \tilde{A}^-_\ell\Psi^0_\ell = C^-_\ell\Psi^0_\ell = \tilde{C}^-_\ell\Psi^0_\ell = B^-_\ell\Psi^0_\ell = \tilde{B}^-_\ell\Psi^0_\ell = 0\,.$$

Thus,

$$\Phi^0_{(l_0=0,l_1=0,l_2)}(\xi,\theta) = N(\cos\theta)^{1/2}(\sin\theta)^{1/2}(\cosh\xi)^{l_2+1/2}\sinh\xi,$$

where $l_2 < -5/2$. In Fig. 2 the point $(l_0 = 0, l_1 = 0, l_2)$ for the cases plotted corresponds to the top vertex of the pyramid $(0,0,-3)$. All the other points can be obtained from this point. All these points correspond to an IUR of $so(4,2)$ that includes IUR's of $su(2,1)$.

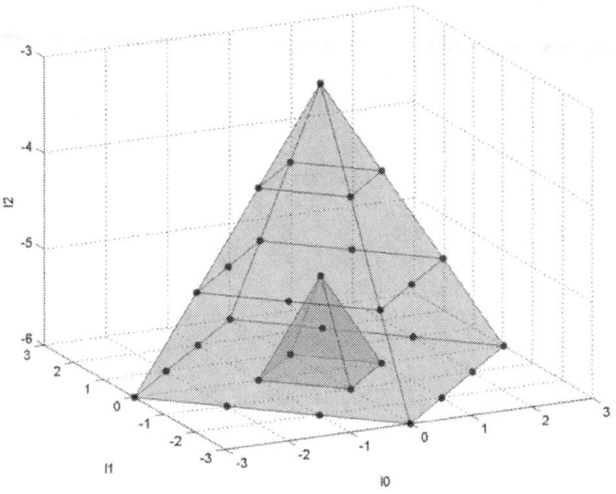

Figure 3. Two pyramids associated to the same IUR of $so(4,2)$. The points of the faces of the exterior pyramid (with vertex $(0,0,-3)$) represent non-degenerated levels. The exterior faces of the inner pyramid (vertex $(0,0,-5)$) are first excited double-degenerated levels.

Acknowledgments

We thank S. Kuru for interesting discussions during the join work which part of this contribution is based. MAO thanks to the organizers of the Workshop on Higher Symmetries in Physics (6–8 November 2008, Universidad Complutense de Madrid, Madrid, Spain) for the invitation to present this work. The financial support of DGES of the Ministerio de Educación y Ciencia of Spain under Projects FIS2005-03989 and FIS2009-09002 and of Junta de Castilla y León (Spain) (Project GR224) is gratefully acknowledged.

References

[1] Evans N W 1990 *Phys. Rev A.* **41** 5666;
1990 *Phys. Lett. A.* **147** 483;
1991 *J. Math. Phys.* **32** 3369.
[2] Grosche C, Pogosyan G S and Sisakian A N 1995 *Fortschr. Phys.* **43** 523;
1996 *Phys. Part. Nucl.* **27** 244;
1997 *Phys. Part. Nucl.* **28** 486.
[3] Ballesteros A, Herranz F J, Santander M and Sanz-Gil T 2003 *J. Phys. A* **36** L93.
[4] Cariñena J F, Rañada M F and Santander M 2005 *J. Math. Phys.* **46** 052702;
2007 *J. Phys. A: Math. Theor.* **40** 13645.
[5] Lakshmanan M and Eswaran K 1975 *J. Phys. A: Math. Gen.* **8** 1658.
[6] Higgs P W 1979 *J. Phys. A: Math. Gen.* **12** 309.
[7] Bonatsos D, Daskaloyannis C and Kokkotas K 1994 *Phys. Rev. A* **50** 3700.
[8] Daskaloyannis C 2001 *J. Math. Phys.* **42** 1100.
[9] Ballesteros A and Ragnisco O 1998 *J. Phys. A: Math. Gen.* **31** 3791.
[10] Ballesteros A, Blasco A, Herranz F J, Musso F and Ragnisco O 2009 "(Super)integrability from coalgebra symmetry: formalism and applications" in this volume; arXiv:0905.1298v1 [math-ph].
[11] Marsden J E and Weinstein A 1974 *Rep. Math. Phys.* **5** 121.
[12] del Olmo M A, Rodríguez M A and Winternitz P 1993 *J. Math. Phys.* **34** 5118.
[13] del Olmo M A, Rodríguez M A and Winternitz P 1996 *Fortschritte der Physik*, **44** 199.
[14] Calzada J A, del Olmo M A and Rodríguez M A 1997 *J. Geom. Phys.* **23** 14.
[15] Calzada J A, del Olmo M A and Rodríguez M A 1999 *J. Math. Phys.* **40**, 188-209.
[16] Calzada J A, Negro J, del Olmo M A and Rodríguez M A 1999 *J. Math. Phys.* **41** 317.
[17] Morse P M 1929 *Phys. Rev.* **34** 57.

[18] Pöschl G and Teller E 1933 *Z. Phys.* **83** 143.
[19] Matveev V B and Salle M A 1991 *Darboux Transformations and Solitons* (Berlin: Springer).
[20] Alhassid Y, Gürsey F and Iachello F, 1983 *Ann. Phys.* **148** 346.
[21] Kuru Ş, Tegmen A and Vercin A 2001 *J. Math. Phys.* **42** (2001), 3344.
[22] Demircioglu B, Kuru Ş, Önder M and Vercin A 2002 *J. Math. Phys.* **43** 2133.
[23] Fernández D J, Negro J and del Olmo M A 1996 *Ann. Phys.* **252** 386.
[24] Infeld L and Hull T E 1951 *Rev. Mod. Phys.* **23** (1951), 21.
[25] Calzada J A, Negro J and del Olmo M A 2006 *J. Math. Phys.* **47** 043511(17pp).
[26] Calzada J A, Kuru Ş, Negro J and del Olmo M A 2008 *J. Phys. A: Math. Theor.* **41** 255201 (11pp).
[27] Calzada J A, Negro J and del Olmo M A 2009 *SIGMA* **5** 039 (23pp).
[28] Kobayashi S and Nomizu K 1969 *Foundations of Differential Geometry*, (New York: Wiley).
[29] del Olmo M A, Rodríguez M A , Winternitz P and Zassenhaus H 1990 *Linear Algebr. Appl.* **135** 79.
[30] Zhedanov A S 1992 *Mod. Phys. Lett. A.* **7** 507.
[31] Bambah B A, Kumar S and Mukku C 2007 *Int. J. Theor. Phys., Group Theor. and Nonlinear Opt.* **11** 265.
[32] Kalnins E G, Miller W Jr. and Pogosyan G S 1996 *J. Math. Phys.* **37** 6439.
[33] Kalnins E G, Miller W Jr. and Pogosyan G S 1997 *J. Math. Phys.* **38** (1997) 5416.

Journal of Physics: Conference Series **175** (2009) 012008 doi:10.1088/1742-6596/175/1/012008

Inequivalent sets of commuting missing label operators for the nuclear surfon model

R. Campoamor-Stursberg†‡

† Dpto. Geometría y Topología
Universidad Complutense de Madrid
Plaza de Ciencias, 3
E-28040 Madrid, Spain
‡ Instituto de Matemática Interdisciplinar U.C.M.

E-mail: rutwig@pdi.ucm.es

Abstract. We exhibit a possible choice for the four available functionally independent labelling operators for the missing label problem associated to the reduction chain $\mathfrak{so}(5) \supset \mathfrak{so}(3)$.

1. Introduction

The labelling of the basis states of irreducible representations (irreps) of a semisimple Lie group G with respect to some subgroup H often leads to the so called missing label problem (MLP), i.e., to the problem of determining subgroup scalars[1] with respect to H that, added to the Casimir operators of G and H, enable to completely classify the states. Basis states for irreps of G are then specified by the common eigenstates of a complete set of labelling operators. The eight reduction chains with one missing label were solved in [1, 2]. Although some general results exist concerning arbitrary reduction chains, mainly about the properties of generating functions and the construction of integrity bases,[2] as well as the development of special techniques like the method of elementary permissible diagrams [1, 3, 4, 5, 6, 7], only for a small number of labelling problems with more than one labelling operator a general solution has been worked out.

In this paper we determine four functionally independent operators depending on a parameter that solve the missing label problem associated to the embedding $\mathfrak{so}(5) \supset \mathfrak{so}(3)$. This reduction chain appears in many applications involving the subalgebra of angular momentum, like the classification of states build up by quadrupole phonon states [8, 9]. The corresponding $n = 2$ missing label problem has been analyzed in [10, 11], where two commuting missing label operators of degrees four and six, respectively, were found. These two operators were shown to constitute, in some sense, the simplest possible choice to solve this labelling problem. A complete set of four independent labelling operators was however not given. In our analysis, it is verified that the pair of commuting operators found in [11] corresponds to the simplest possible solution, and the conjecture on the degree of these operators is confirmed. We moreover show the existence of another pair consisting of operators of degrees four and six that solve the labelling

[1] Polynomials in the enveloping algebra of G that commute with all generators of the subgroup H will be called subgroup scalars.
[2] An integrity basis is a set of elementary subgroup scalars in terms of which all can be expressed by polynomials.

problem. In particular, for special values of the parameter, seven special cases corresponding to the simplest possible solutions to the MLP are found.

2. Missing label operators

For semisimple Lie algebras $\mathfrak{s}$, the rank $l\mathcal{N}(\mathfrak{s}) = l$ specifies the maximal number of (functionally) independent polynomials in the generators such that they commute with the elements of $\mathfrak{s}$ [12]. The eigenvalues of these operators, when applied to representations, provide a first class of suitable labels to distinguish the multiplets and the states. For higher rank algebras, however, degeneracies are found that cannot be solved using the Casimir operators and/or the generators of the Cartan subalgebra. According to Racah, we have to find $f = \frac{1}{2}(\dim \mathfrak{s} - 3l)$ additional operators to distinguish the states. This gives a total number of labels equal to

$$i = \frac{1}{2}(\dim \mathfrak{s} - \mathcal{N}(\mathfrak{s})). \tag{1}$$

Reducing representations of a Lie algebra with respect to some (inner symmetry) subalgebra $\mathfrak{h}$ leads, in the generic case, to a similar situation, where the invariants of the chain do not provide enough labels to separate the states. The reduction chain provides $\frac{1}{2}(\dim \mathfrak{h} + \mathcal{N}(\mathfrak{h})) + l'$ labels, where l' denotes the number of common invariants of $\mathfrak{g}$ and $\mathfrak{h}$.[3] Using this fact, it is easy to see that in addition to the Casimir operators, exactly

$$n = \frac{1}{2}\left(\dim \mathfrak{s} - \mathcal{N}(\mathfrak{s}) - \dim \mathfrak{h} - \mathcal{N}(\mathfrak{h})\right) + l' \tag{2}$$

additional operators are necessary [13]. The scalar $m = 2n$ gives the total number of available operators.

We determine labelling operators by means of the analytical approach [13]. Given the Lie algebra $\mathfrak{s} = \left\{X_1, .., X_n \mid [X_i, X_j] = C_{ij}^k X_k\right\}$ in terms of generators and commutation relations, we realize the generators X_i in the space $C^\infty(\mathfrak{s}^*)$ by means of the differential operators

$$\widehat{X}_i = C_{ij}^k x_k \frac{\partial}{\partial x_j}, \tag{3}$$

where $\{x_1, .., x_n\}$ are the coordinates on a dual basis. In this context, invariants of $\mathfrak{s}$ are the solutions of system of partial differential equations given by:

$$\widehat{X}_i F = 0, \quad 1 \leq i \leq n. \tag{4}$$

For polynomial solutions of (4), the symmetrization map defined by

$$Sym(x_{i_1}..x_{i_p}) = \frac{1}{p!}\sum_{\sigma \in S_p} x_{\sigma(i_1)}..x_{\sigma(i_p)} \tag{5}$$

recover the Casimir operators in their usual form, i.e., as a polynomial in the generators commuting with $\mathfrak{s}$. Observe that the preceding realization shows that the number $\mathcal{N}(\mathfrak{s})$ of independent solutions is given by:

$$\mathcal{N}(\mathfrak{s}) := \dim\ \mathfrak{s} - \text{rank}\left(C_{ij}^k x_k\right), \tag{6}$$

[3] I.e., invariants of $\mathfrak{s}$ that depend only on variables of the subalgebra $\mathfrak{h}$.

Table 1. $\mathfrak{so}(5)$ brackets in a $\mathfrak{so}(3) = \{L_0, L_{\pm 1}\}$ basis.

[,]	Q_3	Q_2	Q_1	Q_0	Q_{-1}	Q_{-2}	Q_{-3}
L_0	$3Q_3$	$2Q_2$	Q_1	0	$-Q_{-1}$	$-2Q_{-2}$	$-3Q_{-3}$
L_1	0	$6Q_3$	Q_2	$2Q_1$	$6Q_0$	$10Q_{-1}$	Q_{-2}
L_{-1}	Q_2	$10Q_1$	$6Q_0$	$2Q_{-1}$	Q_{-2}	$6Q_{-3}$	0
Q_3	0	0	0	Q_3	Q_2	$10Q_1 + 15L_1$	$5Q_0 - 15L_0$
Q_2		0	$-6Q_3$	$-Q_2$	$-15L_1$	$30Q_0 + 60L_0$	$10Q_{-1} - 15L_{-1}$
Q_1			0	$3L_1 - Q_1$	$-3L_0 - 3Q_0$	$15L_{-1}$	Q_{-2}
Q_0				0	$-Q_{-1} - 3L_{-1}$	$-Q_{-2}$	Q_{-3}
Q_{-1}					0	$-6Q_{-3}$	0
Q_{-2}						0	0

where $A(\mathfrak{s}) := \left(C_{ij}^k x_k\right)$ is the matrix associated to the commutator table of $\mathfrak{s}$ over the given basis.

From this perspective, the missing label problem (short MLP) for a reduction chain $\mathfrak{s} \supset \mathfrak{s}'$ constitutes a special application of the analytical ansatz. More specifically, the labelling operators can be seen as those operators that commute with the generators of the subalgebra $\mathfrak{s}'$, thus correspond to a subsystem of (4). The latter has exactly $\mathcal{N}(f(\mathfrak{s}')) = \dim \mathfrak{s} - \dim \mathfrak{s}' - l'$ functionally independent solutions. Transforming this formula using (2), we arrive at the expression

$$\mathcal{N}(f(\mathfrak{s}')) = m + \mathcal{N}(\mathfrak{s}) + \mathcal{N}(\mathfrak{s}') - l'. \tag{7}$$

This shows that the subsystem corresponding to $\mathfrak{s}'$-generators has n more solutions as needed to solve the missing label problem [13]. The ansatz based on differential equations provides an alternative to the usually difficult task of finding integrity bases for the labelling operators [14]. Because of equation (7), if we find m solutions that are independent among each other and from the Casimir operators of the chain, a solution to the missing label problem is given by n polynomials of these solutions that commute mutually. This last step is in practice the most difficult one, and has been completely solved only for a small number of reduction chains.

To be used later, we introduce the counterpart of labelling operators in the space $C^\infty(\mathfrak{s})$, that will be useful to measure the non-commutativity of brackets. Let $P = a^{i_1 \ldots i_p} X_{i_1} \ldots X_{i_p}$ be a labelling operator. Replacing the generators X_{i_j} by the coordinate of the corresponding dual element, we obtain the homogeneous polynomial $P^* = a^{i_1 \ldots i_p} x_{i_1} \ldots x_{i_p}$. Observe that since the variables in $\mathfrak{s}^*$ commute, P^* will usually have less terms than P. It is clear that if $[O_1, O_2] \neq 0$, then the analytical counterpart of the commutator does not vanish, and the number of terms gives an approximate idea of how far the operators O_1, O_2 are from commuting.

3. The nuclear surfon model

For the non-canonical embedding $\mathfrak{so}(5) \supset \mathfrak{so}(3)$, the subalgebra $\mathfrak{so}(3)$ turns out to be the principal simple subalgebra of rank one [3]. We choose the basis of the orthogonal Lie algebra $\mathfrak{so}(5)$ to consist of generators $\{L_0, L_1, L_{-1}\}$ with brackets $[L_0, L_{\pm 1}] = \pm L_{\pm 1}$, $[L_1, L_{-1}] = 2L_0$ together with the irreducible tensor representation Q_μ ($\mu = -3..3$) of dimension seven. The brackets of $\mathfrak{so}(5)$ over this basis are given in Table 1.

In order to describe states when reducing representations of $\mathfrak{so}(5)$ with respect to this $\mathfrak{so}(3)$, we need, according to formula (2), two missing labels among the four available operators. In tems of the differential realization (3), these labelling operators correspond to solutions of the following system of equations:

$$\begin{aligned}
\widehat{L}_0F &:= l_1\frac{\partial F}{\partial l_1} - l_{-1}\frac{\partial F}{\partial l_{-1}} + 3q_3\frac{\partial F}{\partial q_3} + 2q_2\frac{\partial F}{\partial q_2} + q_1\frac{\partial F}{\partial q_1} - q_{-1}\frac{\partial F}{\partial q_{-1}} - 2q_{-2}\frac{\partial F}{\partial q_{-2}} - 3q_{-3}\frac{\partial F}{\partial q_{-3}} = 0,\\
\widehat{L}_1F &:= -l_1\frac{\partial F}{\partial l_0} + 2l_0\frac{\partial F}{\partial l_{-1}} + 6q_3\frac{\partial F}{\partial q_2} + q_2\frac{\partial F}{\partial q_1} + 2q_1\frac{\partial F}{\partial q_0} + 6q_0\frac{\partial F}{\partial q_{-1}} + 10q_{-1}\frac{\partial F}{\partial q_{-2}} + q_{-2}\frac{\partial F}{\partial q_{-3}} = 0,\\
\widehat{L}_{-1}F &:= l_{-1}\frac{\partial F}{\partial l_0} - 2l_0\frac{\partial F}{\partial l_1} + q_2\frac{\partial F}{\partial q_3} + 10q_1\frac{\partial F}{\partial q_2} + 6q_0\frac{\partial F}{\partial q_1} + 2q_{-1}\frac{\partial F}{\partial q_0} + q_{-2}\frac{\partial F}{\partial q_{-1}} + 6q_3\frac{\partial F}{\partial q_{-2}} = 0.
\end{aligned} \tag{8}$$

For labelling purposes, we are only interested on polynomial solutions. These, after the corresponding symmetrization (5) of its terms, provide the classical subgroup scalars. The preceding system (8) has seven functionally independent solutions. It is obvious that the Casimir operators of $\mathfrak{so}(3)$ and $\mathfrak{so}(5)$ satisfy this system, thus can be interpreted as labelling operators of special kind. More specifically, they are composite functions of labelling operators of lower order.[4]

Because of the diagonal action of the generator L_0 of $\mathfrak{so}(3)$ on $L_{\pm1}$ and the irreducible multiplet of dimension seven, polynomials $P = \alpha_{a_{-1}a_0a_1b_3\dots b_{-3}} l_0^{a_0} l_{-1}^{a_{-1}} l_1^{a_1} q_3^{b_3} \dots q_{-3}^{b_{-3}}$ satisfying the system (8) must also satisfy the linear condition

$$-a_{-1} + a_1 + 3b_3 + 2b_2 + b_1 - b_{-1} - 2b_{-2} - 3b_{-3} = 0. \tag{9}$$

Following the notation of [11], we denote by $[k, m]$ a homogeneous polynomial of degree $k + m$ such that its degree in the q_i-variables is k and its degree in $\{l_0, l_1, l_1\}$ is m. Observe that by (9), $k = b_3 + b_2 + b_1 + b_0 + b_{-1} + b_{-2} + b_{-3}$ and $m = a_{-1} + a_0 + a_1$. This notation enables us to rewrite P as a sum of homogeneous polynomials $[k_i, m_j]$. We will say that $[k_i, m_j]$ has bi-degree (k_i, m_i).

To construct the labelling operators, we proceed basing on the degree of polynomial solutions. For any fixed $n \geq 2$, we determine the functionally independent operators $[k, m]$ such that $k + m = n$. Computing such solutions up to a sufficiently high n leads to a set of functions that, once symmetrized, constitute an integrity basis for the MLP. However, here we are only interested in finding two pairs of independent operators such that they solve the MLP for $\mathfrak{so}(5) \supset \mathfrak{so}(3)$ with the additional commutativity condition.

For $n = 2$, only two solutions to (8) exist:

$$[0, 2] = l_0^2 + l_1 l_{-1}, \quad [2, 0] = -\frac{2}{5}(q_3 q_{-3} + q_1 q_{-1}) + \frac{1}{15} q_2 q_{-2} + q_0^2. \tag{10}$$

It follows at once that $[0, 2]$ is the Casimir operator of $\mathfrak{so}(3)$. In addition, the polynomial $C_2 = [0, 2] + [2, 0]$ corresponds to the quadratic Casimir operator of $\mathfrak{so}(5)$.[5]
A generic polynomial of degree 3 in the generators of $\mathfrak{so}(5)$ and satisfying constraint (9) has 26 terms. Due to the latter, it automatically gives a solution of the equation $\widehat{L_0}F = 0$. Inserting such a polynomial into the remaining equations of (8) and solving the corresponding system with respect to the coefficients shows that only the trivial solution is admissible, from which we conclude that system (8) has no polynomial solutions of order three. This provides additional information on the behavior of the quadratic solutions. It is well known that if O_1, O_2 are two

[4] This is more apparent when the Casimir operators are contracted with respect to the natural contraction determined by the embedding of the algebras [15].
[5] In the following, we will identify, for computational purposes, the polynomial function f that solves system (8) with the symmetrized polynomial $Sym(f)$. With an abuse of language, we will call f the labelling operator.

labelling operators, then their commutator $[O_1, O_2]$ also provides a labelling operator [1]. As a consequence of this fact, the two operators $[2,0]$ and $[0,2]$ commute after symmetrization, since their commutator would have terms of order three.

In order four, the three labelling operators $[0,2]^2$, $[0,2]\,[2,0]$ and $[2,0]^2$ are functionally dependent on those of order two, and therefore of no further use for solving the MLP. In the following we will only be interested on solutions that are not of this type. We call a polynomial $[k,m]$ indecomposable if it is not a function of polynomials of lower order. Among the seven linearly independent solutions of (8) of degree four, the only indecomposable ones are the following:

$$\begin{aligned}
[1,3] = \;& \tfrac{1}{4}l_0l_1^2q_{-2} - \tfrac{3}{2}l_0l_1l_{-1}q_0 + \tfrac{1}{4}l_1^3q_{-3} - l_0^2l_{-1}q_1 + l_0^2l_1q_{-1} + l_0^3q_0 + \tfrac{1}{4}l_0l_{-1}^2q_2 - \tfrac{1}{4}l_{-1}^3q_3 \\
& -\tfrac{1}{4}l_1^2l_{-1}q_{-1} + \tfrac{1}{4}l_1l_{-1}^2q_1,
\end{aligned}$$

$$\begin{aligned}
[2,2] = \;& -\tfrac{1}{12}l_{-1}^2q_2q_0 + \tfrac{1}{12}l_0l_{-1}q_2q_{-1} + \tfrac{1}{6}l_0l_1q_0q_{-1} + \tfrac{1}{12}l_1^2q_1q_{-3} - \tfrac{1}{3}l_0^2q_1q_{-1} + \tfrac{7}{12}{l_0}^2q_0^2 \\
& +\tfrac{1}{60}l_1l_{-1}q_2q_{-2} + \tfrac{1}{12}l_{-1}^2q_3q_{-1} - \tfrac{1}{12}l_0l_{-1}q_3q_{-2} + \tfrac{1}{12}l_1l_{-1}q_0^2 - \tfrac{1}{12}l_1^2q_0q_{-2} + \tfrac{1}{12}l_1^2q_{-1}^2 \\
& -\tfrac{1}{6}l_0l_{-1}q_1q_0 - \tfrac{1}{12}l_1l_{-1}q_1q_{-1} + \tfrac{1}{12}l_{-1}^2q_1^2 + \tfrac{1}{15}l_0^2q_3q_{-3} + \tfrac{1}{60}l_0^2q_2q_{-2} - \tfrac{1}{12}l_0l_1q_1q_{-2} \\
& -\tfrac{11}{60}l_1l_{-1}q_3q_{-3} + \tfrac{1}{12}l_0l_1q_2q_{-3},
\end{aligned}$$

$$\begin{aligned}
[3,1] = \;& \tfrac{1}{4}l_1q_2q_0q_{-3} + \tfrac{1}{9}l_0q_2q_0q_{-2} - l_0q_0^3 - \tfrac{1}{2}l_0q_3q_0q_{-3} + \tfrac{17}{18}l_0q_1q_0q_{-1} + \tfrac{1}{36}l_{-1}q_2q_1q_{-2} \\
& -\tfrac{1}{4}l_{-1}q_3q_0q_{-2} - \tfrac{2}{9}l_1q_1^2q_{-3} + \tfrac{1}{36}l_1q_3q_{-2}^2 + \tfrac{1}{9}l_1q_1q_{-1}^2 + \tfrac{1}{18}l_0q_3q_{-1}q_{-2} - \tfrac{1}{9}l_{-1}q_1^2q_{-1} \\
& +\tfrac{1}{36}l_1q_1q_0q_{-2} - \tfrac{1}{6}l_0q_2q_{-1}^2 - \tfrac{1}{3}l_1q_3q_{-1}q_{-3} + \tfrac{2}{9}l_{-1}q_3q_{-1}^2 - \tfrac{1}{36}l_{-1}q_2^2q_{-3} - \tfrac{1}{6}l_0q_1^2q_{-2} \\
& -\tfrac{1}{6}l_1q_0^2q_{-1} + \tfrac{1}{6}l_{-1}q_1q_0^2 - \tfrac{1}{36}l_{-1}q_2q_0q_{-1} - \tfrac{1}{36}l_1q_2q_{-1}q_{-2} + \tfrac{1}{3}l_{-1}q_3q_1q_{-3} + \tfrac{1}{18}l_0q_2q_1q_{-3},
\end{aligned}$$

$$\begin{aligned}
[4,0] = \;& -\tfrac{1}{9}q_1^3q_{-3} - \tfrac{3}{5}q_3q_0^2q_{-3} - \tfrac{1}{36}q_2^2q_{-1}q_{-3} + \tfrac{1}{675}q_2^2q_{-2}^2 + \tfrac{1}{100}q_3q_2q_{-2}q_{-3} - \tfrac{1}{9}q_3q_{-1}^3 \\
& -\tfrac{1}{15}q_2q_0^2q_{-2} - \tfrac{5}{108}q_1^2q_{-1}^2 - \tfrac{1}{540}q_2q_1q_{-1}q_{-2} + \tfrac{1}{18}q_1^2q_0q_{-2} + \tfrac{7}{30}q_3q_1q_{-1}q_{-3} \\
& +\tfrac{1}{18}q_2q_0q_{-1}^2 - \tfrac{3}{100}q_3^2q_{-3}^2 - \tfrac{1}{36}q_3q_1q_{-2}^2 + \tfrac{1}{6}q_2q_1q_0q_{-3} + \tfrac{1}{6}q_3q_0q_{-1}q_{-2}.
\end{aligned}$$

The fourth order Casimir operator of $\mathfrak{so}(5)$ can be recovered from a linear combination of these operators by simply considering $C_4 = [4,0] + [3,1] + [2,2] + [1,3]$.[6] Now, among the operators $[2,0]]\,, C_2, C_4, [1,3]\,, [2,2]\,, [3,1]\,, [4,0]$, at most five are functionally independent. This is easily verified checking a Jacobian. Therefore we can extract two independent linear combinations of the operators $[1,3]\,, [2,2]\,, [3,1]\,, [4,0]$. as labelling operators. However, it can be verified that no such pair of labelling operators commutes, unless one of them coincides with C_4. This follows from the following table, specifying the number of terms in the analytical counterpart of the commutator of the elementary subgroups scalars $[1,3]\,, [2,2]\,, [3,1]\,, [4,0]$:

[,]	[4, 0]	[3, 1]	[2, 2]	[1, 3]
[4, 0]	–	152	282	130
[3, 1]		–	370	218
[2, 2]			–	88

This means that subgroup scalars of higher order have to be considered. Following a reasoning similar to that applied for $n = 3$, it can be verified that (8) does not admit polynomial solutions of order $n = 5$. This further implies that any quadratic labelling operators automatically commutes with the preceding fourth order operators. Observe further that since the latter do not commute, solutions of degree seven must exist. In order six, only five order indecomposable operators $[2,4]$, $[3,3]$ $[4,2]$ $[5,1]$ and $[6,0]$ exist, given in Table 2.

[6] Using the contraction procedure of [16], it follows that the operators of the previous lemma are the components of the decomposition of the Casimir operator C_4 [15].

Table 2. Indecomposable sixth order polynomial solutions to system (8)

$$
\begin{aligned}
[2,4] = \; & (q_{-3}q_3 - \tfrac{1}{4}q_0^2)l_0^4 + \tfrac{1}{2}(q_0(q_1l_{-1} - q_{-1}l_1) + l_1q_2q_{-3} - l_{-1}q_3q_{-2})l_0^3 + l_0^2(l_1l_{-1}q_0^2 - \tfrac{1}{8}(l_{-1}^2q_2 + l_1^2q_{-2})q_0 - \tfrac{1}{4}l_1l_{-1}(-2q_1q_{-1} + q_{-2}q_2 - 2q_{-3}q_3) \\
& + \tfrac{1}{4}\left((5q_3q_{-1} - q_1^2)l_{-1}^2 + (+5q_1q_{-3} - q_{-1}^2)l_1^2)\right) + (\tfrac{3}{16}q_{-1}q_{-3} - \tfrac{1}{64}q_{-2}^2)l_1^4 + \tfrac{1}{32}l_{-1}^2l_1^2(26q_1q_{-1} - q_{-2}q_2 - 50q_0^2 + 2q_{-3}q_3) + (\tfrac{3}{16}q_3q_1 - \tfrac{1}{64}q_2^2)l_{-1}^4 \\
& + \tfrac{1}{8}l_1^3((9q_0q_{-3} - 1q_{-1}q_{-2})l_0 + \tfrac{1}{16}(6q_{-1}^2 + 2q_1q_{-3} - 5q_0q_{-2})l_{-1}) + \tfrac{1}{16}l_{-1}^3((6q_1^2 + 2q_3q_{-1} - 5q_2q_0)l_1 + \tfrac{1}{8}(q_2q_1 - 9q_3q_0)l_0) \\
& + \tfrac{1}{8}l_{-1}^2l_1l_0(4q_2q_{-1} - q_3q_{-2} - 11q_1q_0) + \tfrac{1}{8}l_0l_1^2l_{-1}(+q_2q_{-3} - 4q_1q_{-2} + 11q_0q_{-1}).
\end{aligned}
$$

$$
\begin{aligned}
[3,3] = \; & (q_2q_0q_{-2} - \tfrac{4}{3}\left(q_1^2q_{-2} - q_2q_{-1}^2\right) + 8q_1q_0q_{-1} - 9q_0^3)l_0^3 + \tfrac{1}{12}(6q_0q_{-1}q_{-2} + 12q_1q_{-1}q_{-3} - 27q_0^2q_{-3} - 4q_{-1}^3 - q_1q_{-2}^2)l_1^3 + \tfrac{4}{3}\left(q_1q_{-1}^2 - q_2q_{-1}q_{-2}\right)l_1l_0^2 \\
& + \tfrac{1}{12}(q_2^2q_{-1} - 12q_3q_1q_{-1} + 4q_1^3 + 27q_3q_0^2 - 6q_2q_1q_0)l_{-1}^3 + (3\left(q_2q_0q_{-3} - q_0^2q_{-1}\right) + q_1q_0q_{-2} - 4q_1^2q_{-3})l_1l_0^2 + 12l_0l_1^2\left(q_2q_{-1}q_{-3} - q_0q_{-1}^2\right) \\
& + \tfrac{1}{3}l_{-1}(12q_3q_{-1}^2 + 9\left(q_1q_0^2 - q_3q_0q_{-2}\right) + q_2q_1q_{-2} - 3q_2q_0q_{-1} - 4q_1^2q_{-1})l_0^2 + \tfrac{1}{12}l_0(9q_0^2q_{-2} - 36q_1q_0q_{-3} + 4q_1q_{-1}q_{-2} - q_2q_{-2}^2)l_1^2 \\
& + \tfrac{1}{2}l_0l_1l_{-1}(10q_1q_0q_{-1} - 9q_0^3 - 2q_2q_{-1}^2 + 2q_2q_1q_{-3} + 2q_3q_{-1}q_{-2} - 2q_1^2q_{-2} - 18q_3q_0q_{-3} + q_2q_0q_{-2}) + \tfrac{1}{3}\left(q_2q_1q_{-1} + q_3q_1q_{-2} - q_1^2q_0\right)l_{-1}^2l_0 \\
& + \tfrac{1}{12}(9q_2q_0^2 - 36q_3q_0q_{-1} - q_2^2q_{-2})l_{-1}^2l_0 - \tfrac{1}{12}l_1l_{-1}(((8q_1^2q_{-1} - 9q_1q_0^2 - 36q_3q_1q_{-3} + 18q_3q_0q_{-2} - 2q_2q_1q_{-2} - 12q_3q_{-1}^2 + 3q_2^2q_{-3})l_{-1}) \\
& + ((36q_3q_{-1}q_{-3} + 9q_0^2q_{-1} + 2q_2q_{-1}q_{-2} - 3q_3q_{-2}^2 - 8q_1q_{-1}^2 - 18q_2q_0q_{-3} + 12q_1^2q_{-3})l_1))
\end{aligned}
$$

$$
\begin{aligned}
[4,2] = \; & \tfrac{1}{2}(27l_1l_{-1} - 135l_0^2)q_0^4 + (27\left(l_0l_{-1}q_1 - l_0l_1q_{-1}\right) - \tfrac{9}{2}\left(l_1^2q_{-2} + l_{-1}^2q_2\right))q_0^3 + q_0^2(l_0(\tfrac{3}{2}(l_1q_1q_{-2} - l_{-1}q_2q_{-1}) + \tfrac{27}{2}(l_{-1}q_3q_{-2} - l_1q_2q_{-3})) \\
& + (27q_3q_{-3} + 81q_1q_{-1} + 9q_2q_{-2})l_0^2 + \tfrac{27}{2}l_{-1}(5q_3q_{-3} - q_1q_{-1})l_1 + \tfrac{1}{2}(3q_1^2 + 9q_3q_{-1})l_{-1}^2 + \tfrac{1}{2}(3q_{-1}^2 + 9q_1q_{-3})l_1^2) + 16l_0^2\left(q_1^3q_{-3} + q_3q_{-1}^3\right) \\
& - \tfrac{1}{6}l_0^2(108(q_3q_0q_{-1}q_{-2} + q_2q_1q_0q_{-3}) - 6(q_3q_1q_{-2}^2 + q_2^2q_{-1}q_{-3}) + 84(q_1^2q_0q_{-2} + q_2q_0q_{-1}^2) - 20q_2q_1q_{-1}q_{-2} + q_2^2q_{-2}^2 + 64q_1^2q_{-1}^2) + 9l_1^2q_3q_{-3}q_{-1}^2 \\
& - \tfrac{1}{2}l_1^2q_2q_{-2}q_{-1}^2 - l_1^2\left((q_{-1}q_{-3} + \tfrac{3}{4}q_{-2}^2)q_1^2 + (4q_1q_{-1}q_{-2} - 3q_2q_{-1}q_{-3} - 9q_3q_{-2}q_{-3} + \tfrac{1}{2}q_2q_{-2}^2)q_0 - q_1q_{-1}^3 + 9q_3q_1q_{-3}^2 + q_2q_1q_{-2}q_{-3} - \tfrac{3}{4}q_2^2q_{-3}^2\right) \\
& + l_{-1}^2((-9q_3q_2q_{-3} - 3q_3q_1q_{-2} + \tfrac{1}{2}q_2^2q_{-2} + 4q_2q_1q_{-1})q_0 + (9q_3q_{-3} - \tfrac{1}{2}q_2q_{-2})q_1^2 - (q_3q_1 + \tfrac{3}{4}q_2^2)q_{-1}^2 + 9q_3^2q_{-1}q_{-3} - q_1^3q_{-1} + q_3q_2q_{-1}q_{-2} - \tfrac{3}{4}q_3^2q_{-2}^2) \\
& + (-6l_0l_1q_2 + 4l_1l_{-1}q_3)q_{-1}^3 + (6l_0l_{-1}q_{-2} + 4l_1l_{-1}q_{-3})q_1^3 + q_1^2(l_0((-28q_{-1}l_{-1} + 12l_1q_{-3})q_0 - (6l_{-1}q_2q_{-3} + 8l_1q_{-1}q_{-2})) + \tfrac{1}{3}l_{-1}l_1(3q_0q_{-2} + 4q_{-1}^2)) \\
& + q_{-1}^2((-12l_0l_{-1}q_3 + l_1l_{-1}q_2 + 28l_0l_1q_1)q_0 + (8l_{-1}q_2q_1 + 6q_{-2}q_3l_1)l_0) + l_1l_{-1}(\tfrac{5}{2}q_3q_1q_{-2}^2 - 9q_3q_0q_{-1}q_{-2} + \tfrac{1}{3}q_2q_1q_{-1}q_{-2} + \tfrac{5}{2}q_2^2q_{-1}q_{-3} - \tfrac{1}{6}q_2^2q_{-2}^2 \\
& - 36q_3q_1q_{-1}q_{-3} - 9q_2q_1q_0q_{-3}) + l_0l_1((6l_{-1}q_2^2q_{-3} + 5l_1q_2q_{-1}q_{-2} - 6l_1q_3q_{-2}^2 - 5l_{-1}q_2q_1q_{-2})q_0 + (6q_3q_1q_{-2}q_{-3} + \tfrac{1}{2}\left(q_2q_1q_{-2}^2 - q_2^2q_{-2}q_{-3}\right)) \\
& + 2l_0l_1q_2q_1q_{-1}q_{-3} + l_0l_{-1}(\tfrac{1}{2}\left(q_3q_2q_{-2}^2 - q_2^2q_{-1}q_{-2}\right) - 6q_3q_2q_{-1}q_{-3} - 2q_3q_1q_{-1}q_{-2}).
\end{aligned}
$$

$$
\begin{aligned}
[5,1] = \; & -8l_0q_0^5 - \tfrac{4}{3}(l_1q_{-1} - l_{-1}q_1)q_0^4 + \tfrac{2}{9}q_0^3((2q_2q_{-2} - 9q_3q_{-3} + 58q_1q_{-1})l_0 + (6q_2q_{-3} + q_1q_{-2})l_1 - (q_2q_{-1} + 6q_3q_{-2})l_{-1}) + q_2^2q_{-3}l_1(\tfrac{1}{12}q_3q_{-3} - \tfrac{1}{81}q_2q_{-2}) \\
& + \tfrac{43}{81}(l_{-1}q_1^3q_{-1}^2 - l_1q_1^2q_{-1}^3) - \tfrac{1}{108}(l_0q_3^2q_{-2}^3 + l_0q_2^3q_{-3}^2) + q_0^2(\tfrac{5}{27}\left(l_{-1}q_2q_1q_{-2} - l_1q_2q_{-1}q_{-2}\right) + \tfrac{1}{3}\left(l_1q_3q_{-2}^2 - l_{-1}q_2^2q_{-3} + l_0q_2q_1q_{-3} + l_0q_3q_{-1}q_{-2}\right)) \\
& + \left(\tfrac{11}{9}\left(l_{-1}q_3q_{-1}^2 - l_1q_1^2q_{-3}\right) + \tfrac{5}{3}\left(l_{-1}q_3q_1q_{-3} - l_1q_3q_{-1}q_{-3}\right) + \tfrac{16}{9}\left(l_1q_1q_{-1}^2 - l_{-1}q_1^2q_{-1}\right) - \tfrac{38}{27}\left(l_0q_2q_{-1}^2 + l_0q_1^2q_{-2}\right)\right)q_0^2 + \tfrac{1}{36}(l_1q_1^3q_{-2}^2 - l_{-1}q_2^2q_{-1}^3) \\
& - \tfrac{1}{18}(l_1q_3q_2q_{-1}q_{-2}q_{-3} - l_{-1}q_3q_2q_1q_{-2}q_{-3}) + \tfrac{4}{27}(l_1q_3q_{-1}^4 - l_{-1}q_1^4q_{-3}) + l_0q_0(\tfrac{4}{27}\left(q_1^3q_{-3} + q_3q_{-1}^3\right) + \tfrac{1}{18}\left(q_3q_1q_{-2}^2 + q_2^2q_{-1}q_{-3}\right) + q_3^2q_{-3}^2(l_1q_{-1} - l_{-1}q_1) \\
& + \tfrac{10}{81}q_1q_{-1}(l_{-1}q_2^2q_{-3} - l_1q_3q_{-2}^2) + \tfrac{1}{18}l_0q_0(20q_1q_{-1} - 11q_2q_{-2} + 18q_3q_{-3})q_3q_{-3} + \tfrac{4}{81}q_2^2q_{-2}^2l_0q_0 - \left(\tfrac{403}{81}q_1^2q_{-1}^2 + \tfrac{5}{54}q_2q_1q_{-1}q_{-2}\right)l_0q_0 + \tfrac{68}{81}q_1^3q_{-3}q_{-1}l_1 \\
& + (\tfrac{8}{9}q_1q_{-1} - \tfrac{1}{12}q_{-2}q_2 + \tfrac{8}{27}q_3q_{-3})(q_{-1}^2q_2 + q_1^2q_{-2})l_0 + (\tfrac{7}{162}q_{-2}q_2 - \tfrac{32}{81}q_1q_{-1})(q_3q_{-1}q_{-2} + q_1q_{-3}q_2)l_0 + \tfrac{19}{162}q_2q_{-2}(l_{-1}q_3q_{-1}^2 - l_1q_{-3}q_1^2) \\
& + q_0(\tfrac{41}{162}q_1q_{-1} + \tfrac{7}{18}q_3q_{-3} + \tfrac{2}{81}q_{-2}q_2)(q_2l_{-1}q_{-1} - l_1q_{-2}q_1) + q_0(\tfrac{4}{27}q_{-2}q_2 - \tfrac{47}{54}q_1q_{-1} - \tfrac{5}{6}q_3q_{-3})(q_{-3}q_2l_1 - l_{-1}q_3q_{-2}) - \tfrac{2}{27}l_0l_1q_{-1}^2(q_{-1}q_2 + 4q_{-2}q_3) \\
& + \tfrac{2}{27}q_0q_1^2l_{-1}(q_1q_{-2} + 4q_{-3}q_2) - (\tfrac{38}{27}q_{-3}q_3q_{-1}q_1 + \tfrac{17}{81}q_1q_{-1}q_{-2}q_2 - \tfrac{1}{81}q_2^2q_{-2}^2)(q_1l_{-1} - l_1q_{-1}) + l_1q_1q_{-3}^2(\tfrac{8}{9}q_3q_1 - \tfrac{1}{108}q_2^2) + l_1q_{-3}q_{-1}(\tfrac{68}{81}q_1^3 - \tfrac{5}{108}q_2^2q_{-1}) \\
& - \tfrac{5}{108}l_1q_{-3}q_{-1}^2q_2^2 + (\tfrac{1}{81}q_2q_{-2} - \tfrac{1}{12}q_3q_{-3})q_3q_{-2}^2l_1 + l_{-1}q_3^2q_{-1}(\tfrac{1}{108}q_{-2}^2 - \tfrac{8}{9}q_{-1}q_{-3}) + l_{-1}q_1q_3(\tfrac{5}{108}q_1q_{-2}^2 - \tfrac{68}{81}q_{-1}^3)
\end{aligned}
$$

$$
\begin{aligned}
[6,0] = \; & -729q_0^6 - 54q_1^4q_{-2}^2 + 54q_3q_{-3}\left(9q_2q_0^2q_{-2} + 162q_1q_0^2q_{-1} - 32q_1^2q_{-1}^2 + 6q_2q_1q_{-1}q_{-2}\right) + 6q_2q_{-2}\left(6q_3q_{-1}^3 - 10q_1^2q_{-1}^2 + 6q_{-3}q_1^3 - 63q_1q_0^2q_{-1}\right) \\
& -162q_0^2\left(q_{-2}^2q_3q_1 + q_2^2q_{-3}q_{-1}\right) + 54\left(q_0^2\left(27q_3^2q_{-3}^2 - 8q_{-3}q_1^3 - 8q_3q_{-1}^3 - 13q_1^2q_{-1}^2\right) - q_3^2\left(-q_0q_{-2}^3 + q_{-1}^2q_{-2}^2\right)\right) - 64q_1^3q_{-1}^3 + q_2^3q_0q_{-3}^2 + q_2^3q_{-2}^3 \\
& -54\left(q_1^2q_{-3}^2 + q_{-1}^4\right)q_2^2 - 3q_2^2q_{-2}^2\left(4q_1q_{-1} + 9q_0^2\right) - 324q_0^3\left(q_1^2q_{-2} + q_2q_{-1}^2\right) - 18q_{-2}q_2\left(q_{-2}^2q_3q_1 + q_2^2q_{-3}q_{-1}\right) - 756q_0q_1q_{-1}\left(q_3q_{-1}q_{-2} + q_2q_1q_{-3}\right) \\
& +972\left(q_0^3\left(q_3q_{-1}q_{-2} + q_2q_1q_{-3}\right) - \left(q_3^2q_{-1}q_{-2}q_{-3} + \left(q_2q_1q_{-3}^2 + q_1^2q_{-2}q_{-3}\right)q_3\right)q_0\right) + 243\left(6q_1q_{-1} - 30q_3q_{-3} + q_2q_{-2}\right)q_0^4 + 288q_{-1}q_1\left(q_{-3}q_1^3 + q_3q_{-1}^3\right) \\
& +864q_{-3}^2q_3q_1^3 + 972q_2q_{-1}^2q_{-3}q_3q_0 + 90q_{-2}q_2\left(q_1^2q_{-2} + q_2q_{-1}^2\right)q_0 + 396q_{-1}q_0q_1\left(q_1^2q_{-2} + q_2q_{-1}^2\right) + 180q_1q_{-1}\left(q_{-2}^2q_3q_1 + q_2^2q_{-3}q_{-1}\right) + 864q_{-3}q_3^2q_{-1}^3
\end{aligned}
$$

There is a first interesting fact concerning these labelling operators. Although their symmetrized expressions belong to an integrity basis for this MLP [14], they cannot be deduced neither from the reduction chain $\mathfrak{so}(5) \supset \mathfrak{so}(3)$ itself nor from the associated inhomogeneous contraction. This is due essentially to the fact that the Lie $\mathfrak{so}(5)$ has no primitive Casimir operator of degree six. According to [15], these labelling operators are purely formal, and do not inherit an obvious physical meaning. A routine computation shows that these elementary scalars do not commute mutually, as shown in the following table:[7]

$[\,,\,]$	$[2,4]$	$[3,3]$	$[4,2]$	$[5,1]$	$[6,0]$
$[2,4]$	–	1070	2112	2980	2112
$[3,3]$		–	2862	3562	2490
$[4,2]$			–	3840	2602
$[5,1]$				–	2154

Clearly all these operators commute with $[0,2]$ and the quadratic Casimir operator C_2 of $\mathfrak{so}(5)$, thus with $[2,0]$. The next case to be analyzed corresponds to the commutator of an operator of degree four with another of degree six. Here, in accordance with [11], we find the first nontrivial pairs of commuting labelling operators. To this extent, we analyze the commutator of an arbitrary linear combination of the scalars $[4,0]$, $[3,1]$, $[2,2]$ and $[1,3]$ with an operator of degree six formed by $[2,4]$, $[3,3]$, $[4,2]$, $[5,1]$ and $[6,0]$ and products of lower order operators.[8] Proceeding in this way, we find the two following pairs of linearly independent operators $\{X_1^1, X_2^1\}$ and $\{X_1^2, X_2^2\}$, where

$$\begin{aligned} X_1^1 =\; & [4,0]+[3,1]+(4-3\alpha)[2,2]+\alpha[1,3], \quad \alpha\neq 1,\\ X_2^1 =\; & -\tfrac{27}{5}[6,0]-162[5,1]+[4,2]-216[3,3]-[2,0]\left(5310[2,2]+\tfrac{2025}{2}[4,0]\right)\\ & +[0,2](2124[3,1]+528[1,3])+768[2,4] \end{aligned} \tag{11}$$

and

$$\begin{aligned} X_1^2 =\; & \left(\tfrac{4}{3}-\beta\right)[4,0]+\tfrac{3}{2}(1-\beta)[3,1]+3\beta[2,2]+[1,3], \quad \beta\neq\tfrac{1}{3}\\ X_2^2 =\; & -\tfrac{12}{5}[6,0]+108[5,1]+[4,2]+324[3,3]-[2,0](180[2,2]-1035[3,1])\\ & -[0,2](17172[2,2]+1728[2,4]+3998[1,3]). \end{aligned} \tag{12}$$

The corresponding symmetrized operators $O_i^j = Sym(X_i^j)$ in the enveloping algebra of $\mathfrak{so}(5)$ satisfy the requirements

$$\begin{aligned} \left[O_1^1, O_2^1\right] = 0, \quad \left[O_1^1, O_2^2\right] \neq 0,\\ \left[O_1^2, O_2^1\right] \neq 0, \quad \left[O_1^2, O_2^2\right] = 0. \end{aligned} \tag{13}$$

Observe that for the excluded values of the parameters α and β, the fourth order labelling operator is reduced to the Casimir operator C_4 of $\mathfrak{so}(5)$. Equation (13) further confirms that the sets $\mathcal{F}_{1,\alpha} = \{O_1^1, O_2^1\}$ and $\mathcal{F}_{2,\beta} = \{O_1^2, O_2^2\}$ are inequivalent sets, since no element of $\mathcal{F}_{1,\alpha}$ commutes with an element of $\mathcal{F}_{2,\beta}$.

We claim that the two preceding pairs of labelling operators can be taken as a possible choice for the four available operators that solve this MLP. This is equivalent to show that the seven operators $[0,2], [2,0], C_4, X_1^1, X_1^2, X_2^1, X_2^2$ are functionally independent. To prove this assertion, it suffices to find a set of seven independent variables such that the Jacobian

[7] As before, the table specifies the number of terms for the analytical counterpart of the commutator.
[8] Observe that for the four dimensional operator, the products and powers of quadratic labelling operators need not to be considered, since they commute with the sixth order solutions.

of these operators with respect to these variables does not vanish. We take the variables $\{l_0, l_{-1}, q_0, q_{-1}, q_{-2}, q_{-3}, q_3\}$ and consider

$$\frac{\partial\left([0,2],[2,0],C_4,X_1^1,X_1^2,X_2^1,X_2^2\right)}{\partial\left(l_0,l_{-1},q_0,q_{-1},q_{-2},q_{-3},q_3\right)} \neq 0. \tag{14}$$

Since the operators are independent, they can be taken as a fundamental set of solutions to system (8). We remark that this can also be proved using an indirect argument based on equation (13).

It is clear that any linear combination $X_1^1 + \mu C_4$ or $X_2^1 + \mu C_4$ is also a possible fourth order labelling operator that commutes with X_1^2 and X_2^2, respectively. In view of these possibilities, it is natural to look for commuting pairs of operators with the lowest possible number of components. Since the found operators of degree six cannot be modified, this leads to look for operators of degree four having four or less components. Having in mind that $\alpha \neq 1$ and $\beta \neq \frac{1}{3}$, it can be shown that, up to scalars, only seven non-equivalent operators with less of four components exist:

(i) $X_1^1(0) = [4,0] + [3,1] + 4\,[2,2]$,
(ii) $X_1^1(\frac{4}{3}) = [4,0] + [3,1] + \frac{4}{3}\,[1,3]$,
(iii) $X_1^1 - C_4 = 3\,[2,2] - 1\,[1,3]$,
(iv) $X_2^1(\frac{4}{3}) = -\frac{1}{2}\,[3,1] + 4\,[2,2] + [1,3]$,
(v) $X_2^1(1) = [4,0] + 9\,[2,2] + [1,3]$,
(vi) $X_2^1(0) = \frac{4}{3}\,[4,0] + \frac{3}{2}\,[3,1] + [1,3]$,
(vii) $X_2^1 - C_4 = [4,0] + \frac{3}{2}\,[3,1] - 3\,[2,2]$.

The two component solution found in [11] is equivalent to the symmetrized operators obtained from $\{X_1^1 - C_4, X_1^2\}$. The discrepancy in the coefficients of the scalars [2,2] and [1,3] is due to a different normalization factor to that used in [11]. Moreover, it follows from the previous list that this is the only solution with two terms. In this sense, the pair proposed in [11] is actually the simplest possible choice for solving the missing label problem. The pair $\{X_2^1, X_2^2\}$ constitutes a new solution and has no analogue in the previous analysis. The non-equivalence of these sets of labelling operators refers to their independence and to the fact they do not mutually commute. Thus the class of labelling operators is divided into two incompatible sets with respect to the commutativity requirement. Starting from either the operators of $\mathcal{F}_{1,\alpha}$ or $\mathcal{F}_{2,\beta}$, labelling operators of higher even order may be constructed. In particular, for arbitrary constants a, b, the pairs $\{(a\,[2,0] + b\,[0,2])\,X_1^1, X_1^2\}$ and $\{(a\,[2,0] + b\,[0,2])\,X_2^1, X_2^2\}$ are solutions to the MLP consisting of two operators of degree six, although one of them is decomposable. It was further verified that there do not exist two independent of operators of the form $O = a_1\,[6,0] + a_2\,[5,1] + a_3\,[4,2] + a_4\,[3,3] + a_5\,[2,4]$ that commute.

Looking for higher order solutions to system (8), we found that it has only three solutions of degree seven, which correspond to operators of bi-degree [5,2], [4,3] and [3,4] respectively.[9] These operators can be shown to appear in the commutator of fourth order elementary scalars:

$$[[4,0],[3,1]] = [5,2]\,,\ [[4,0],[1,3]] = [4,3]\,,\ [[2,2],[3,1]] = [3,4]\,. \tag{15}$$

As a consequence of the non-existence of elementary scalars of degree three and five, it follows that no linear combination of [5,2], [4,3] and [3,4] commutes with a fourth order labelling

[9] Their symmetrized expressions coincide with those found by other authors for the integrity basis. The computation of indecomposable solutions up to order nine also agreed completely with the results of [14].

operator. In agreement with the results of [14], we found also seven indecomposable scalars of degree nine among the 13 solutions of (8) having this degree. However, no lineal combination of the latter commutes with an operator of degree four. Due to the computational complications for these higher order operators, we did not further extended the search of commuting pairs supplementary to those already found. A question left open in this analysis concerns the existence of a pair of commuting operators consisting of an operator of degree four and an operator of odd degree.

Final remarks

We have obtained one possible choice for the four functionally independent label operators available among the 28 subgroups scalars computed in [14] for the reduction chain $\mathfrak{so}(5) \supset \mathfrak{so}(3)$. Two inequivalent sets of solutions have been found, each consisting of operators of degree four and six, respectively. While the fourth operator can be interpreted as a linear combination of "broken" Casimir operators with respect to the contraction defined by the embedding of $\mathfrak{so}(3)$ into $\mathfrak{so}(5)$ [15, 16], the operators of order six have no obvious interpretation in terms of the initial data. As observed in [16], it is unlikely that such operators, which have to be determined by pure formal means, have really a significant physical meaning other than that conferred by the labelling of basis states. As shown in [17, 18], for degenerate representations the number of missing label operators needed is lower than that given by the generic formula (2). This actually happens for the applications of this chain to the description of quadrupole vibrations of the nuclear surface about a spherical equilibrium shape and their coupling to giant dipole resonances [19, 20], where the totally symmetric representations are considered. In these situations, linear combinations of the operators X_1^1, X_1^2 constitute the natural choice, and allow to determine the most general form of a fourth order labelling operator. The numerical part, left untouched here, concerns the determination of the complete eigenvalue spectrum for these missing label operators.

Finally, one striking point about the missing label problem $\mathfrak{so}(5) \supset \mathfrak{so}(3)$ is its difficulty when compared to similar problems with more than one labelling operator, like the Wigner's scheme $\mathfrak{su}(4) \supset \mathfrak{su}(2) \times \mathfrak{su}(2)$ [21]. Although the decomposition of the fourth order Casimir operator provides a sufficient number of labelling operators, the commutativity requirement is not fulfilled, regardless of the linear combinations considered. This forces to determine higher order operators with no apparent link to the information provided by the embedding. One may wonder whether this difficulty is related to the fact that the subalgebra $\mathfrak{so}(5)$ is principal, and therefore possesses special characteristics not shared by other types of subalgebras. Another obstruction to find suitable pairs of commuting operators seems to be the non-existence of solutions of orders three and five, and the fact that the seventh order labelling operators arise as commutators of operators of order four.

Acknowledgments

The author expresses his gratitude to J. Van der Jeugt and S. De Baerdemacker for useful discussions and reference [9]. During the preparation of this work, the author was financially supported by the research projects MTM2006-09152 of the M.I.C.I.N.N. and the GR58/4120818-920920 of the UCM-BSCH.

References

[1] Sharp R T 1975 *J. Math. Phys.* **16**, 2050
[2] Louck J D 1976 *Proceedings of the International Symposium on Mathematical Physics*, Mexico City, Mexico
[3] Judd B R, Miller W, Patera J and Winternitz P 1974 *J. Math. Phys.* **15** 1787
[4] Syamela Devi V and Venkatarayayudu T 1968 *J. Math. Phys.* **9** 1057
[5] Syamela Devi V 1970 *Nouvo Cimento* **A68** 1

[6] Wybourne B G 1972 *J. Math. Phys.* **13** 455
[7] Jarvis P D and Zhang R B 2005 *J. Phys. A: Math. Gen.* **38** L219
[8] Kerman A K 1953 *Phys. Rev.* **92** 1176
[9] De Baerdemacker S, Heyde K and Hellemans V 2007 *J. Phys. A: Math. Gen.* **40** 2733
[10] Vanden Berghe G and De Meyer H 1980 *J. Math. Phys.* **21** 1902
[11] De Meyer H, Vanden Berghe G, Van der Jeugt J and De Wilde P 1985 *J. Math. Phys.* **26** 2124
[12] Racah G 1951 *Group Theory and Spectroscopy*, (Princeton Univ. Press, N. J.)
[13] Peccia A and Sharp R T 1976 *J. Math. Phys.* **17** 1313
[14] Gaskell R, Peccia A and Sharp R T 1987 *J. Math. Phys.* **19** 727
[15] Campoamor-Stursberg R 2008 *J. Phys. Conf. Ser.* **128** 012052
[16] Campoamor-Stursberg R 2007 *J. Phys. A: Math. Gen.* **40** (2007), 14773
[17] Giroux Y and Sharp R T 1987 *J. Math. Phys.* **28** 1671
[18] Gilmore R and Draayer J P 1985 *J. Math. Phys.* **26** 3053
[19] Hecht K T 1965 *Nucl. Phys.* **63** 177
[20] Frank A and van Isacker P 1994 *Algebraic Methods in Molecular and Nuclear Structure Physics* (Wiley: New York)
[21] Quesne Ch 1977 *J. Math. Phys.* **18** 1210

A geometric approach to higher-order Riccati chain: Darboux polynomials and constants of the motion

José F. Cariñena[1], Partha Guha[§2] and Manuel F. Rañada[3]

Departamento de Física Teórica, Facultad de Ciencias, Universidad de Zaragoza, 50009 Zaragoza, Spain
[§] Max Planck Institute for Mathematics in the Sciences
Inselstrasse 22, D-04103 Leipzig, Germany
[§] S.N. Bose National Centre for Basic Sciences, JD Block
Sector-3, Salt Lake, Calcutta-700098, India

E-mail: [1]`jfc@unizar.es`, [2]`partha@bose.res.in`, [3]`mfran@unizar.es`

Abstract. The properties of higher-order Riccati equations are investigated. The second-order equation is a Lagrangian system and can be studied by using the symplectic formalism. The second-, third- and fourth-order cases are studied by proving the existence of Darboux functions. The corresponding cofactors are obtained and some related properties are discussed. The existence of generators of t-dependent constants of motion is also proved and then the expressions of the associated time-dependent first integrals are explicitly obtained. The connection of these time-dependent first integrals with the so-called master symmetries, characterizing some particular Hamiltonian systems, is also discussed. Finally the general n-th-order case is analyzed.

1. Introduction

Symmetry analysis and the corresponding reduction process is a fundamental approach in the solution of many physical problems. The introduction by Lie of the concept of the now called Lie groups and algebras has played a relevant rôle in the solution of systems of differential equations. The knowledge of infinitesimal symmetries for a given system of differential equations allows us to introduced adapted coordinates which reduce the system to one of a simpler form, and in particular differential equations to lower order ones. For instance linear second order differential equations are invariant under dilations and the corresponding reduction equation turns out to be a first order Riccati equation. This is one of the reasons for the relevance in physics of such an equation.

The second-order Riccati equation has been studied in [1] from a geometric perspective and it has been proved to admit two alternative Lagrangian formulations, both Lagrangians being of a non-natural class (neither potential nor kinetic term). Our aim is to use analogous geometric techniques to deal with the higher-order terms of the so called Riccati sequence [2, 3, 4]. The symmetry properties of the Riccati sequence (also known as Riccati hierarchy or chain) has been recently studied by Euler *et al* [5]. They showed that the members of the Riccati hierarchy admit the maximal number of Lie point symmetries for an equation of order no lower than two. The first two members of the hierarchy are special cases of the standard Riccati equation and the second-order Riccati equation (SORE) (also known as Painlevé-Ince equation).

With this aim we find some Darboux polynomials (see [6, 7] for introduction) which may be helpful in the reduction or even solution of the given problem. In 1878 Darboux [8] showed how one can construct the first integrals of planar polynomial ordinary differential equations (ODEs) possessing sufficient invariant algebraic curves. Nowadays Darboux method of integrability is considered to be one of the best methods for finding first integrals of polynomial ODEs. Recall that Prelle and Singer [9] proposed a few years ago a method for solving first-order ODEs that presents their solutions in terms of elementary functions when such a solution exists. Recently Duarte *et al.* [10] modified the Prelle-Singer technique and applied it to second-order ODEs. The method was based on a conjecture that if an elementary solution exists for the given second-order ODE then there exists at least one elementary first integral $I(t, x, \dot{x})$ whose derivatives are all rational functions of t, x and $\dot{x}$. More recently Chandrasekar *et al.* [11] discussed a method for solving n-th-order scalar ODEs by extending some ideas based on the Prelle-Singer method for second-order ODEs; however there is not so much work done on higher-order operators.

Our first objective is the study of higher-order Riccati equations, as $\ddot{x}+3k\,x\,\dot{x}+k^2x^3 = 0$, which was recently studied in [1] within a Lagrangian formulation, from a more geometric perspective and using their relations with the theory of Darboux and the extended Prelle-Singer methods. We give a new development of a deeper analysis of these equations using presymplectic forms, Darboux polynomials and Jacobi multipliers. This suggests us how to deal with higher-order Riccati equations in a similar approach.

The plan of this article is as follows: In Sec. 2 we recall the Riccati chain and its relation with the series of differential equations $y^{(n)} = 0$ from which the elements of the Riccati sequence appear by using the Lie reduction method. Section 3 is devoted to review some results of [1] for the second-order Riccati equation and other mathematical tools as Darboux polynomials and Jacobi last multipliers which enable us with a better understanding of the Lagrangian theory developed in [1]. We develop in Section 4 an analogous formalism for the third-order Riccati equation (TORE). In particular, we find a Darboux polynomial and a presymplectic form for the TORE. Section 5 is devoted to an analogous formalism for the fourth-order Riccati equation. Finally the generic n-th-order case is analysed in Section 6 and the results are summarized in the last section.

2. The Riccati chain

The Riccati sequence can be introduced as follows. Let us define, for a given real number $k \in \mathbb{R}$, the differential operator

$$\mathbb{D}_k = \frac{d}{dt} + k\,x\,, \tag{1}$$

and then consider the sequence obtained by applying such a differential operator to the function $x(t)$ in an iterative way so that we obtain

$$\begin{aligned}
n&=0 \qquad \mathbb{D}_k^0 x = x\\
n&=1 \qquad \mathbb{D}_k x = \left(\frac{d}{dt}+kx\right) x = \dot{x}+k\,x^2\\
n&=2 \qquad \mathbb{D}_k^2 x = \left(\frac{d}{dt}+kx\right)^2 x = \ddot{x}+3k\,x\,\dot{x}+k^2\,x^3\\
n&=3 \qquad \mathbb{D}_k^3 x = \left(\frac{d}{dt}+kx\right)^3 x = \dddot{x}+4\,k\,x\ddot{x}+6k^2\,x^2\dot{x}+3k\,\dot{x}^2+k^3\,x^4\\
n&=4 \qquad \mathbb{D}_k^4 x = \left(\frac{d}{dt}+kx\right)^3 x = \ddddot{x}+5kx\,\dddot{x}+10k\,\dot{x}\ddot{x}+15k^2\,x\,\dot{x}^2+10k^2x^2\ddot{x}\\
&\qquad\qquad + 10k^3x^3\dot{x}+k^4\,x^5
\end{aligned} \tag{2}$$

and analogous expressions for higher values of n.

The equation

$$R^{(j)}(x,\ldots,x^{(j)}) = \mathbb{D}_k^j x = 0\,, \quad j=0,1,\ldots,$$

is a Riccati equation of order j. Note that $R^{(0)}(x)=x$, the particular case $R^{(1)}(x,\dot{x})=0$, obtained for $j=1$, is the standard Riccati equation with constant coefficients in its reduced form:

$$\dot{x}+k\,x^2=0\,, \tag{3}$$

and the particular case $R^{(2)}(x,\dot{x},\ddot{x})=0$, obtained for $j=2$, is the second-order Riccati equation

$$\ddot{x}+3k\,x\,\dot{x}+k^2\,x^3=0\,, \tag{4}$$

recently studied in [1]. Our aim here is to first study the Riccati equation of third-order

$$\dddot{x}+4\,k\,x\ddot{x}+6k^2\,x^2\dot{x}+3k\,\dot{x}^2+k^3\,x^4=0\,, \tag{5}$$

which appears in the same way as the first- (3) and second-order Riccati (4) equations; then the study proceeds with other higher-order equations. Of course, for any j we have $R^{(j+1)}=\mathbb{D}_k R^{(j)}$, by definition.

The relevance of usual first-order Riccati equation in classical and quantum physics is mainly due to its appearance in the Lie reduction process when taking into account invariance under infinitesimal dilations of linear second-order differential equations [12] and it was proved in [1] that the second-order Riccati equations appear in a similar way from linear third-order differential equations and in general the j-order Riccati equation appears as a reduction of a linear $(j+1)$-order differential equation. More specifically, given a differential equation $y^{(n)}=0$, its invariance under dilations suggests, according to Lie recipe, to look for a new variable z such that the dilation vector field $y\,\partial/\partial y$ becomes $(1/k)\partial/\partial z$, the factor k being purely conventional. Then $y=e^{kz}$, up to an irrelevant factor, and with this change of variable, $y^{(n)}=0$, for $n>1$, becomes $R^{(n-1)}(x)=0$ with $x=\dot{z}$, because, when $D=d/dt$, we have that $e^{-kz}De^{kz}=D+k\,x$, and then we can prove by complete induction that $y^{(n)}=k\,e^{kz}R^{(n-1)}(x)$. In fact, the property is true for $n=1$, because $y'=k\,\dot{z}\,e^{kz}=e^{kz}\,k\,x=k\,e^{kz}\,R^{(0)}(x)$, and if the property is true for n, it also holds for $n+1$, because

$$D^{n+1}y = D(y^{(n)}) = D(k\,e^{kz}R^{(n-1)}(x)) = k\,e^{kz}(D+k\,x)R^{(n-1)}(x) = k\,e^{kz}\,R^{(n)}(x).$$

Therefore, Lie recipe applied to the invariant under dilation equation $y^{(n)} = 0$ transforms such equation into $R^{(n-1)}(x) = 0$. Of course the most general Riccati equation of order n can be obtained from a linear combination $\sum_{j=0}^{n+1} a_j(t) y^{(j)}$ which gives rise to

$$a_0(t) + k \sum_{j=0}^{n} a_{j+1}(t)\, R^{(j)}(x, \ldots, x^{(j)}) = 0\,.$$

There is an identity generalising the chain rule to higher derivatives which provides us the element of order n of the Riccati chain. This identity, usually known as Faá di Bruno formula [13, 14], is given by

$$\frac{d^n}{dt^n} g(f(t)) = \sum \frac{n!}{b_1! \cdots b_n!} g^{(b_1+\cdots+b_n)}(f(t)) \left(\frac{\dot f(t)}{1!}\right)^{b_1} \cdots \left(\frac{f^{(n)}(t)}{n!}\right)^{b_n}, \tag{6}$$

where the sum is over all natural numbers such that $b_1 + 2b_2 + \cdots + n\, b_n = n$.

In our case $f(t) = z(t)$ and $g(t) = e^{kt}$, and taking into account that $g^{(l)} = k^l\, g$ and that $\dot z = x$:

$$y^{(n)} = \frac{d^n}{dt^n} e^z = e^z \sum_{b_1+\cdots+nb_n=n} \frac{n!}{b_1! \cdots b_n!} k^{b_1+\cdots+b_n} x^{b_1} \left(\frac{\dot x}{2!}\right)^{b_2} \cdots \left(\frac{x^{(n-1)}}{n!}\right)^{b_n}, \tag{7}$$

which can be used to find the nth-element of the Riccati sequence:

$$\sum_{b_1+\cdots+nb_n=n} \frac{n!}{b_1! \cdots b_n!} k^{b_1+\cdots+b_n} x^{b_1} \left(\frac{\dot x}{2!}\right)^{b_2} \cdots \left(\frac{x^{(n-1)}}{n!}\right)^{b_n} = 0\,. \tag{8}$$

Note that the coefficient of the highest-order derivative in (8), which corresponds to $b_n = 1$, is 1.

3. The Lagrangian form of the second-order Riccati equation

We denote $\widetilde{R}^{(j)}(x, v^{(1)}, \ldots, v^{(j)})$ the expression obtained from $R^{(j)}$ by a simple substitution of $v = v^{(1)}$ by $\dot x$, $a = v^{(2)}$ by $\ddot x$, and so on. For instance,

$$\widetilde{R}^{(0)}(x) = x, \quad \widetilde{R}^{(1)}(x, v) = v + kx^2, \quad \widetilde{R}^{(2)}(x, v, a) = a + 3kxv + k^2x^3,$$

$$\widetilde{R}^{(3)}(x, v, a, w) = w + 4kxa + 6k^2x^2\, v + 3kv^2 + k^3x^4,$$

$$\widetilde{R}^{(4)}(x, v, a, w, u) = u + 5kxw + 10kva + 15k^2xv^2 + 10k^2x^2a + 10k^3x^3v + k^4x^5\,.$$

In the geometric approach to differential equations the element of order j in our Riccati sequence is given by the zero level set of the function $\widetilde{R}^{(j)}$ defined in $T^j\mathbb{R}$ (the j-th-order tangent bundle, see e.g. [15, 16]), which can also be seen as a section for the natural map $\tau_{j,j-1} : T^j\mathbb{R} \to T^{j-1}\mathbb{R}$.

Journal of Physics: Conference Series **175** (2009) 012009 doi:10.1088/1742-6596/175/1/012009

There is another geometric interpretation of a differential equation of order j as given by a vector field in $T^{j-1}\mathbb{R}$. For instance, recall that an autonomous system of second-order differential equations like $\ddot{x} = F(x, \dot{x})$ has associated the system of first-order differential equations

$$\left\{ \begin{array}{rcl} \dfrac{dx}{dt} & = & v \\ \dfrac{dv}{dt} & = & F(x,v) \end{array} \right. , \tag{9}$$

whose solutions are the integral curves of the vector field in $T\mathbb{R}$

$$\Gamma_F^{(2)} = v\,\frac{\partial}{\partial x} + F(x,v)\frac{\partial}{\partial v}\,.$$

In particular, for the second-order Riccati equation (4), we have

$$\Gamma^{(2)} = v\,\frac{\partial}{\partial x} - (3kxv + k^2x^3)\frac{\partial}{\partial v}\,. \tag{10}$$

In the alternative geometric approach, the second-order differential equation can be seen as a section for $\tau_{2,1} : T^2\mathbb{R} \to T\mathbb{R}$, $\sigma(x,v) = (x, v, F(x,v))$, which in the second-order Riccati equation (4) reduces to $\sigma(x,v) = (x, v, -(3kxv + k^2x^3))$.

Similarly, an autonomous third-order differential equation $\dddot{x} = F(x, \dot{x}, \ddot{x})$ is studied either by means of the system of first-order differential equations

$$\left\{ \begin{array}{rcl} \dfrac{dx}{dt} & = & v \\ \dfrac{dv}{dt} & = & a \\ \dfrac{da}{dt} & = & F(x,v,a) \end{array} \right. \tag{11}$$

whose solutions are the integral curves of the vector field in $T^2\mathbb{R}$

$$\Gamma_F^{(3)} = v\frac{\partial}{\partial x} + a\frac{\partial}{\partial v} + F(x,v,a)\frac{\partial}{\partial a}\,,$$

or by the section σ for $\tau_{3,2} : T^3\mathbb{R} \to T^2\mathbb{R}$ given by $\sigma(x,v,a) = (x, v, a, F(x,v,a))$.

In particular, (5) is described either by the vector field

$$\Gamma^{(3)} = v\frac{\partial}{\partial x} + a\frac{\partial}{\partial v} - (4\,k\,x\,a + 6k^2\,x^2\,v + 3k\,v^2 + k^3\,x^4)\frac{\partial}{\partial a}\,, \tag{12}$$

or by the section $\sigma(x,v,a) = (x, v, a, -(4\,k\,x\,a + 6k^2\,x^2\,v + 3k\,v^2 + k^3\,x^4))$.

In the general case of a generic $\Gamma^{(j)}$ the connection between the vector field and the section describing the system is given by the $\mathbf{T}^{(j-1)}$ operator introduced in [15] or the canonical immersion $i_{j-1,1}$ used in [16].

In the search for constants of the motion allowing us to reduce a given system, the Jacobi multipliers [17] play a relevant rôle (see later on for the definition of Jacobi multiplier). A very useful ingredient for the determination of a Jacobi multiplier for a vector field, or even a Lagrangian in the case of a second-order differential equation, is the concept of Darboux polynomial we recall next.

Journal of Physics: Conference Series **175** (2009) 012009 doi:10.1088/1742-6596/175/1/012009

Definition 1 *Let U be an open subset of $\mathbb{R}^n$. We say that a polynomial function $\mathcal{D} : U \to \mathbb{R}$ is a Darboux polynomial for the polynomial vector field X if there is a polynomial function f defined in U such that $X\mathcal{D} = f\mathcal{D}$. The function f is said to be the cofactor corresponding to such Darboux polynomial.*

Note that when $\mathcal{D}$ is a Darboux polynomial for the polynomial vector field X, the zero level set of $\mathcal{D}$ is an algebraic curve invariant under X, and that given two Darboux polynomials for X, $\mathcal{D}_i : U_i \to \mathbb{R}$, $i = 1, 2$, with cofactors f_1 and f_2 respectively, its product $\mathcal{D}_1\mathcal{D}_2 : U_1 \cap U_2 \to \mathbb{R}$ is a Darboux polynomial with cofactor $f_1 + f_2$. Consequently, the power $\mathcal{D}^n$, $n \in \mathbb{N}$, of a Darboux polynomial $\mathcal{D}$ with cofactor f is a Darboux polynomial with cofactor $n\,f$. Moreover, for any integer number p, $X\mathcal{D}^p = p\,f\,\mathcal{D}^p$, and $X(\mathcal{D}_1/\mathcal{D}_2) = (f_1 - f_2)(\mathcal{D}_1/\mathcal{D}_2)$. Consequently, if $\mathcal{D}_1$ and $\mathcal{D}_2$ are Darboux polynomials with the same cofactor, the quotient $\mathcal{D}_1/\mathcal{D}_2$ is a first integral.

The rational functions $\mathcal{F}$ such that $X\mathcal{F} = f\,\mathcal{F}$, with f a polynomial function are also known as second-integrals for X, because they are first-integrals when $f = 0$. More generally, a set of C^1-functions $I_i = I_i(x)$, $i = 1, \ldots, r$, is considered to be a set of higher order integrals if there exist r^2 C^1-functions β_{ij} such that

$$XI_i = \sum_{j=1}^{r} \beta_{ij} I_j, \qquad i = 1, \cdots, r.$$

In this case, if $dI_1 \wedge \cdots \wedge dI_r \neq 0$ the set $\bigcap_{k=1}^{r} I_k^{-1}(0)$ is an invariant $(n-r)$-dimensional manifold. If for an index i we have that $\beta_{i1} = \beta_{i2} = \cdots = \beta_{ir} = 0$, the function I_i is called first integral, otherwise, if some β_{ij} are not zero, the system can be used to define second, third or higher order integrals.

In this paper we often use products of powers of Darboux polynomials as playing a rôle similar to that of the Lagrangian. In fact, it can be seen that for second-order Riccati equations the Lagrangian is the inverse of a Darboux polynomial, because as it has been proved in [1], a general second-order Riccati equation admits a Lagrangian formulation with a Lagrangian of a non-mechanical type

$$L(x, v) = \frac{1}{v + k\,U(x, t)},$$

where $U(x, t) = c_0(t) + c_1(t)\,x^2$ for a convenient choice of functions $c_i(t)$. For instance, (4) is described by the Lagrangian function

$$L_1(x, v) = \frac{1}{v + k\,x^2} = (\widetilde{R}^{(1)}(x, v))^{-1}. \tag{13}$$

Note that $\widetilde{R}^{(1)}(x, v)$ is a Darboux polynomial for $\Gamma^{(2)}$ with cofactor $-kx$, because $(\Gamma^{(2)}\widetilde{R}^{(1)})(x, v) = -kxv - k^2x^2 = -kx\,R^{(1)}(x, v)$, and when $L_1(x, v) = (R^{(1)}(x, v))^{-1}$, the vector field $\Gamma^{(2)}$ given by (10) is such that $\Gamma^{(2)}L_1 = k\,x\,L_1$. Moreover, the associated Liouville 1-form θ_{L_1} and the corresponding symplectic 2-form $\omega_{L_1} = -d\theta_{L_1}$,

$$\theta_{L_1} = \frac{\partial L_1}{\partial v}\,dx = -\frac{1}{(v + kx^2)^2}\,dx = -{L_1}^2\,dx\,, \tag{14}$$

$$\omega_{L_1} = -d\theta_{L_1} = \frac{2\,dx \wedge dv}{(v + kx^2)^3} = d{L_1}^2 \wedge dx\,, \tag{15}$$

together with the corresponding energy function

$$E_{L_1}(x,v) = v\,\frac{\partial L_1}{\partial v} - L_1 = \frac{-(2v+kx^2)}{(v+kx^2)^2}\,, \tag{16}$$

show that $\Gamma^{(2)}$ is such that $i(\Gamma^{(2)})\omega_{L_1} = dE_{L_1}$, i.e. $\Gamma^{(2)}$ is the dynamical vector field defined by the Lagrangian L_1.

We recall that given a vector field X in an oriented manifold (M,Ω), a function R such that $R\,i(X)\Omega$ is closed is said to be a Jacobi multiplier for X. Note that this means that RX is a divergenceless vector field and then

$$\mathcal{L}_{RX}\Omega = \operatorname{div}(RX)\;\Omega = [X(R) + R\operatorname{div}X]\Omega = 0\,,$$

and therefore we see that R is a last multiplier for X if and only if

$$X(R) + R\operatorname{div}X = 0\,. \tag{17}$$

Note that if R is a never vanishing Jacobi multiplier, then $f\,R$ is a Jacobi multiplier too if and only if f is a constant of motion, i.e. $Xf = 0$, because if (17) is true, then for any function f,

$$X(f\,R) + f\,R\operatorname{div}X = (Xf)\,R + f(X(R) + R\operatorname{div}X) = (Xf)\,R\,,$$

and consequently $f\,R$ is a multiplier too if and only if $Xf = 0$.

The remarkable point is that if $\mathcal{D}_1,\ldots,\mathcal{D}_k$ are Darboux polynomials with corresponding cofactors f_i, $i=1,\ldots,k$, one can look for multiplier factors of the form

$$R = \prod_{i=1}^{k}\mathcal{D}_i^{\nu_i} \tag{18}$$

and then

$$\frac{X(R)}{R} = \sum_{i=1}^{k}\nu_i\frac{X(\mathcal{D}_i)}{\mathcal{D}_i} = \sum_{i=1}^{k}\nu_i\,f_i\,,$$

and therefore, if the coefficients ν_i can be chosen such that

$$\sum_{i=1}^{k}\nu_i\,f_i = -\operatorname{div}X \tag{19}$$

holds, then we arrive to

$$\frac{X(R)}{R} = \sum_{i=1}^{k}\nu_i\,f_i = -\operatorname{div}X\,,$$

and consequently R is a Jacobi last multiplier for X.

In the case of $X = \Gamma^{(2)}$ we are considering we see that as $\operatorname{div}\Gamma^{(2)} = -3kx$, a Jacobi multiplier obtained as a power of the Darboux polynomial $\mathcal{D}_1 = \widetilde{R}^{(1)}$ is $R(x,v) = (\widetilde{R}^{(1)}(x,v))^{-3}$.

The main point is that if R is a Jacobi multiplier for a vector field which corresponds to a second-order differential equation, then there is an essentially unique Lagrangian (up to addition of a gauge term) such that $R = \partial^2 L/\partial v^2$ [18, 19].

Journal of Physics: Conference Series **175** (2009) 012009 doi:10.1088/1742-6596/175/1/012009

Actually, the inverse problem for the existence of a Lagrangian was analysed by Helmholtz and he found a set of conditions that a multiplier matrix $g_{ij}(x,\dot{x})$ must satisfy in order for a given system of second-order equations

$$\ddot{x}^j = F^j(x,\dot{x})\,,\quad j=1,2,\ldots,n,$$

or the corresponding system

$$\left\{\begin{array}{rcl}\dfrac{dx^i}{d\tau} & = & v^i\\ \dfrac{dv^i}{dt} & = & F^i(x,v)\end{array}\right.$$

which provides us the integral curves for the vector field

$$X = v^i\frac{\partial}{\partial x^i} + F^i(x,v)\frac{\partial}{\partial v^i}\,,$$

when written of the form

$$g_{ij}(\ddot{x}^j - F^j(x,\dot{x})) = 0\,,\quad i,j=1,2,\ldots,n,$$

to be the set of Euler–Lagrange equations for a certain Lagrangian L [20, 21, 22, 23] (the summation convention on repeated indices is assumed). If a matrix solution g_{ij} is obtained then it can be identified with the Hessian matrix of L, that is $g_{ij} = \partial L/\partial v^i \partial v^j$, and a Lagrangian L can be obtained by direct integration of the g_{ij} functions. The two first conditions just impose regularity and symmetry of the matrix g_{ij}; the two other ones are equations introducing relations between the derivatives of g_{ij} and the derivatives of the functions F^i. Here we only write the fourth set of conditions that determine the time-evolution of the g_{ij}

$$X(g_{ij}) = g_{ik}A^k{}_j + g_{jk}A^k{}_i\,,\quad A^i{}_j = -\frac{1}{2}\frac{\partial F^i}{\partial v^j}\,.$$

When the system is one-dimensional we have $i=j=k=1$ and then the three first set of conditions become trivial and the fourth one reduces to one single first-order P.D.E.

$$X(g) + g\,\frac{\partial F}{\partial v} \equiv v\,\frac{\partial g}{\partial x} + F\,\frac{\partial g}{\partial v} + g\,\frac{\partial F}{\partial v} = 0\,, \qquad (20)$$

which is nothing but the equation (17) defining the Jacobi multipliers, because $\mathrm{divX} = \partial F/\partial v$.

Therefore, the generalised inverse problem reduces to find the function g as a solution of this equation, i.e. a Jacobi multiplier. Once a solution g is known a Lagrangian L is obtained by integrating the function g two times with respect to velocities. The funtion L so obtained from g is unique up to addition of a gauge term. For instance in the case of the equation (4) we find, up to a gauge term, the Lagrangian (13).

It is also noteworthy that if a system with one degree of freedom admits two different regular Lagrangins then the function F definida por

$$\frac{\partial^2 L_1}{\partial v^2} = F\,\frac{\partial L_2}{\partial v^2}$$

is a constant of the motion, because such a function F is the quotient of two different Jacobi multipliers. This result is usually attributed to Currie and Saletan [24], but actually data back to Jacobi's time.

Journal of Physics: Conference Series **175** (2009) 012009 doi:10.1088/1742-6596/175/1/012009

Note however that $\mathcal{D}_2(x,v) = 2v - kx^2$ is also a Darboux polynomial with cofactor $f_2(x,v) = -2kx$, because $(\Gamma^{(2)}\mathcal{D}_2)(x,v) = v(2kx) - 2(3kxv + k^2x^3) = -2kx(2v + kx^2)$. One can therefore look for a new Jacobi multiplier that is a power of $\mathcal{D}_2$ and we obtain $R'(x,v) = (2v + kx^2)^{-3/2}$, defining the new Lagrangian function

$$L'_1(x,v) = \sqrt{2v + kx^2}$$

Note that the quotient of both Jacobi multipliers is but a function of the energy, which is a constant of the motion.

A function T that satisfies the following property

$$\frac{d}{dt}T \neq 0 \quad , \ldots , \quad \frac{d^m}{dt^m}T \neq 0\,, \quad \frac{d^{m+1}}{dt^{m+1}}T = 0\,,$$

is called a generator of integrals of motion of degree m. Notice that this means that the function T is a non-constant function generating a constant of motion by time derivation and also that T can be used for generating a t-dependent constant of the motion. This function, to be denoted by J_t, turns out to be a polynomial of order m in the time t. In the Hamiltonian (symplectic) case this property is directly related with the existence of "master symmetries" [25, 26, 27, 28]. If the dynamics is represented by a Hamiltonian vector field Γ_H, then a vector field Z that satisfies the following two properties

$$[Z, \Gamma_H] = \widetilde{Z} \neq 0\,, \qquad [\widetilde{Z}, \Gamma_H] = 0\,,$$

is called a "master symmetry" of degree $m = 1$ for Γ_H. If Z is such that

$$[Z, \Gamma_H] = \widetilde{Z} \neq 0\,,\ [\widetilde{Z}, \Gamma_H] \neq 0 \quad \text{and} \quad [[\widetilde{Z}, \Gamma_H], \Gamma_H] = 0\,,$$

then it is called a "master symmetry" of degree $m = 2$, and similarly for $m = 3, 4, \ldots$.

In the second-order Riccati case (4), we have that the Lagrangian $T_1 = L_1$ is a generator of degree three of first-integrals for $\Gamma^{(2)}$, as corresponding to a master symmetry of the dynamics [1]. More explicitly, we have

$$\Gamma^{(2)}T_1 = T_2 = \frac{k\,x}{v + k\,x^2}\,,$$

$$\Gamma^{(2)}(\Gamma^{(2)}T_1) = \Gamma^{(2)}T_2 = k\,,$$

$$\Gamma^{(2)}(\Gamma^{(2)}(\Gamma^{(2)}T_1)) = 0\,.$$

Hence T_1 and T_2 are generators of t-dependent constants of motion for the Lagrangian. The two functions

$$J^{(2)}_{t1} = T_2 - k\,t\,, \quad J^{(2)}_{t2} = T_1 - t\,T_2 + \frac{1}{2}\,k\,t^2\,,$$

are the corresponding constants of the motion obtained from T_1 and T_2.

Journal of Physics: Conference Series **175** (2009) 012009 doi:10.1088/1742-6596/175/1/012009

4. A geometric approach to third-order Riccati equation

For the case of the third-order Riccati equation in the Riccati sequence one can proceed in a similar way but of course we cannot have a (regular) Lagrangian formulation.

First, when applying the vector field (12) to $\widetilde{R}^{(2)}(x,v,a)$ we see that

$$(\Gamma^{(3)}\widetilde{R}^{(2)})(x,v,a) = -kx\,\widetilde{R}^{(2)}(x,v,a)\,,$$

and therefore the polynomial function $\mathcal{D}_1 = \widetilde{R}^{(2)}$ is a Darboux polynomial with cofactor equal to $-k\,x$. Consequently, we can consider the corresponding generalisation of the function L_1 given by (13):

$$L_2(x,v,a) = \frac{1}{a+3k\,x\,v+k^2\,x^3} = (\widetilde{R}^{(2)}(x,v,a))^{-1}\,, \tag{21}$$

which is not a Lagrangian for (5) but note that the function L_2 is such that

$$\Gamma^{(3)}(L_2) = k\,x\,L_2\,. \tag{22}$$

This means that the function L_2 is not a constant of motion for the dynamics $\Gamma^{(3)}$ and the foliation of level surfaces of L_2 is not invariant under the vector field (12), but its zero level set is invariant. Furthermore, as $\operatorname{div}\Gamma^{(3)} = -4\,k\,x$, we can find a Jacobi multiplier as a power $(\mathcal{D}_1)^{-4}$ of the Darboux polynomial $\mathcal{D}_1$, i.e. we find a solution of (19) with only one term.

Note also that

$$dL_2 = -\frac{3k(kx^2+v)dx + 3kx\,dv + da}{(a+3k\,x\,v+k^2\,x^3)^2}\,,$$

and therefore we can define a 2-form ω_{L_2} in $T^2\mathbb{R}$ in full similarity with the second relation in (15) by

$$\omega_{L_2} = dL_2^2 \wedge dx, \tag{23}$$

whose coordinate expression turns out to be given by

$$\omega_{L_2} = 2\frac{3kx\,dx\wedge dv + dx\wedge da}{(a+3k\,x\,v+k^2\,x^3)^3}\,.$$

Such 2-form ω_{L_2} is exact and therefore $d\omega_{L_2} = 0$, but as $T^2\mathbb{R}$ is three-dimensional, ω_{L_2} is degenerate: the rank of ω_L is two and therefore its kernel is one-dimensional. More explicitly, one can see that such kernel is generated by the vector field in $T^2\mathbb{R}$

$$K = \frac{\partial}{\partial v} - 3kx\frac{\partial}{\partial a}\,,$$

because such vector field is such that $K(L_2) = K(x) = 0$.

The presymplectic form ω_{L_2} gives rise to a foliation determined by $\ker\omega_{L_2}$, i.e. spanned by the vector field K, and induces a symplectic form in the quotient, the set of leaves. Moreover, both functions L_2 and x are K-invariant and therefore projectable functions. This points out the convenience of using instead of coordinates (x,v,a) three different coordinates (x,f,L_2) with f being a function such that $Kf = 1$. In such coordinates the projected 2-form looks like in (23).

The function L_2^2 is such that there exist vector fields whose contraction with ω_{L_2} is dL_2^2. Such vector fields are those of the form

$$Y = -\frac{\partial}{\partial x} + 3k(v + kx^2)\frac{\partial}{\partial a} + b(x, v, a)K\,.$$

In particular for $b(x, v, a) = kx$ we obtain the vector field $X_1^{(2)}$ given by

$$X_1^{(2)} = -\frac{\partial}{\partial x} + kx\,\frac{\partial}{\partial v} + 3kv\,\frac{\partial}{\partial a}\,.$$

In fact, note that as L_2 is given by (13), we have

$$X_1^{(2)}(x) = -1\,, \qquad X_1^{(2)}(L_2) = -\frac{X_1^{(2)}(a + 3k\,x\,v + k^2\,x^3)}{(a + 3k\,x\,v + k^2\,x^3)^2} = 0\,,$$

and consequently, from the definition (23) one easily check that

$$i(X_1^{(2)})\omega_{L_2} = (X_1^{(2)}L_2^2)\;dx - (X_1^{(2)}x)\;dL_2^2 = dL_2^2\,.$$

Let us explore the existence of t-dependent constants of motion. The function $T_1 = L_2$ is such that

$$\Gamma^{(3)}(T_1) = T_2 = \frac{kx}{a + 3k\,x\,v + k^2\,x^3}\,,$$

$$\Gamma^{(3)}(T_2) = T_3 = \frac{k(v + kx^2)}{a + 3k\,x\,v + k^2\,x^3}\,,$$

$$\Gamma^{(3)}(T_3) = T_4 = k\,.$$

Hence T_1, T_2 and T_3 are generators of integrals of motion for the vector field $\Gamma^{(3)}$ and the t-dependent functions

$$J_{t1}^{(3)} = T_3 - t\,, \qquad J_{t2}^{(3)} = T_2 - t\,T_3 + \frac{1}{2}\,kt^2\,, \qquad J_{t3}^{(3)} = T_1 - t\,T_2 + \frac{1}{2}\,t^2T_3 - \frac{1}{6}\,kt^3\,,$$

are constants of the motion for the dynamical vector field $\Gamma^{(3)}$.

5. Darboux function, closed two form and fourth-order Riccati equation

Let us now consider the fourth-order case of Riccati equation $R^{(4)}(x, \dot{x}, \ddot{x}, \dddot{x}, \ddddot{x}) = 0$, i.e.

$$\ddddot{x} + 5kx\,\dddot{x} + 10k\,\dot{x}\ddot{x} + 15k^2\,x\,\dot{x}^2 + 10k^2x^2\ddot{x} + 10k^3x^3\dot{x} + k^4\,x^5 = 0\,, \tag{24}$$

with the following associated system of first-order differential equations:

$$\left\{\begin{aligned} \frac{dx}{dt} &= v \\ \frac{dv}{dt} &= a \\ \frac{da}{dt} &= w \\ \frac{dw}{dt} &= -(5\,kxw + 10k\,v\,a + 15k^2\,x\,v^2 + 10k^2x^2\,a + 10k^3x^3\,v + k^4\,x^5) \end{aligned}\right. \tag{25}$$

whose solutions are the integral curves of the vector field

$$\Gamma^{(4)} = v\frac{\partial}{\partial x} + a\frac{\partial}{\partial v} + w\frac{\partial}{\partial a} - (5kxw + 10k\,v\,a + 15k^2\,x\,v^2 + 10k^2x^2\,a + 10k^3x^3\,v + k^4\,x^5)\,\frac{\partial}{\partial w}\,. \quad (26)$$

Note that now in full similarity with the preceding cases

$$(\Gamma^{(4)}\widetilde{R}^{(3)})(x,v,a,w) = -kx\,\widetilde{R}^{(3)}(x,v,a,w)\,,$$

i.e. $\widetilde{R}^{(3)}(x,v,a,w)$ is a Darboux polynomial for $\Gamma^{(4)}$ with respective cofactor $-k\,x$.

Moreover, as $\operatorname{div}\Gamma^{(4)} = -5\,k\,x$, we can find a Jacobi multiplier as a power $R = (\mathcal{D}_1)^{-5}$ of the Darboux polynomial $\mathcal{D}_1$, as a solution of (19) with only one term.

By analogy with the preceding cases we can define the function L_3 given by

$$L_3(x,v,a,w) = \frac{1}{w + 4\,k\,x\,a + 6k^2\,x^2\,v + 3k\,v^2 + k^3\,x^4} = (\widetilde{R}^{(3)}(x,v,a,w))^{-1} \quad (27)$$

which satisfies the condition analogous to (22) with the new vector field $\Gamma^{(4)}$ and the function L_3. Define the 2-form ω_{L_3} as in the preceding example by an expression analogous to (15) and (23) but with L_3 instead of L_1 and L_2 respectively, which is written in this case as

$$\omega_{L_3} = -2\,\frac{6k(v+kx^2)\,dv\wedge dx + 4k\,x\;da\wedge dx + dw\wedge dx}{(w + 4\,k\,x\,a + 6k^2\,x^2\,v + 3k\,v^2 + k^3\,x^4)^3}\,.$$

Then the 2-form ω_{L_3} is presymplectic, its kernel defining a foliation generated by the two commuting vector fields

$$K_1 = \frac{\partial}{\partial v} - 6k(v+kx^2)\frac{\partial}{\partial w}\,, \qquad K_2 = \frac{\partial}{\partial a} - 4kx\frac{\partial}{\partial w}\,,$$

and therefore the functions x and L_3 are projectable because

$$K_1(x) = K_1(L_3) = 0\,, \qquad K_2(x) = K_2(L_3) = 0\,.$$

Consequently, the symplectic form induced in the quotient space is given by the same expression $dL_3^2 \wedge dx$.

This fourth-order Riccati equation also admits a set of t-dependent constants of the motion. Starting with the function $T_1 = L_3$ it is easy to check that

$$\Gamma^{(4)}(T_1) = T_2 = \frac{kx}{w + 4kxa + 3kv^2 + 6k^2x^2v + k^3x^4}$$

$$\Gamma^{(4)}(T_2) = T_3 = \frac{k(v+kx^2)}{w + 4kxa + 3kv^2 + 6k^2x^2v + k^3x^4}\,,$$

$$\Gamma^{(4)}(T_3) = T_4 = \frac{k(a + 3kvx + k^2x^3)}{w + 4kxa + 3kv^2 + 6k^2x^2v + k^3x^4}\,,$$

$$\Gamma^{(4)}(T_4) = T_5 = k\,,$$

and the functions $J_{t1}^{(4)}$, $J_{t2}^{(4)}$, $J_{t3}^{(4)}$ and $J_{t4}^{(4)}$ given by

$$\begin{aligned} J_{t1}^{(4)} &= T_4 - kt \\ J_{t2}^{(4)} &= T_3 - t\,T_4 + \tfrac{1}{2}\,kt^2 \\ J_{t3}^{(4)} &= T_2 - t\,T_3 + \tfrac{1}{2}\,t^2T_4 - \tfrac{1}{6}\,kt^3 \\ J_{t4}^{(4)} &= T_1 - t\,T_2 + \tfrac{1}{2}\,t^2T_3 - \tfrac{1}{6}\,t^3T_4 + \tfrac{1}{24}\,kt^4 \end{aligned}$$

are t-dependent constants of the motion.

6. The n-th-order Riccati equation of the sequence

The theory can be extended to the n-th-order Riccati equation of the above Riccati sequence for a generic value of n, which is obtained from an iterated action of the operator $\mathbb{D}_k$,

$$R^{(n)}(x, \dot{x}, \dots, x^{(n)}) = \mathbb{D}_k^n x = 0\,,$$

we can write the n-th-order Riccati equation of the Riccati sequence both as a section for $\tau_{n,n-1} : T^n\mathbb{R} \to T^{n-1}\mathbb{R}$ defined by the zeros of the map $\widetilde{R}^{(n)}(x, v^{(1)}, \dots, v^{(n)})$ or as vector field $\Gamma^{(n)}(x, v^{(1)}, \dots, v^{(n-1)}) \in \mathfrak{X}(T^{n-1}\mathbb{R})$. We remark that $\Gamma^{(n+k)}$ acting on (the pull-back of) a function $\widetilde{F}(x, v^{(1)}, \dots, v^{(n)})$ on $T^n\mathbb{R}$ obtained from the function $F(x, \dot{x}, \dots, x^{(n)})$ when replacing $\dot{x}$ by $v^{(1)}$, $\ddot{x}$ by $v^{(2)}$, and so on, coincides with the function obtained from the total time derivative dF/dt with the same replacements. Therefore $(\Gamma^{(n+1)} + kx)\widetilde{R}^{(n)} = \widetilde{R}^{(n+1)}$, as a consequence of the definition of the Riccati sequence, and therefore we obtain

$$(\Gamma^{(n+1)} + kx)\widetilde{R}^{(n)} = 0\,,$$

because $T^n\mathbb{R}$ is identified as a submanifold of $T^{n+1}\mathbb{R}$ given by the vanishing of the function $\widetilde{R}^{(n+1)}$. Consequently, $\mathcal{D} = \widetilde{R}^{(n)}$ is a Darboux polynomial for $\Gamma^{(n+1)}$ with a cofactor $-k\,x$.

On the other side, one can see form the explicit expression (8) that

$$\operatorname{div}\Gamma^{(n)} = -(n+1)\,k\,x$$

and therefore we find a Jacobi multiplier for $\Gamma^{(n)}$ as a power of $\mathcal{D}$,

$$R = (\mathcal{D})^{-(n+1)}\,.$$

Note that if a function $\mathcal{F}$ is such that $X\mathcal{F} = f\,\mathcal{F}$ for a vector field X and a polynomial function f, then

$$X^2\mathcal{F} = X(f\,\mathcal{F}) = (Xf + f^2)\,\mathcal{F} = ((X+f)f)\mathcal{F}\,,$$

and iterating the process

$$X^{n+1}\mathcal{F} = ((X+f)^n f)\;\mathcal{F}\,.$$

In our case $X = \Gamma^{(n)}$, and $\mathcal{F} = L_1 = (\widetilde{R}^{(n-1)})^{-1}$, and as we started with the cofactor $f = kx$ we find from the relation

$$(\Gamma^{(n)} + kx)^n x = \widetilde{R}^{(n)}(x) = 0\,,$$

that if $L_{n-1} = (\widetilde{R}^{(n-1)})^{-1}$

$$(\Gamma^{(n)})^{n+1}L_1 = ((\Gamma^{(n)} + kx)^n kx)L_1 = 0\,,$$

which shows that $T_1 = L_1$ is a generator of degree $n+1$ of first-integrals for $\Gamma^{(n)}$ and there will be n t-dependent constants of motion.

7. Conclusions

It has been proved that the use of geometrical techniques to deal with the elements of the Riccati equation is very efficient to unveil some previously hidden aspects of such equations. So this communication has raised many interesting problems, viz., the use of Darboux factors for finding Jacobi multipliers for higher order ODEs via Prelle-Singer method, a deeper understanding t-dependent constants of the motion and their relation with master symmetries and the Lagrangian formalism in the second order case and many other problems. The application of this differential geometric Prelle-Singer method to multicomponent systems and their integrability is a very interesting subject to be studied. In fact there are several interesting issues connected to this paper – we have only hit the tip of the iceberg. For instance the method here developed can also be used to study higher-order Abel equations [29]. We hope to answer some of these questions in forthcoming papers.

Acknowledgements.

JFC and MFR acknowledge support from research projects MTM-2006-10531, E24/1 (DGA), and MTM-2005-09183. PG thanks the Departamento de Física Teórica de la Universidad de Zaragoza for its hospitality and acknowledges support from Max Planck Institute for Mathematics in the Sciences, Leipzig.

References

[1] J.F. Cariñena, M.F. Rañada and M. Santander, "Lagrangian formalism for nonlinear second-order Riccati systems: one-dimensional integrability and two-dimensional superintegrability", J. Math. Phys. **46**, no. 6, 062703 (2005).
[2] A.M. Grundland and D. Levi, "On higher-order Riccati equations as Bäcklund transformations", J. Phys. A: Math. Gen. **32**, 3931–3937 (1999).
[3] P. Guha and D. Mayer, "Riccati Chain, Ramanujan's Differential Equations For Eisenstein Series and Chazy Flows", to appear in: *Recent development in Nonlinear Dynamics*, edited by M. Daniel *et al*, Narosa (2009).
[4] K. Andriopoulos, P.G.L. Leach and A. Maharaj, "On Differential Sequences", arXiv:0704.3243v1 (2007).
[5] M. Euler, N. Euler and P.G.L. Leach, "The Riccati and Ermakov-Pinney hierarchies", J. Nonlinear Math. Phys. **14**, 290–310 (2007).
[6] F. Dumortier, J.S. Llibre and J.C. Artes, *Qualitative Theory of Planar Differential Systems*, Springer Verlag, Berlin 2006.
[7] A. Goriely, *Integrability and nonintegrability of dynamical systems*, Advanced Series in Nonlinear Dynamics, vol. 19. World Scientific, River Edge, NJ, 2001.
[8] G. Darboux, "Mémoire sur les équations différentielles algébriques du premier ordre et du premier degré", Bull. Sci. Math. (2) **2**, 60–96, 123–144, 151–200 (1878).
[9] M.J. Prelle and M.F. Singer, "Elementary first integrals of differential equations", Trans. Amer. Math. Soc. **279**, 215–229 (1983).
[10] L.G.S. Duarte, S.E.S. Duarte, L.A.C. da Mota and J.E.F. Skea, "Solving second-order ordinary differential equations by extending the Prelle-Singer method", J. Phys. A: Math. Gen. **34**, 3015–3024 (2001).
[11] V.K. Chandrasekar, M. Senthilvelan and M. Lakshmanan, "Extended Prelle-Singer method and integrability/solvability of a class of nonlinear nth-order ordinary differential equations", J. Nonlinear Math. Phys. **12**, suppl. 1, 184–201 (2005).
[12] J.F. Cariñena, G. Marmo and J. Nasarre, "The non-linear superposition principle and the Wei–Norman method", Int. J. Mod. Phys. A **13**, 3601–3627 (1998).
[13] W.P. Johnson, "The curious history of Faà di Bruno formula", Amer. Math. Month. **109**, 217–234 (2002).
[14] S. Roman, "The formula of Faà di Bruno", Amer. Math. Month. **87**, 805–809 (1980).
[15] J.F. Cariñena, C. López and E. Martínez, "A new approach to the converse of Noether's theorem", J. Phys. A: Math. Gen. **22**, 4777–4787 (1989).

[16] J.F. Cariñena, C. López and E. Martínez, "Sections along a map applied to higher-order Lagrangian Mechanics. Noether's theorem", Acta Appl. Math. **25**, 127–151 (1991).

[17] E.T. Whittaker, *A Treatise on the Analytical Dynamics of Particles and Rigid Bodies*, Cambridge U.P., 1999.

[18] M.C. Nucci, "Jacobi Last Multiplier and Lie Symmetries: A Novel Application of an Old Relationship", J. Nonl. Math. Phys. **12**, 284–304 (2005).

[19] M.C. Nucci and P.G.L. Leach, "The Jacobi Last Multiplier and its applications in mechanics", Phys. Scr. **78**, 065011 (2008).

[20] W. Sarlet, "The Helmholtz conditions revisited. A new approach to the inverse problem of Lagrangian dynamics", J. Phys. A **15**, 1503–1517 (1982).

[21] M. Crampin, G. Prince and G. Thompson, "A geometrical version of the Helmholtz conditions in time-dependent Lagrangian dynamics", J. Phys. A **17**, 1437–1447 (1984).

[22] J. Lopuszanski, *The inverse variational problem in classical mechanics* (World Scientific Publishing, 1999).

[23] J.F. Cariñena and M.F. Rañada, "Helmholtz conditions and alternative Lagrangians: Study of an integrable Henon-Heiles system", Intern. J. of Theor. Phys. **38**, 2049–2061 (1999).

[24] D.G. Currie, E.J. Saletan, "q-Equivalent Particle Hamiltonians I. The Classical one-dimensional case", J. Math. Phys. **7**, 967–74 (1966).

[25] W. Sarlet and F. Cantrijn, "Higher-order Noether symmetries and constants of motion", J. Phys. A: Math. Gen. **14**, 479–492 (1981).

[26] P.A. Damianou, "Symmetries of Toda equations", J. Phys. A: Math. Gen. **26**, 3791–3796 (1993).

[27] R.L. Fernandes, "On the master symmetries and bi-Hamiltonian structure of the Toda Lattice", J. Phys. A: Math. Gen. **26**, 3797–3803 (1993).

[28] M.F. Rañada, "Superintegrability of the Calogero-Moser system: constants of motion, master symmetries, and time-dependent symmetries", J. Math. Phys. **40**, 236–247 (1999).

[29] J.F. Cariñena, P. Guha and M.F. Rañada, "Higher-order Abel equations: Lagrangian formalism, Darboux polynomials and constants of the motion", Preprint, Univ. Zaragoza (2009).

Journal of Physics: Conference Series **175** (2009) 012010 doi:10.1088/1742-6596/175/1/012010

Functional Description of $\mathbb{S}^1 \times \mathbb{S}^2$ and $\mathbb{S}^3$ Gowdy Cosmologies

Daniel Gómez Vergel[1] **and Eduardo J S Villaseñor**[1,2]

[1] Instituto de Estructura de la Materia, CSIC, Serrano 121, 28006 Madrid, Spain
[2] Grupo de Modelización y Simulación Numérica, Universidad Carlos III de Madrid, Avda. de la Universidad 30, 28911 Leganés, Spain

E-mail: dgvergel@iem.cfmac.csic.es, ejsanche@math.uc3m.es

Abstract. We will briefly review the classical formulation of the 3-handle $\mathbb{S}^1 \times \mathbb{S}^2$ and 3-sphere $\mathbb{S}^3$ Gowdy cosmological models coupled with massless scalar fields and their exact (non-perturbative) quantization by defining suitable Schrödinger functional representations in terms of appropriate probability spaces. We will pay special attention to the construction of closed expressions for the corresponding quantum time evolution propagators.

1. Introduction

Two-Killing symmetry reductions of General Relativity are appealing testing grounds to probe the features of gravity in its quantum regime, and also to deepen our knowledge in the axiomatic structure of quantum field theory in curved spacetimes. This is due to their exact solvability –both classical and quantum mechanical–, the existence of local degrees of freedom, and the invariance under a certain (restricted) class of diffeomorphisms. These models differ from each other in the action of the biparametric and abelian isometry group $G^{(2)}$ and the topologies of the spatial sections Σ compatible with the group action (see *Table 1* and [1]).

The so-called Gowdy cosmological models correspond to the group $G^{(2)} = U(1) \times U(1)$ acting on symmetric compact slices [2]. The possible spatial topologies are those of the 3-torus $\mathbb{T}^3$, the 3-handle $\mathbb{S}^1 \times \mathbb{S}^2$, the 3-sphere $\mathbb{S}^3$, or the lens spaces $L(p,q)$. The 3-torus topology has been profusely analyzed in the past and is, by far, the preferred choice to discuss quantization issues [3, 4, 5, 6]. Here, we will restrict our attention on the remaining spatial topologies, the 3-handle and the 3-sphere, which are less known than the 3-torus one but equally relevant in cosmology and quantum gravity due to the fact that they have both initial and final singularities. The lens spaces $L(p,q)$ can be studied by imposing discrete symmetries on the 3-sphere case and, in fact, the general arguments presented for $\mathbb{S}^3$ remain valid for them. Among other interesting issues, these models are privileged frameworks to discuss the canonical quantization of cyclic universes, some fundamental aspects of quantum field theory in curved spacetimes –particularly, the unitary implementability of the dynamics–, and also the application of modern differential geometry techniques to the Hamiltonian formulation of field theories.

We will closely rely on references [7, 8, 9, 10], where the reader will find the rigorous classical and quantum treatments for these models coupled to massless scalar fields.

Group $G^{(2)}$	Manifold Σ	Action	Name of the model
$U(1)\times U(1)$	$\mathbb{R}^2\times\mathbb{S}^1$	Not free	
	$\mathbb{R}\times\mathbb{T}^2$	Free	Schmidt model
	$\mathbb{T}^3$	Free	$\mathbb{T}^3$ Gowdy model
	$\mathbb{S}^2\times\mathbb{S}^1$	Not free	$\mathbb{S}^1\times\mathbb{S}^2$ Gowdy model
	$\mathbb{S}^3$	Not free	$\mathbb{S}^3$ Gowdy model
$\mathbb{R}\times U(1)$	$\mathbb{R}^3$	Not free	Cylindrical gravitational waves
	$\mathbb{R}\times\mathbb{S}^2$	Free	Cylindrical wormhole
	$\mathbb{R}\times\mathbb{T}^2$	Free	
$\mathbb{R}^2$	$\mathbb{R}^3$	Free	
	$\mathbb{R}^2\times\mathbb{S}^1$	Free	

Table 1. Spatial topologies compatible with the abelian biparametric Lie group $G^{(2)}$, with a smooth, effective and proper action on the spatial sections of a globally hyperbolic spacetime $(M^4\simeq\mathbb{R}\times\Sigma,g^{(4)})$. The action of the group, unique up to automorphisms of $G^{(2)}$ and diffeomorphisms of Σ, can be free or have degenerate orbits.

2. Group action

Consider a smooth, effective, and proper action of the biparametric Lie Group $U(1)\times U(1)$ on a compact, connected, and oriented 3-manifold Σ. As we explained in the *Introduction*, the spatial manifold Σ is then restricted to have the topology of a 3-torus $\mathbb{T}^3$, a 3-handle $\mathbb{S}^1\times\mathbb{S}^2$, a 3-sphere $\mathbb{S}^3$, or the lens spaces $L(p,q)$. Moreover, the action of the group is unique up to automorphisms of $G^{(2)}$ and diffeomorphisms of Σ [11, 12]. We will focus our attention on the 3-handle and 3-sphere Gowdy models. Let start by defining the smooth, effective, and proper action ϱ of $U(1)\times U(1)=\{(e^{x_1},e^{x_2})\,|\,x_\alpha\in\mathbb{R}(\mathrm{mod}\,2\pi),\alpha=1,2\}$ on $\Sigma=\mathbb{S}^1\times\mathbb{S}^2=\{(e^{i\kappa},e^{i\sigma}\sin\theta,\cos\theta)\,|\,\theta\in[0,\pi],\ \kappa,\sigma\in\mathbb{R}(\mathrm{mod}\,2\pi)\}$ by

$$\varrho((e^{x_1},e^{x_2}),(e^{i\kappa},e^{i\sigma}\sin\theta,\cos\theta)):=(e^{i(x_1+\kappa)},e^{i(x_2+\sigma)}\sin\theta,\cos\theta)\,.$$

The tangent vectors corresponding to the orbits of the two $U(1)$ commuting subgroup factors of $U(1)\times U(1)$ are

$$(ie^{i\kappa},0,0)\,,\quad(0,ie^{i\sigma}\sin\theta,0)\,.$$

Both fields commute. The first one is never zero but the latter vanishes at $\theta=0$ and $\theta=\pi$. For the 3-sphere case, we take Hopf coordinates $\Sigma=\mathbb{S}^3=\{(z_1,z_2)\in\mathbb{C}^2:|z_1|^2+|z_2|^2=1\}$, $z_1=e^{i\sigma}\sin(\theta/2)$, $z_2=e^{i\kappa}\cos(\theta/2)$, with $\theta\in[0,\pi]$ and $\kappa,\sigma\in\mathbb{R}(\mathrm{mod}\,2\pi)$, and define the action

$$\varrho((g_1,g_2),(z_1,z_2)):=(e^{i(x_1+\sigma)}\sin(\theta/2),e^{i(x_2+\kappa)}\cos(\theta/2))\,.$$

Here, the commuting tangent vectors corresponding to the action of the two $U(1)$ subgroup factors are $(iz_1,0)$ and $(0,iz_2)$. They alternatively vanish at the circles $(0,e^{i\kappa})$ and $(e^{i\sigma},0)$.

3. Einstein field equations

Here, we briefly review the construction of the 3-handle and 3-sphere Gowdy models coupled to massless scalar fields. We will assume that all fields are smooth. We consider 4-manifolds M^4 diffeomorphic to $\mathbb{R}\times\Sigma$, endowed with Lorentzian metrics $g^{(4)}$. The spacetimes $(M^4,g^{(4)})$ are considered to be globally hyperbolic. We further require $U(1)\times U(1)$ to act by isometries on the spatial slices Σ of M^4, obtaining in this way the so-called $\mathbb{S}^1 \times \mathbb{S}^2$ and $\mathbb{S}^3$ Gowdy cosmological models. We will focus on the *linearly polarized* cases, where the isometry group is generated by the pair of mutually orthogonal, commuting, spacelike, and globally defined, hypersurface-orthogonal Killing vector fields $\boldsymbol{\kappa} := (\partial/\partial\kappa)$ and $\boldsymbol{\sigma} := (\partial/\partial\sigma)$.

The 3-handle and 3-sphere Gowdy models coupled to massless scalar fields are determined by solving the Einstein-Klein-Gordon equations

$$\mathrm{Ric}(g^{(4)}) = \mathrm{d}\phi \otimes \mathrm{d}\phi\,, \quad \Box^{(4)}\phi = 0\,, \tag{1}$$

corresponding to (1+3)-dimensional gravity minimally coupled to a massless scalar field, $\phi : M^4 \to \mathbb{R}$, symmetric under the diffeomorphisms generated by $U(1) \times U(1)$. Here, $\mathrm{Ric}(g^{(4)})$ and $\Box^{(4)}$ denote the Ricci tensor and the d'Alembertian associated with $g^{(4)}$, respectively. As usual, the exterior derivative and the tensor product are denoted by d and $\otimes$, and we use units such that $8\pi G_N = 1$, where G_N is the Newton constant. In order to get a simplified, lower dimensional description we perform a Geroch reduction [13] with respect to one of the Killing fields, say $\boldsymbol{\kappa}$. Let $\tilde{M}^4$ denote the set of points of M in which $\boldsymbol{\kappa}$ does not vanish. Consider the space of orbits $M^3 = \tilde{M}^4/U(1)$. In the present situation, hypersurface orthogonality allows us to view M^3 as an embedded submanifold of $\tilde{M}^4$, everywhere orthogonal to the closed orbits of $\boldsymbol{\kappa}$, endowed with the induced metric

$$g^{(3)} := g^{(4)} - \frac{1}{\lambda_{\boldsymbol{\kappa}}} K_{\boldsymbol{\kappa}} \otimes K_{\boldsymbol{\kappa}}\,,$$

where $\lambda_{\boldsymbol{\kappa}} := g^{(4)}(\boldsymbol{\kappa}, \boldsymbol{\kappa}) > 0$ and $K_{\boldsymbol{\kappa}}$ is the one-form field defined by $K_{\boldsymbol{\kappa}}(\cdot) = g^{(4)}(\boldsymbol{\kappa}, \cdot)$. We can reintroduce the previously removed points $M^4 \setminus \tilde{M}^4$, that correspond to the *symmetry axis*, as a boundary where the fields must satisfy certain regularity conditions. In the so called linearly polarized case, the twist of the Killing fields vanishes, i.e.

$$K_{\boldsymbol{\kappa}} \wedge \mathrm{d}K_{\boldsymbol{\kappa}} = 0\,,$$

and similarly for $\boldsymbol{\sigma}$. As a consequence, the field equations can be written as those corresponding to a set of massless scalar fields coupled to (1+2)-gravity by performing the conformal transformation $g := \lambda_{\boldsymbol{\kappa}}\, g^{(3)}$. The system (1) is then equivalent to

$$\mathrm{Ric}(g) = \frac{1}{2}\sum_i \mathrm{d}\phi_i \otimes \mathrm{d}\phi_i\,, \quad \Box\phi_i = 0\,, \quad L_{\boldsymbol{\sigma}} g = 0\,, \quad L_{\boldsymbol{\sigma}}\phi_i = 0\,, \quad i = 1, 2\,, \tag{2}$$

where $\mathrm{Ric}(g)$ and $\Box$ are, respectively, the Ricci tensor and the d'Alembertian associated with g, all of them are 3-dimensional objects defined on M^3, and $L_{\boldsymbol{\sigma}}$ denotes the Lie derivative with respect $\boldsymbol{\sigma}$. Here, we have defined the scalar fields $\phi_1 := \log\lambda_{\boldsymbol{\kappa}}$ and $\phi_2 := \sqrt{2}\phi$, and we must remember that we have the additional symmetry generated by the remaining Killing vector field $\boldsymbol{\sigma}$. Note that the equations (2) are formally symmetric under the exchange of the gravitational and matter scalars. However, it is important to realize that these fields are subject to different regularity conditions in the gravitational and matter sectors that effectively break the symmetry among them. These are the so-called *polar constraints* [7], which are necessary ingredients to ensure the consistency of the models as they guarantee the differentiability of the other constraints.

4. Hamiltonian formalism and reduced phase space

Now, we can proceed to obtain the Hamiltonian formalisms of these models within the (1+2)-dimensional scheme provided by (2). These equations can be derived from the standard (1+2)

Einstein-Hilbert Lagrangian coupled to massless scalar fields [7]. The singular nature of the Lagrangian function forces us to carefully apply the Dirac-Bergmann theory of constrained systems. As expected for closed universes, one finds that the physical dynamical variables are restricted to belong to first class (i.e., coisotropic) constrained submanifolds of the phase spaces, where the Hamiltonian functions identically vanish. In order to isolate the true physical degrees of freedom and recover a dynamics for these models we perform a deparameterization (partial gauge fixation process) by imposing appropriate extra conditions on the phase space variables. This allows us to describe its reduced phase space by quadratic nonautonomous Hamiltonian systems.

An interesting feature of both the 3-handle and 3-sphere cases is the fact that after the deparameterization process there are no Dirac constraints left. This is in contrast with the situation for the 3-torus topology, where in addition to the dynamics generated by the time-dependent Hamiltonian there is an additional constraint in the system that must be appropriately taken into account.

Apart from some global modes present in these models, whose quantization follows readily in terms of standard position and momentum operators with dense domains in $L^2(\mathbb{R})$, the dynamics of both the gravitational and matter local degrees of freedom in the linearly polarized 3-handle and 3-sphere Gowdy models obeys the same field equations. These are the Klein-Gordon equations corresponding to massless scalar fields evolving in the same fixed background spacetime $((0,\pi)\times\mathbb{S}^2,\mathring{g})$ conformally equivalent to the (1+2)-dimensional Einstein universe. Explicitly,

$$\mathring{g} \;=\; \sin^2 t\Big(-\mathrm{d}t\otimes\mathrm{d}t+\gamma\Big)\,.$$

where γ is the unit round metric in the 2-sphere that, using spherical coordinates $(\theta,\sigma)\in(0,\pi)\times(0,2\pi)$ on $\mathbb{S}^2$, has the form $\gamma=\mathrm{d}\theta\otimes\mathrm{d}\theta+\sin^2\theta\,\mathrm{d}\sigma\otimes\mathrm{d}\sigma$. The field equations are then given by

$$\ddot{\varphi}_i+\cot t\,\dot{\varphi}_i-\Delta\varphi_i=0\,,\quad L_{\boldsymbol{\sigma}}\varphi_i=0$$

where, in the Hamiltonian formalism, the scalar fields φ_i are related to the original fields ϕ_i through nontrivial symplectic transformations [7] and Δ denotes the Laplace-Beltrami operator on the round 2-sphere $(\mathbb{S}^2,\gamma)$. We have to impose an extra symmetry condition: invariance under the diffeomorphisms generated by the Killing vector field $\boldsymbol{\sigma}=(\partial/\partial\sigma)$. Concerning the exact (non-perturbative) quantization of these models, there are subtleties associated with the unitary implementability of the quantum dynamics. First, it is shown in [8] that it is not possible to encode the quantum time evolution in a unitary operator for any $SO(3)$-invariant Fock/Schrödinger representation when the systems are written in terms of the dynamical variables encoded in the fields φ_i that naturally appear in their Hamiltonian formalisms. This is in conflict with the axiomatic structure of the quantum theory, and makes it difficult to properly perform a probabilistic interpretation for these models. However, it is possible to overcome this problem by performing a redefinition of the fields at the classical level involving precisely the conformal factor $\sin t$ mentioned above [8]. This type of field redefinitions was used for the first time in [5] to deal with the problem of the unitary implementability of the quantum dynamics for the vacuum 3-torus model, and it was generalized and reinterpreted later on geometric grounds for the remaining topologies [8]. Specifically, we define the rescaled fields

$$\xi_i:=\sqrt{\sin t}\,\varphi_i$$

satisfying

$$-\ddot{\xi}_i+\Delta\xi_i=\frac{1}{4}(1+\csc^2 t)\xi_i\,,\quad L_{\boldsymbol{\sigma}}\xi_i=0\,. \tag{3}$$

A way to understand what is going on is to realize that the singular behavior introduced by the conformal factors is translated, in terms of the redefined fields, into the behavior of a singular, time-dependent, potential term for the rescaled fields. The time evolution can now be implemented unitarily as a direct consequence of the fact that, in spite of being singular at some instants of time, these potentials are well behaved enough as functions of the time variable in a definite sense that will be clarified below. There is another difficulty in the quantization process: The quantum counterpart of the classical Hamiltonian does not contain the vacuum state in its domain. As a consequence, the action of the operator is not defined on the dense subspace of states with a finite number of particles. This difficulty can be overcome right from the start by describing the classical dynamics through a positive definite Hamiltonian that is related to the original one by a symplectic transformation. Taking the above prescriptions into account, both gravitational and matter local degrees of freedom of the systems under consideration, whose dynamics is governed by (3), can be described by the same nonautonomous Hamiltonian system $((0,\pi)\times\Gamma, \mathrm{d}t, \boldsymbol{\omega}, H(t))$, mathematically described by an infinite-dimensional cosymplectic space. Here, $\Gamma := \mathscr{C}\times\mathscr{C}$ is the space of Cauchy data, where the classical configuration space $\mathscr{C}$ is identified with the Fréchet space of rapidly decreasing real sequences $\mathbf{q} := (q_\ell : \ell\in\mathbb{N}_0)$. The space Γ is endowed with the natural symplectic structure

$$\boldsymbol{\omega}((\mathbf{q}_1,\mathbf{p}_1),(\mathbf{q}_2,\mathbf{p}_2)) := \sum_{\ell\in\mathbb{N}_0}(p_{1\ell}q_{2\ell}-p_{2\ell}q_{1\ell}), \quad (\mathbf{q}_1,\mathbf{p}_1),(\mathbf{q}_2,\mathbf{p}_2)\in\Gamma.$$

The nonautonomous Hamiltonian $H(t):\Gamma\to\mathbb{R}_+$ derived from (3) is given by a positive-definite quadratic form on Γ that can be diagonalized as a sum of time-dependent harmonic oscillator Hamiltonians. Explicitly,

$$H(t,\mathbf{q},\mathbf{p}) := \frac{1}{2}\sum_{\ell\in\mathbb{N}_0}(p_\ell^2+\omega_\ell^2(t)q_\ell^2), \quad t\in(0,\pi), \tag{4}$$

where the time-dependent squared frequencies $\omega_\ell^2(t)$ are given by

$$\omega_\ell^2(t) = \ell(\ell+1)+(1+\csc^2 t)/4, \quad \ell\in\mathbb{N}_0. \tag{5}$$

The time evolution is implemented by the symplectic transformations

$$\begin{pmatrix} q_{H\ell}(t,t_0) \\ p_{H\ell}(t,t_0)\end{pmatrix} = \mathcal{T}^{(\ell)}_{(t,t_0)}\cdot\begin{pmatrix} q_\ell \\ p_\ell\end{pmatrix}, \quad \mathcal{T}^{(\ell)}_{(t,t_0)} := \begin{pmatrix} c_\ell(t,t_0) & s_\ell(t,t_0) \\ \dot c_\ell(t,t_0) & \dot s_\ell(t,t_0)\end{pmatrix}, \quad \ell\in\mathbb{N}_0,$$

where, for each $\ell\in\mathbb{N}_0$, $c_\ell(\cdot,t_0)$ and $s_\ell(\cdot,t_0)$ are the solutions to the time-dependent harmonic oscillator equation

$$\ddot u + \omega_\ell^2(t)\,u = 0$$

such that $c_\ell(t_0,t_0)=1=\dot s_\ell(t_0,t_0)$, $s_\ell(t_0,t_0)=0=\dot c_\ell(t_0,t_0)$. Here, the dot denotes partial derivative with respect to the first argument. Explicitly,

$$\begin{aligned} c_\ell(t,t_0) &= \frac{1}{2}\sqrt{\sin t/\sin t_0}\,\mathscr{Q}_{(\omega'_\ell-1)/2}(\cos t)((1+\omega')\mathscr{P}_{(1+\omega'_\ell)/2}(\cos t_0)-\omega'_\ell\mathscr{P}_{(\omega'_\ell-1)/2}(\cos t_0)) \\ &- \frac{1}{2}\sqrt{\sin t/\sin t_0}\,\mathscr{P}_{(\omega'_\ell-1)/2}(\cos t)((1+\omega')\mathscr{Q}_{(1+\omega'_\ell)/2}(\cos t_0)-\omega'_\ell\mathscr{Q}_{(\omega'_\ell-1)/2}(\cos t_0)), \\ s_\ell(t,t_0) &= \sqrt{\sin t\sin t_0}(\mathscr{Q}_{(\omega'_\ell-1)/2}(\cos t_0)\mathscr{P}_{(\omega'_\ell-1)/2}(\cos t)-\mathscr{P}_{(\omega'_\ell-1)/2}(\cos t_0)\mathscr{Q}_{(\omega'_\ell-1)/2}(\cos t)), \end{aligned} \tag{6}$$

where $\omega'_\ell := \sqrt{1+4\ell(\ell+1)}$.

5. Schrödinger functional representation

In order to exactly quantize the systems under consideration, we introduce Schrödinger representations [14, 15, 16, 9], where quantum state vectors act as functionals $\Psi : \overline{\mathscr{C}} \to \mathbb{C}$ belonging to suitable L^2-spaces $\mathscr{H}_{\boldsymbol{\alpha}} = L^2(\overline{\mathscr{C}}, \sigma(\mathrm{Cyl}(\overline{\mathscr{C}})), \mathrm{d}\mu_{\boldsymbol{\alpha}})$ endowed with Gaussian measures.[1] The quantum configuration space is given by a proper distributional extension of the classical configuration space $\mathscr{C}$. In this case, it suffices to consider the dual $\overline{\mathscr{C}}$, i.e., the linear space of slowly increasing real sequences. The Gaussian measure $\mu_{\boldsymbol{\alpha}}$ is defined on the cylinder sets σ-algebra $\sigma(\mathrm{Cyl}(\overline{\mathscr{C}}))$ on $\overline{\mathscr{C}}$. Finally, the configuration observables act as multiplication operators, whereas the canonically conjugate momenta differ from the usual ones in terms of derivatives by the appearance of multiplicative terms which are necessary to ensure their self-adjointness; specifically, for cylindrical functions Ψ,

$$(Q_\ell \Psi)(\mathbf{q}) = q_\ell \Psi(\mathbf{q}), \quad (P_\ell \Psi)(\mathbf{q}) = -i\frac{\partial \Psi}{\partial q_\ell}(\mathbf{q}) + \frac{\bar{\beta}_\ell}{\bar{\alpha}_\ell} q_\ell \Psi(\mathbf{q}).$$

The complex sequences $\boldsymbol{\alpha}$ and $\boldsymbol{\beta}$ must satisfy

$$\alpha_\ell \bar{\beta}_\ell - \beta_\ell \bar{\alpha}_\ell = i, \quad \forall \ell \in \mathbb{N}_0,$$

by virtue of the CCR, $[Q_\ell, P_{\ell'}] = i\delta(\ell, \ell')\mathbf{1}$ and $[Q_\ell, Q_{\ell'}] = 0 = [P_\ell, P_{\ell'}]$. The functional form of $\boldsymbol{\alpha}$ and $\boldsymbol{\beta}$ is crucial to ensure the unitary implementation of the dynamics. Specifically, this is guaranteed for $\boldsymbol{\alpha}$ and $\boldsymbol{\beta}$ sequences with the asymptotic expansions [8]

$$\alpha_\ell = \frac{1}{\sqrt{2(|\ell|+1)}} \exp(i\eta_\ell) + O(|\ell|^{-3/2}), \quad \beta_\ell = -i\sqrt{\frac{(|\ell|+1)}{2}} \exp(i\eta_\ell) + O(|\ell|^{-1/2}),$$

where $\boldsymbol{\eta}$ is an arbitrary real-valued sequence. Moreover, all $SO(3)$-invariant representations for which the dynamics is well-defined and unitary are unitarily equivalent [8]. In what follows, we will simply take

$$\alpha_\ell = \frac{1}{\sqrt{2(|\ell|+1)}}, \quad \beta_\ell = -i\sqrt{\frac{(|\ell|+1)}{2}}. \tag{7}$$

We can rewrite the configuration and momentum operators in terms of annihilation and creation operators a_ℓ and a_ℓ^*, with $[a_\ell, a_{\ell'}^*] = \delta(\ell, \ell')\mathbf{1}$ and $[a_\ell, a_{\ell'}] = 0 = [a_\ell^*, a_{\ell'}^*]$, such that

$$Q_\ell = \alpha_\ell a_\ell + \bar{\alpha}_\ell a_\ell^*, \; P_\ell = \beta_\ell a_\ell + \bar{\beta}_\ell a_\ell^* \Leftrightarrow a_\ell = -i\bar{\beta}_\ell Q_\ell + i\bar{\alpha}_\ell P_\ell, \; a_\ell^* = i\beta_\ell Q_\ell - i\alpha_\ell P_\ell.$$

Note that the annihilation operators are given by the derivatives

$$a_\ell = \bar{\alpha}_\ell \frac{\partial}{\partial q_\ell}$$

and, hence, the vacuum state $\Psi_0 \in \mathscr{H}_{\boldsymbol{\alpha}}$ satisfying $a_\ell \Psi_0 = 0$ for all $\ell \in \mathbb{N}_0$ is given by the unit constant functional $\Psi_0(\mathbf{q}) = 1$ up to multiplicative phase. The quantum time evolution, characterized by a unitary operator $U(t, t_0)$, is univocally determined *up to phase* by the evolution of these creation and annihilation operators

$$\begin{aligned} U^{-1}(t,t_0)\, a_\ell\, U(t,t_0) &= A_\ell(t,t_0) a_\ell + B_\ell(t,t_0) a_\ell^*, \\ U^{-1}(t,t_0)\, a_\ell^*\, U(t,t_0) &= \bar{B}_\ell(t,t_0) a_\ell + \bar{A}_\ell(t,t_0) a_\ell^*, \end{aligned}$$

[1] Given the infinite-dimensionality of the classical configuration space, it is not possible to define nontrivial Lebesgue-type translation invariant measures $\mu_{\boldsymbol{\alpha}}$, but rather probability ones.

and the evolution of the vacuum state

$$U(t,t_0)\Psi_0 = \langle \Psi_0 \,|\, U(t,t_0)\Psi_0\rangle \exp\left(-\frac{1}{2}\sum_{\ell\in\mathbb{N}_0}\frac{B_\ell(t_0,t)}{A_\ell(t_0,t)}a_\ell^{*2}\right)\Psi_0\,,$$

with

$$|\langle \Psi_0 \,|\, U(t,t_0)\Psi_0\rangle| = \prod_{\ell\in\mathbb{N}_0}\frac{1}{\sqrt{|A_\ell(t_0,t)|}}\,. \tag{8}$$

Here, $A_\ell(t,t_0)$ and $B_\ell(t,t_0)$ are the Bogoliubov coefficients given by

$$\begin{aligned} A_\ell(t,t_0) &:= i\Big(\dot s_\ell(t,t_0)\bar\alpha_\ell\beta_\ell - c_\ell(t,t_0)\bar\beta_\ell\alpha_\ell + \dot c_\ell(t,t_0)|\alpha_\ell|^2 - s_\ell(t,t_0)|\beta_\ell|^2\Big)\,,\\ B_\ell(t,t_0) &:= i\Big((\dot s_\ell(t,t_0) - c_\ell(t,t_0))\bar\alpha_\ell\bar\beta_\ell + \dot c_\ell(t,t_0)\bar\alpha_\ell^2 - s_\ell(t,t_0)\bar\beta_\ell^2\Big)\,,\end{aligned}$$

satisfying $A_\ell(t_0,t) = \overline{A}_\ell(t,t_0)$, $B_\ell(t_0,t) = -B_\ell(t,t_0)$, and $|A_\ell(t,t_0)|^2 - |B_\ell(t,t_0)|^2 = 1$, for all $\ell\in\mathbb{N}_0$. In particular, $A_\ell(t,t_0)\neq 0$ for all $t,t_0\in(0,\pi)$. Due to the unitary implementability of the dynamics, the square summability of the sequence $(B_\ell(t_0,t) : \ell\in\mathbb{N}_0)$ is guaranteed [17] and, hence, the action of $U(t,t_0)$ is well defined over states with finite number of particles. The phase of the expectation value (8), though being irrelevant to answer most of the physical questions, can be explicitly calculated once a quantum Hamiltonian has been fixed. Given the quadratic nature of the classical Hamiltonian (4), its quantum counterpart coincides with the operator directly promoted from the classical expression modulo an arbitrary t-dependent real term proportional to the identity which encodes the choice of $U(t,t_0)$, i.e.,

$$H(t) = \frac{1}{2}\sum_{\ell\in\mathbb{N}_0}\left(P_\ell^2 + \omega_\ell^2(t)Q_\ell^2 + 2\vartheta_\ell(t)\mathbf{1}\right),$$

where the sequence $\vartheta_\ell(t)\in C^0(I)$, $\ell\in\mathbb{N}_0$, may be employed to avoid the appearance of infinite phases. Analogously to the one-dimensional case, when the dynamics is unitarily implementable, we define the time evolution propagator through the relation

$$(U(t,t_0)\Psi)(\mathbf{q}) = \int_{\overline{\mathscr{C}}} K_{\boldsymbol{\alpha\beta}}(\mathbf{q},t;\mathbf{q}_0,t_0)\Psi(\mathbf{q}_0)\,\mathrm{d}\mu_{\boldsymbol{\alpha}}(\mathbf{q}_0)\,,$$

where a straightforward calculation formally provides

$$K_{\boldsymbol{\alpha\beta}}(\mathbf{q},t;\mathbf{q}_0,t_0) = \prod_{\ell\in\mathbb{N}_0}\sqrt{2\pi}|\alpha_\ell|\exp\left(\frac{i}{2}\left(\frac{\beta_\ell}{\alpha_\ell}q_{0\ell}^2 - \frac{\bar\beta_\ell}{\bar\alpha_\ell}q_\ell^2\right)\right)K_\ell(q_\ell,t;q_{0\ell},t_0)\exp\left(-i\int_{t_0}^t \mathrm{d}\tau\,\vartheta_\ell(\tau)\right),$$

with K_ℓ denoting the propagator in the standard Schrödinger representation corresponding to the one-dimensional oscillator of squared frequency $\omega_\ell^2(t)$ defined in (5), whose construction is analyzed in [10]. For the sequences (7), this equation provides the expectation value

$$\langle \Psi_0 \,|\, U(t,t_0)\Psi_0\rangle = \prod_{\ell\in\mathbb{X}}\frac{1}{\sqrt{|A_\ell(t_0,t)|}}\exp\left(i\Big(\sigma_\ell(t,t_0) - \int_{t_0}^t \mathrm{d}\tau\,\vartheta_\ell(\tau)\Big)\right),$$

with

$$\sigma_\ell(t,t_0) = -\frac{1}{2}\arctan\left(\frac{(|\ell|+1)s_\ell(t,t_0) - (|\ell|+1)^{-1}\dot c_\ell(t,t_0)}{c_\ell(t,t_0)+\dot s_\ell(t,t_0)}\right)$$

for times t close to t_0. Taking the asymptotic behavior of the c_ℓ and s_ℓ functions defined in (6) into account, it is easy to check that normal ordering allows only the cancelation of the phases at high frequencies, for which $\vartheta_\ell(t)\sim -|\ell|/2$ as $\ell\to+\infty$. This is in contrast with the situation for the Minkowskian quantum field theory generalized to a spacetime $\mathbb{R}\times\mathbb{T}^3$ with closed spatial sections, where normal ordering allows the cancelation of *all* phases.

6. Conclusions

We have performed the classical Hamiltonian formulation and the exact canonical quantization of the 3-handle and 3-sphere Gowdy models in mathematically rigorous terms. At the classical level, we have carefully applied the Dirac-Bergmann theory of constrained systems and modern symplectic geometry techniques in order to attain appealing reduced phase space descriptions for these systems. Here, the dynamics is governed by nonautonomous Hamiltonians, where the time dependence reflects the appearance of both initial and final singularities. Concerning the quantization process, we have analyzed the construction of Schrödinger functional representations in terms of appropriate probability spaces, the unitarity of the time evolution, and the uniqueness of the representation, obtaining closed expressions for the evolution operators of these models for the first time. These systems provide a privileged framework to discuss quantization issues related to cyclic universes, and they are expected to have interesting applications in quantum cosmology, particularly, within the loop quantum cosmology program.

Acknowledgements

The authors want to thank Rutwig Campoamor for his kind invitation to participate in the Higher Symmetries in Physics Workshop. D. G. Vergel acknowledges the support of the Spanish Research Council (CSIC) through an I3P research assistantship. This work is also supported by the Spanish MICINN research grant FIS2008-03221.

References

[1] B. K. Berger, P. T. Chruściel and V. Moncrief, Ann. Phys. **237**, 322 (1995).
[2] R. H. Gowdy, Phys. Rev. Lett. **27**, 826 (1971); Ann. Phys. **83**, 203 (1974).
[3] G. A. Mena Marugán, Phys. Rev. D **56**, 907 (1997).
[4] M. Pierri, Int. J. Mod. Phys. D **11**, 135 (2002).
[5] A. Corichi, J. Cortez, and G. A. Mena Marugán, Phys. Rev. D **73**, 084020 (2006).
[6] J. F. Barbero G., D. G. Vergel, and E. J. S. Villaseñor, Phys. Rev. D **74**, 024003 (2006).
[7] J. F. Barbero G., D. G. Vergel, and E. J. S. Villaseñor, Class. Quant. Grav. **24**, 5945 (2007).
[8] J. F. Barbero G., D. G. Vergel, and E. J. S. Villaseñor, Class. Quant. Grav. **25**, 085002 (2008).
[9] D. G. Vergel, Class. Quant. Grav. **25**, 175016 (2008).
[10] D. G. Vergel and E. J. S. Villaseñor, Ann. Phys. **324**, 1360-1385 (2009).
[11] P. S. Mostert, Ann. of Math. **65**, 447 (1957); **66**, 589 (1957).
[12] P. T. Chruściel, Ann. Phys. **202**, 100 (1990).
[13] R Geroch, J. Math. Phys. **12**, 918-924 (1971).
[14] A. Corichi, J. Cortez, and H. Quevedo, Phys. Rev. **D66**, 085025 (2002); Class. Quant. Grav. **20**, L83 (2003); Annals Phys. **313**, 446-478 (2004).
[15] C. G. Torre, Class. Quant. Grav. **24**, 1 (2007).
[16] A. Corichi, J. Cortez, G. A. Mena Marugán, and J. M. Velhinho, Phys. Rev. D **76**, 124031 (2007).
[17] D. Shale, Trans. Am. Math. Soc. **103**, 149 (1962).

Journal of Physics: Conference Series **175** (2009) 012011 doi:10.1088/1742-6596/175/1/012011

Symmetries and singularities in Hamiltonian systems

Eva Miranda

Departament de Matemàtiques, Universitat Autònoma de Barcelona, E-08193 Bellaterra, Spain

E-mail: `miranda@mat.uab.cat`

Abstract. This paper contains several results concerning the role of symmetries and singularities in the mathematical formulation of many physical systems. We concentrate in systems which find their mathematical model on a symplectic or Poisson manifold and we present old and new results from a global perspective.

1. Introduction

Many physical problems can be formulated as mathematical problems in the language of symplectic 2-forms or, more generally, in the language of Poisson bivector fields. It was probably Jacobi who first formulated the equations of motion of a particle in terms of position and momenta of a cotangent bundle. He also observed that the set of functions in this manifold is endowed naturally with a bracket, called later the Poisson bracket after Denis Simon Poisson.

A cotangent bundle is naturally endowed with a symplectic structure for which Hamilton's equations correspond simply to the differential equation of the vector field associated to a privileged function called Hamiltonian function (which is a conserved quantity).

The Hamiltonian function of a Hamiltonian system is an integral of motion and the flow of the Hamiltonian vector field (and therefore, the trajectory of the particle) lies on the set $\{H = ct.\}$. This fact is know as Noether's theorem or conservation of energy. Noether's principle for detecting symmetries is based in the preservation of extra functions associated to our system. The set of first integrals have to satisfy special rules; those of complete integrability in the commutative or non-commutative case.

A first example of optimal application of Noether's principle is that of completely integrable systems (in the commutative case). Assume that we have extra $n-1$ functions (where $2n$ is the dimension of the manifold) which pairwise Poisson commute and are independent; then there is a generalization of Noether's principle to the set of n functions $F = (f_1, \dots, f_n)$ with $f_1 = H$ which says that indeed the particle moves on the invariant set $\{p \in M^{2n}, F(p) = (c_1, \dots, c_n)\}$. Under regularity assumptions, this set is an open submanifold of a cylinder.

In the case the flows of the Hamiltonian vector fields X_{f_i} are complete, the commutation of the flow gives a natural action of $\mathbb{R}^n$. By identification of the orbit with the quotient of $\mathbb{R}^n$ with its isotropy group we can prove that his manifold is an n-dimensional manifold which indeed is a generalized cylinder $\mathbb{R}^{n-k} \times \mathbb{T}^k$. When the orbit is compact, we obtain the classical Liouville tori which foliate a neighbourhood of the orbit. As a consequence completely integrable systems carry a lot of symmetry (there is an induced action of the cylinder $\mathbb{R}^{n-k} \times \mathbb{T}^k$ acting as a translation on each tori) . The orbits of the system lie on a foliation by tori of our symplectic manifold.

But not only that, there is also a Darboux theorem for those systems which tells us that also the symplectic structure can be expressed in a Darboux form with respect to special coordinates associated to our system (action-angle coordinate). This theorem is classically known as Arnold-Liouville theorem but indeed the formulas for the action functions based on integration along the cycles of the Liouville tori can already be found in the works of Henri Mineur [34], [35] and [36]. [1]

Observe that in the case the manifold is compact the mapping $F = (f_1, \ldots, f_n)$ will have singularities because of the maximum principle on this manifold. This giving non-independence of all the functions on M^{2n}. Thus, we generally impose "generic" independence of the functions which means that the rank of the differential of F is maximal on an open dense set.

Compactness can be seen as a weird topological condition for a physicist but indeed many mechanical systems are shown naturally on a compact manifold after symplectic reduction of a cotangent bundle. This is the case of the Euler equations on the sphere (see for instance [10]).

But what happens with singularities? Physically, singularities of the system correpond to relative equilibria of the system. From a mathematical point of view this answer is not satisfactory enough. These questions immediately arise:

(i) What is the Geometry hidden in these singularities?
(ii) How are the invariant sets (orbits)?
(iii) Do we still get normal form results like the one of Arnold-Liouville-Minneur for those?
(iv) Can we get applications to stability like in the regular case?

These are the main questions that this review paper addresses. In the first section of this paper we give an overlook over the regular results for integrable systems (in the commutative case) and recall some equivariant versions for general symplectic actions (obtained in [41]). In the second section of this paper we will review some results obtained for non-degenerate singularities. Grosso modo, there is a symplectic Morse-bott result that also concerns the neighbourhood of a whole orbit. There is also an equivariant version of this result for compact group actions preserving the system. Most of these results have been obtained by the author in her thesis [37] or together with Nguyen Tien Zung in [38]. In this section, we also review some examples with Morse singularities showing up in physical examples. We finally give applications of these normal form results that have been recently obtained by the author in [39], [40] and [24].

Another way singularities show up in mechanical systems is in the geometrical form associated to the problem. For instance, if the system is considered on a "union" of symplectic manifolds that foliate a Poisson manifold, then singularities may also arise as singularities of the symplectic foliation (non-regular leaves). This is the case of Gelfand-Ceitlin systems on $\mathfrak{u}(n)^*$. The integrable system lies on each coadjoint orbit of $\mathfrak{u}(n)^*$ but the collection of integrable systems can be, indeed, considered as an integrable system on a Poisson manifold. Another context in which the Poisson formulation seems effective is in the magnetic flow in an homogeneous space). We will also describe these examples here. Equivariant aspects in Poisson geometry are interesting field of study since singularities may show up either as singular symplectic leaves of the symplectic foliation or as fixed points of the Poisson/Hamiltonian actions (in the case of integrable systems as singular points of the moment mapping F).

The existence of action-angle coordinates for integrable systems (commutative and non-commutative systems) has been a recent field of study by the author together with Camille Laurent-Gengoux and Pol Vanhaecke in [27]. We will not spell out these results here instead we invite the reader to consult these references.

[1] Henri Mineur is best known for his works in astronomy (a crater on the moon has his name) and for his participation in the French resistance. However, the action formulas that we can find in his paper related to astronomy have constituted a major contribution in classical mechanics.

For compact group actions (not necessarily abelian), the closer result to "Darboux theorem" that we can get in the Poisson context is an equivariant version of Weinstein's splitting theorem that asserts that locally a Poisson manifolds splits as the Poisson product of a symplectic manifold with dimension equal to the rank of the Poisson bivector field at that point, together with a Poisson transverse structure that vanishes at that point. The proof of this result is tricky and needs extra assumptions mainly because the path method which is so popular in symplectic geometry ([43]) does not seem to work so well in the Poisson context.

2. The symplectic regular case

Assume that we are given a Hamiltonian system given by a Hamiltonian function H on a symplectic manifold (M^{2n}, ω). Hamilton's equations correspond to the equations of the flow of the Hamiltonian vector field (denoted in the sequel by X_H) associated to H in such a way that the following equation is satisfied,

$$i_{X_H}\omega = -dH.$$

The main philosophy beyond Noether's principle [44] is that conserved quantities give rise to symmetries in physical systems and can be encoded as group actions on these manifolds.

For instance, the existence of symmetries allows reduction of degrees of freedom of differential equations given by Hamilton's equations above.

This reduction of degrees of freedom can be formulated in geometrical terms using "Symplectic Reduction" (Marsden-Weinstein Theorem) for Hamiltonian group actions (see for instance [20].)

We are then given a *Moment Map* $F : M^{2n} \longrightarrow \mathfrak{g}^*$ and the reduced space at $\mu \in \mathfrak{g}^*$ is

$$F^{-1}(\mu)/G_\mu[2].$$

From an algebraic-geometric point of view those quotients are studied in Geometric Invariant Theory [25]. The study of singularities of these quotients is specially relevant from an algebro-geometric point of view. Some of these quotients can be interpreted as moduli spaces. For instance in [5] a symplectic structure is given in the moduli space of flat connections on a Riemann surface using reduction. A more recent generalization of this result to the case of punctured Riemann surfaces is contained in [2].

Recall that given a symplectic structure we can associate to it a Poisson bracket in a natural way via the formula:

$$\{f, g\} = \omega(X_f, X_g).$$

In the case there is a set of n (where n is half of the dimension of the symplectic manifolds) functions commuting with respect to this Poisson bracket and the orbits are compact, the group associated to these set of functions is a torus $\mathbb{T}^n$ which acts by translation on the sets $F = cte$ and therefore the reduced space is just a point. This is a consequence of Arnold-Liouville-Mineur theorem that we explain in the first subsection of this section.

In the second subsection we plunge into the non-abelian group case and recall some equivariant Darboux results and give some rigidity results for Hamiltonian and general symplectic group actions.

[2] There is a Poisson counterpart of this reduction procedure.

2.1. Abelian symmetries

In June 29*th* of 1853 Joseph Liouville presented a communication entitled "Sur l'intégration des equations différentielles de la Dynamique" at the "Bureau des longitudes". In the resulting note [26] he relates the notion of integrability of the system to the existence of n integrals in involution with respect to the Poisson bracket attached to the symplectic form. These systems come to the scene with the classical denomination of "completely integrable systems". In another language, a particular choice of n-first integrals in involution determines the n components of a moment map $\mathbf{F} : M^{2n} \longrightarrow \mathbb{R}^n$.

A lot of work has been done in the subject after Liouville.

Consider a completely integrable Hamiltonian system. The symplectic gradients of the components of the moment map define an involutive distribution. Assume that the moment map is proper (or instead that the invariant sets are compact). Let L be a regular orbit of this distribution then this orbit is a Lagrangian submanifold. Moreover, it is a torus and the neighbouring orbits are also tori. Those tori are called Liouville tori. This is the topological contribution of a theorem which has been known in the literature as Arnold-Liouville theorem. The geometrical contribution of the above-mentioned theorem ensures the existence of symplectic normal forms in the neighbourhood of a compact regular orbit. The formulae for the action formulas can already be found in the work of Mineur. That is why we will refer to the classical Arnold-Liouville theorem as Liouville-Mineur-Arnold theorem. Let us state the theorem below,

Theorem 2.1 (Liouville-Mineur-Arnold Theorem)
Let (M^{2n}, ω) be a symplectic manifold and let $F : M^{2n} \longrightarrow \mathbb{R}^n$ be a proper moment map. Assume that the components f_i of F are pairwise in involution with respect to the Poisson bracket associated to ω and that $df_1 \wedge \cdots \wedge df_n \neq 0$ on a dense set. Let $N = F^{-1}(c)$, $c \in \mathbb{R}^n$ be a connected levelset. Then there exists a neighbourhood $U(N)$ of N and a diffeomorphism $\phi : U(N) \longrightarrow D^n \times \mathbb{T}^n$ such that,

(i) $\phi(N) = \{0\} \times \mathbb{T}^n$.

(ii) A set of coordinates μ_i in D^n and a set of coordinates β_i in $\mathbb{T}^n$ for which, $\phi^*(\sum_{i=1}^n d\mu_i \wedge d\beta_i) = \omega$.

(iii) F depends only on $\phi^*(\mu_i) = p_i$ and it does not depend on $\phi^*(\beta_i) = \theta_i$.

The new coordinates p_i obtained are called action coordinates. The coordinates θ_i are called angle coordinates.

Remark 2.2 *The orbits of an integrable system in a neighbourhood of a compact orbit are tori. In action-angle coordinates (p_i, θ_i) the foliation is given by the fibration $\{p_i = c_i\}$ and the symplectic structure is Darboux $\omega = \sum_{i=1}^n dp_i \wedge d\theta_i$.*

Remark 2.3 *Another consequence of this theorem is that there is a natural group action associated to a completely integrable systems in a neighbourhood of a compact regular orbit. It is the action of a torus $\mathbb{T}^n$ which acts by translations on the orbits (invariant manifolds). This action is naturally induced from the action of $\mathbb{R}^n$ given by the complete joint flow. The isotropy group of this action is a discrete lattice $\mathbb{Z}^n$. Therefore the action of $\mathbb{R}^n$ gives a toric action when restricted to each orbit. But not only this, the periods of this action can be uniformized in such a way that the action really extends to an action by translations on neighbouring tori.*

The existence of action-angle coordinates in a neighbourhood of a compact orbit provides a symplectic model for the Lagrangian foliation $\mathcal{F}$ determined by the symplectic gradients of the n component functions f_i of the moment map F. In fact, Liouville-Mineur-Arnold theorem entails a "uniqueness" result for the symplectic structures making $\mathcal{F}$ into a Lagrangian foliation.

Journal of Physics: Conference Series **175** (2009) 012011 doi:10.1088/1742-6596/175/1/012011

In other words, if ω_1 and ω_2 are two symplectic structures defined in a neighbourhood of N for which $\mathcal{F}$ is Lagrangian then there exists a symplectomorphism preserving the foliation, fixing N and carrying ω_1 to ω_2. This is due to the following observation: Let X_{f_i} be the symplectic gradients of the functions f_i for any $1 \leq i \leq n$, then the Lagrangian condition implies that in fact $\mathcal{F} =< X_{f_1}, \ldots, X_{f_n} >$, further $\{f_j, f_k\}_i = 0$ where $\{.,.\}_i$ stands for the Poisson bracket attached to ω_i , $i = 1, 2$. Then by virtue of Liouville-Mineur-Arnold theorem there exists a foliation-preserving symplectomorphism ϕ_i taking ω_i to $\omega_0 = \sum_{i=1}^n dp_i \wedge d\theta_i$. In all, the diffeomorphism $\phi_2^{-1} \circ \phi_1$ does the job. It takes ω_1 to ω_2, it fixes N and it is foliation preserving.

So if the orbit is regular the existence of action-angle coordinates allows to classify the symplectic germs, up to foliation-preserving symplectomorphism, for which $\mathcal{F}$ is Lagrangian in a neighbourhood of a compact orbit. There is just one class of symplectic germs for which the foliation is Lagrangian.

One could look at the problem of existence of action-angle coordinates on the whole manifold. There are topological obstructions to the existence of global action-angle coordinates as it was shown by Duistermaat in [15].

2.2. *General symmetries*

Symmetries in physical systems can often be encoded in a Lie group action.

Given an action of a compact Lie group G on a smooth manifold M, Bochner's theorem says that we can linearize it in a neighbourhood of a fixed point for the action. Namely, in a neighbourhood of a fixed point the action of a compact Lie group is locally equivalent to the linear action on the tangent space at this point. Equivalence here means that there exists a diffeomorphism intertwining both actions. Bochner's theorem entails local rigidity for compact group actions.

What is the global counterpart to linearization? Global rigidity.

Rigidity is a concept in geometrical analysis that establishes that in some cases "close" objects are equivalent. When we use the expression close it means that we are using a topology in the set of objects that we are comparing. John Mather [30, 31, 32] was the first to study of rigidity of spaces of functions. Any action α, can be considered as a mapping from $G \times M$ on M with additional properties. This is why the topology considered here is the compact open topology in the set of C^k mappings from $G \times M$ on M, $Map(G \times M, M)$.

In the case of global group actions on compact manifolds, Palais [45] proved that two C^1-close group actions are equivalent. This result can be understood as a result of equivalence of close compact Lie group actions.

The result of Bochner and Palais, can be easily exported to the symplectic case.

Let us recall these results:

Theorem 2.4 (Equivariant Darboux Theorem) *Let $\rho : G \times M \longrightarrow M$ stand for a symplectic group action on (M, ω) and let p be a fixed point for the action. Then there exists coordinates $(x_1, y_1, \ldots, x_n, y_n)$ in a neighbourhood of p such that the group action is linear and the symplectic form can be written as $\omega_0 = \sum_i dx_i \wedge dy_i$.*

The method to produce equivariant normal forms from normal forms was first mentioned by Weinstein in [54]. For a complete proof of Darboux equivariant we refer to the work of Chaperon [8]. The main idea of the proof is to use Moser's path method and the average with respect to a Haar mesure of the compact group G.

The result of rigidity for symplectic actions was not found in the literature by the author.

The author proved it in [41]. For a more detailed proof look at [42].

Theorem 2.5 *Let ρ_0 and ρ_1 be two C^2-close symplectic actions of a compact Lie group G on a compact symplectic manifold (M, ω) then there exists a symplectomorphism ϕ satisfying $\rho(g) \circ \phi = \phi \circ \rho_1(g), \forall g \in G$.*

3. The symplectic Morse singular case

What can be said about action-angle coordinates if the completely integrable systems has singularities?

This question is quite natural because singularities are present in many well-known examples of integrable systems. In fact, if the completely integrable system is defined on a compact manifold then the singularities cannot be avoided.

In the singular case, the problem can be posed at three different levels:

(i) At the orbit level: In the neighbourhood of a compact singular orbit.
(ii) At a semi-local level: In the neighbourhood of a compact singular leaf.
(iii) At a global level.

The problem of topological classification of integrable Hamiltonian systems began with Fomenko [18] in some particular cases. Nguyen Tien Zung [57] studied the general case for the semi-local problem for non-degenerate singularities. It turns out that from a topological point of view we have a product-like description of the singularities in terms of the Williamson type.

The condition of non-degeneracy is always present in the works cited above.

But what is "non-degenerate" in this context?

The singularity of the orbit can be described in terms of the singularity of the functions f_i.

Case I. Let us start with the case L is reduced to a point.

Observe that the Poisson bracket induces a Lie algebra structure in the set of functions. Since the functions f_i are in involution with respect to the Poisson bracket, the quadratic parts of the functions f_i commute defining in this way an abelian subalgebra of $Q(2n, \mathbb{R})$ (the set of quadratic forms on $2n$-variables). In the case the singularity of the functions f_i is of Morse type this subalgebra is indeed a Cartan subalgebra. We call these singularities of non-degenerate type.

The problem of classification of singularities for the quadratic parts of the functions f_i can be therefore converted into the problem of classification of Cartan subalgebras of $Q(2n, \mathbb{R})$. The singularities for the quadratic parts are well-understood thanks to a result of Williamson [56] where Cartan subalgebras of $Q(2n, \mathbb{R})$ are classified. Let us recall its precise statement,

Theorem 3.1 (Williamson)
For any Cartan subalgebra $\mathcal{C}$ of $Q(2n, \mathbb{R})$ there is a symplectic system of coordinates $(x_1, \ldots, x_n, y_1, \ldots, y_n)$ in $\mathbb{R}^{2n}$ and a basis $f_1, \ldots, f_n$ of $\mathcal{C}$ such that each f_i is one of the following:

$$\begin{array}{lll} f_i = x_i^2 + y_i^2 & \text{for } 1 \le i \le k_e\,, & \text{(elliptic)} \\ f_i = x_i y_i & \text{for } k_e + 1 \le i \le k_e + k_h\,, & \text{(hyperbolic)} \\ \begin{cases} f_i = x_i y_{i+1} - x_{i+1} y_i, \\ f_{i+1} = x_i y_i + x_{i+1} y_{i+1} \end{cases} & \begin{array}{l}\text{(focus-focus pair)} \\ \text{for } i = k_e + k_h + 2j - 1,\ 1 \le j \le k_f \end{array} & \end{array} \tag{3.1}$$

The linear system given by the quadratic parts of the f_i is called the linear model for a singularity.

We may attach a triple of natural numbers (k_e, k_h, k_f) to a non-degenerate singularity p of F, where k_e stand for the number of elliptic components in the linear model, k_h and k_f the number of hyperbolic and focus-focus components in the linear model respectively.

By virtue of Williamson theorem this triple is an invariant of the linear system. That is why this triple is often called the Williamson type of the singularity.

The result of Eliasson in [16] says that an integrable system that has a non-degenerate singularity can be linearized symplectically in a neighbourhood of this singularity.

This means that we can find coordinates in which we can simultaneously write

- The integrable system in a Williamson basis.
- The symplectic form in Darboux form.

Namely,

Theorem 3.2 (Eliasson) *The integrable system at a point of Williamson type (k_e, k_h, k_f) is locally equivalent to its linear model.*

This theorem was stated for all Williamson's type in Eliasson's thesis [16]. Though a complete proof was only found in the completely elliptic case (Williamson type $(k_e, 0, 0)$). In [37], we give a complete proof for the other cases. The proof included in my thesis is based on an argument that uses symplectically orthogonal decomposition via Moser's path method [43].

Remark 3.3 *This result can be seen as a symplectic Morse result. Because we obtain a Morse theorem for all the components of the moment map in coordinates in which the symplectic structure has a Darboux form.*

Observe that Eliasson's theorem can be seen as a symplectic linearization result which ensures that the initial completely integrable system can be taken to the linear system and that the symplectic form can be taken to the standard one. As a byproduct we obtain a multiple differentiable linearization result for n commuting vector fields with singularities of non-degenerate type.

Case II. L is a compact orbit.

What happens now if we have a compact orbit?

We can define a notion of non-degenerate orbits using reduction.

For these orbits we have the following result which determines existence of normal forms for the symplectic form and the integrable system in the neighbourhood of a nondegenerate orbit but first we need to define a *linear model for higher-dimensional orbits in a covering* as follows,

Denote by $(p_1, ..., p_m)$ a linear coordinate system of a small ball D^m, $(q_1(mod1), ..., q_m(mod1))$ a standard periodic coordinate system of the torus $\mathbb{T}^m$, and $(x_1, y_1, ..., x_{n-m}, y_{n-m})$ a linear coordinate system of a small ball $D^{2(n-m)}$. Consider the manifold

$$V = D^m \times \mathbb{T}^m \times D^{2(n-m)} \tag{3.2}$$

with the standard symplectic form $\sum dp_i \wedge dq_i + \sum dx_j \wedge dy_j$, and the following moment map:

$$(\mathbf{p}, \mathbf{h}) = (p_1, ..., p_m, h_1, ..., h_{n-m}) : V \to \mathbb{R}^n \tag{3.3}$$

where

$$\begin{array}{l} h_i = x_i^2 + y_i^2 \ \text{ for } \ 1 \le i \le k_e \ , \\ h_i = x_i y_i \ \text{ for } \ k_e + 1 \le i \le k_e + k_h \ , \\ h_i = x_i y_{i+1} - x_{i+1} y_i \ \text{ and } \\ h_{i+1} = x_i y_i + x_{i+1} y_{i+1} \ \text{ for } \ i = k_e + k_h + 2j - 1, \ 1 \le j \le k_f \end{array} \tag{3.4}$$

We will refer to this model as the linear model in a covering of the initial neighbourhood. Indeed, we can define the linearized model on the initial neighbourhood of a point when we do not have hyperbolic components.

For each hyperbolic component, we may have an "hyperbolic twist" giving a generator of $\mathbb{Z}/2\mathbb{Z}$ (for details look at the author's paper with N.T Zung [40].

The linear model is V/Γ, with an integrable system on it given by the same moment map:

$$(\mathbf{p}, \mathbf{h}) = (p_1, ..., p_m, h_1, ..., h_{n-m}) : V/\Gamma \to \mathbb{R}^n \tag{3.5}$$

(Γ is a finite group that acts freely and symplectically on the linear model in a covering and comes from the hyperbolic twists associated to hyperbolic components).

The proof of the following theorem is contained in [40],

Theorem 3.4 (Eliasson-Miranda-Zung) *The integrable Hamiltonian system is equivalent to the linear model for Williamson type (k_e, k_h, k_f) in a neighbourhood of non-degenerate compact orbit.*

Remark 3.5 *This theorem is telling us that a neighbourhood of a compact non-degenerate orbit of rank k and Williamson type (k_e, k_h, k_f) of an integrable system is equivalent in a finite covering of it to a symplectic product of different types of cells.*

(i) A Liouville-Mineur-Arnold cell which consists of $D^k \times \mathbb{T}^k$ with a symplectic form in action-angle coordinates, $\omega = \sum_i dp_i \wedge d\theta_i$.

(ii) A symplectic product of exactly k_e two dimensional cells with elliptic singularities with induced integrable system with moment map $f = x^2 + y^2$ and symplectic form $\omega = dx \wedge dy$.

(iii) A symplectic product of exactly k_h two dimensional cells with elliptic singularities with induced integrable system with moment map $f = xy$ and symplectic form $\omega = dx \wedge dy$.

(iv) A symplectic product of exactly k_f four dimensional cells with focus-focus singularities with induced integrable system with components of the moment map $h_i = x_i y_{i+1} - x_{i+1} y_i$ and $h_{i+1} = x_i y_i + x_{i+1} y_{i+1}$ and symplectic form $\omega = dx_i \wedge dy_i + dx_{i+1} \wedge dy_{i+1}$.

In the case there are no hyperbolic components, it is exactly equivalent to the symplectic product.

In [40] we ended up proving the equivariant version of this theorem result in a neighbourhood of a singular compact orbit.

This result is really relevant because it encodes extra symmetries of the integrable system. All the symmetries considered are compact.

A nice consequence is the abelianity of the connected component of the identity of the group of symplectomorphisms preserving the moment map. In particular, in the case the action of the group is effective then this group is Abelian, in all, since it is also compact it is a product of a torus with a finite group. In the end, in the case the group is connected we recover actions by tori in the spirit of the theorem of Delzant.

Namely in [40] we proved,

Theorem 3.6 (Miranda-Zung) *The integrable system is equivariantly equivalent in a neighbourhood of a compact orbit of dimension k and Williamson type (k_e, k_h, k_f) to the integrable system given by the linear model endowed with the linear group action.*

Journal of Physics: Conference Series **175** (2009) 012011 doi:10.1088/1742-6596/175/1/012011

Remark 3.7 *The linear group action means transversally linear and acting by translations on Liouville tori. For details look at the paper [40].*

Remark 3.8 *This result can be seen as a symplectic Morse-Bott result. Because we obtain a Morse-Bott theorem (transversally Morse) for all the components of the moment map in coordinates in which the symplectic structure has a Darboux form.*

Indeed we already needed the equivariant version to obtain symplectic equivalence in the initial neighbourhood applied to the discrete group of order two.

3.1. Examples showing up in physical systems and construction of integrable systems with singularities

The above results will have much more value if these kind of singularities are present in simple physical systems. Here we present a list of examples that contain non-degenerate singularities (most of them just non-degenerate). In the inspiring book by Larry Bates and Richard Cushman [10] we can find many more and a close inspection of the symmetries naturally associated to these problems.

The Rotation of a sphere

Consider a 2-sphere spinning around the z-axis. This defines a Hamiltonian system associated to the height function of the sphere. Indeed the natural $\mathbb{R}$-action is periodic and gives an S^1-action.

The fixed points for the S^1-action correspond to singularities of the Hamiltonian function. The singular values are the boundary of the interval $[-1, 1]$ which is the image under the height function. The singular points are the north and the south pole.

Indeed the north and the south poles are really elliptic singularities. (This construction recalls the one of Morse theory [33]).

The harmonic oscillator

The harmonic oscillator correspond to the movement of a particle folowing the solution of the differential equations.

$$\ddot{x} = -ax$$

The equation above is known as Hooke's law and corresponds to the physical situation of a spring-mass system.

The symplectic structure is the one of the cotangent bundle of $\mathbb{R}$.

By conservation of energy the Hamiltonian of the system is the sum of Kinetic and potential energy which in position and momenta coordinates reads as,

$$H = \frac{1}{2}(x^2 + ay^2).$$

The origin is the only singular point and the singularity is of elliptic type. Indeed, for $a = 1$ we get exactly Eliasson's normal form.

The Euler's equations on the sphere

The Euler's equations on the sphere correspond to the movement of the Euler top.

Let us quickly review this example. Take our initial symplectic manifold to be $T^*(SO(3,\mathbb{R}))$. The action of $SO(3,\mathbb{R})$ on itself by left translations can be lifted to the cotangent bundle. Recall that the lift of any action of a manifold to its cotangent bundle is Hamiltonian.

In the sequel, we closely follow [10] and notation therein.

The reduced symplectic spaces $F^{-1}(\mu)/SO(3,\mathbb{R})_\mu$ coincide with the coadjoint orbits on $\mathfrak{so}(3)^*$ endowed with the Kostant-Kirillov-Souriau symplectic form. Identifying $\mathfrak{so}(3)$ with $(\mathbb{R}^3,\wedge)$, the coadjoint orbits are spheres and the Kostant-Kirillow-Souriau form is the standard area form.

The reduced Euler equations are the Hamiltonian equations on these reduced symplectic spaces coming from the following Hamiltonian on $T^*(SO(3))$:

$$\begin{array}{rcl} H: T^*(SO(3)) & \longrightarrow & \mathbb{R} \\ \alpha_A & \mapsto & \frac{1}{2}\rho(A)(\alpha_A,\alpha_A) \end{array}$$

(where α_A denotes an element in $T^*(SO(3))$ according to the standard trivialization of the cotangent bundle of a Lie group and $\rho(A)$ is a left invariant metric on $T^*(SO(3))$.)

In [10], Cushman and Bates use this trivialization of $T^*(SO(3))$ to write down the Hamiltonian in terms of the *moments of inertia*, and find the reduced Euler equations in the reduced space. In section III.4 in their book, they show that this system has exactly two hyperbolic singularities and four elliptic singularities.

To see a beautiful picture of this system and their singularities look at the cover of the book by Marsden and Ratiu where we can appreciate very clearly the dynamics of these singularities [29].

The spherical pendulum

Consider the movement of a point that represents a ball attached to a fixed point. This gives the simple pendulum.

If we consider the movement of this ball constrained to a sphere, we obtain the spherical pendulum.

The phase space is $T^*(S^2)$. The manifold being four-dimensional in order to prove integrability we need to show that there exists two first integrals in involution.

One of the first integrals is given by the complete energy of the system and the second one is given by the existence of an additional symmetry given by rotation around the z-axis.

This system can be shown to be completely integrable see for instance [10] and it has two critical points one of hyperbolic type and one of focus-focus type.

Construction of integrable systems

Bolsinov and Fomenko give in their book [19] a procedure to include singularities of different type of non-degenerate via surgery adapted to the foliation given by the integrable systems.

In the case of compact surface their method is very simple to describe and can be done by changing strips with regular foliations by strips containing one hyperbolic and one elliptic singularity. This contribution does not change the Euler characteristic as expected.

Together with Mark Hamilton, we gave given a precise description of this procedure in [24].

Journal of Physics: Conference Series **175** (2009) 012011 doi:10.1088/1742-6596/175/1/012011

4. Applications of Normal forms

The theorems listed above in the regular and singular case are normal form theorems for the symplectic form and the integrable system.

As such it has many applications. We explicit a pair of applications here. Another beautiful application of these normal form results is the computation of topological entropy of Hamiltonian system. This is a joint work with Jean-Pierre Marco [28] but we will not spell it out here.

4.1. Stability of integrable systems

The stability consequences of the Liouville-Mineur-Arnold theorem have been largely explored.

The normal form results for non-degenerate singularities (points or orbits) implies stability in the set of moment maps.

Indeed we also obtain infinitesimal stability. This is a result that we have obtained together with San Vu Ngoc [39].

An infinitesimal deformation of an integrable system $F = (f_1, \dots, f_n)$ is given by a set of n functions g_i that modulo ε^2 satisfy ,

$$\{f_i + \varepsilon g_i, f_j + \varepsilon g_j\} \equiv 0.$$

We can view these infinitesimal deformations as 1-cocycles of a complex (deformation complex) associated to the integrable system [39].

Given integrable system F we can introduce a deformation complex by considering the adjoint representation of $\mathbb{R}^n$ on $C^\infty(M)$ given by the Poisson bracket. This complex is defined via the Chevalley-Eilenberg cohomology.

Together with San Vu Ngoc, we proved in [39] an infinitesimally stability result which corresponds to the vanishing of the first cohomology group of this complex.

Indeed the vanishing of the first cohomology group is given via a singular Poincaré lemma which proves that every 1-cocycle is a coboundary.

Theorem 4.1 (Miranda, San Vu Ngoc) *Let $q_1, \dots, q_n$ be a standard basis (in the sense of Williamson) of a Cartan subalgebra of $\mathcal{Q}(2n, \mathbb{R})$. Then the corresponding completely integrable is smoothly infinitesimally stable at $m = 0$: that is,*

$$H^1(\mathbf{q}) = 0.$$

4.2. Applications to geometric quantization

Geometric quantization associated to a real polarization on a symplectic manifold (M^{2n}, ω) with a prequantum line bundle can be given via the cohomology groups with coefficients in the sheaf of flat section $\mathcal{J}$. This geometric quantization is often called Kostant's quantization.

Real polarizations are just given by Lagrangian foliations on the symplectic manifold. Leaves that admit sections that are flat along them are called Bohr-Sommerfeld leaves.

In the case the real polarization is fibrating these computations can be carried using sheaf cohomology. Sniatycki proved in [46] the following result:

Theorem 4.2 *In the case the polarization is fibrating and the fibers are compact all the cohomology groups $H^i(M, \mathcal{J})$ vanish except for the one in dimension n. This cohomology group is determined by the Bohr-Sommerfeld leaves.*

In the paper [21] by Guillemin and Sternberg this result is reinterpreted in terms of action-angle coordinates. More precisely, they construct action coordinates on the basis of the fibration and identify the Bohr-Sommerfeld leaves with the ones whose action functions are integer-valued.

Together with Mark Hamilton, we have considered a real polarization given by an integrable systems on a compact manifolds with non-degenerate singularities. We have used the singular normal forms explained before to compute the Geometric Quantization in the surface case [24].

5. Symmetries in Poisson manifolds

Many Hamiltonian systems have their natural context in the Poisson setting. The study the symmetries for these systems becomes a challenging thing since we have extra contribution of singularities which comes from the singularities of the symplectic foliation associated to the Poisson vector field.

Some of them are integrable (in the commutative or non-commutative sense). We will show some examples in the second subsection of this section.

5.1. General symmetries

In this section we describe the local structure of the Poisson manifolds under the existence of symmetries.

There is no Darboux theorem for Poisson bivector fields. The best that we can achieve is a splitting theorem due to Weinstein.

Theorem 5.1 (Weinstein) *Let (P^n, Π) be a smooth Poisson manifold and let p be a point of P of rank $2k$, then there is a smooth local coordinate system $(x_1, y_1, \ldots, x_{2k}, y_{2k}, z_1, \ldots, z_{n-2k})$ near p, in which the Poisson structure Π can be written as*

$$\Pi = \sum_{i=1}^{k} \frac{\partial}{\partial x_i} \wedge \frac{\partial}{\partial y_i} + \sum_{ij} f_{ij}(z) \frac{\partial}{\partial z_i} \wedge \frac{\partial}{\partial z_j},$$

where f_{ij} vanish at the origin.

This result is telling us that he Poisson manifold is locally a product of a symplectic manifold with a Poisson manifold with vanishing Poisson structure at the point.

$$(P^n, \Pi, p) \approx (M^{2k}, \omega, p_1) \times (P_0^{n-2k}, \Pi_0, p_2)$$

With Nguyen Tien Zung we have proved the equivariant version of this result under some mild assumptions ("tameness") on the Poisson structure.

The result is contained in [38].

Theorem 5.2 (Miranda-Zung) *Any tame Poisson manifold admits an equivariant splitting for an action of a compact Lie group.*

Remark 5.3 *This equivariant splitting is obtained using Vorobjev coupling [50] associated to a Poisson fibration. This method extends the minimal coupling method of Sternberg [47] and geometric data associated to Symplectic fibration to the Poisson context.*

In the case the transverse Poisson structure has linear part of semisimple type we can combine the linearization result of Conn [9] together with this equivariant splitting to obtain an equivariant linearization result contained in [38].

We have also obtained some rigidity results similar to those of Palais in the Poisson context. These are contained in the recent work [42].

In the case of abelian symmetries or symmetries given by an integrable system, we have obtained an extension of the action-angle theorem to the Poisson context together with Camille Laurent-Gengoux and Pol Vanhaecke in [27]. The reader is invited to consult this reference for further information.

5.2. Examples of integrable system in the Poisson context

In this section we present two examples that justify the interest of integrable systems in the Poisson context. With Alain Albouy we have been recently looking at another wide source of examples which come from Projective Dynamics and use Appel transformations. Other "more classical" examples include,

The Gelfand-Ceitlin system

The Gelfand-Cetlin system has been classically [20] considered as an integrable system on each coadjoint orbit $\mathcal{O}$ of $\mathfrak{u}(n)^*$.

On the other hand the collection of coadjoint orbits gives the Poisson manifold $\mathfrak{u}(n)^*$ and the Gelfand-Ceitlin system can be seen as an integrable system in the Poisson sense.

This example is interesting because the dual of a Lie algebra's constitute one of the basic examples of Poisson manifolds.

Let us recover this system in another way.

(i) Consider the increasing sequence of Lie algebra inclusions which comes from considering the Lie algebra $\mathfrak{u}(k)$ as the left-upper diagonal block of $\mathfrak{u}(k+1)$,

$$\mathfrak{u}(1) \subset \cdots \subset \mathfrak{u}(n-1) \subset \mathfrak{u}(n)$$

(ii) We can dualize the inclusions to get a sequence of surjective mappings (the projections) which are indeed Poisson maps:

$$\mathfrak{u}(n)^* \longmapsto \mathfrak{u}(n-1)^* \longmapsto \cdots \longmapsto \mathfrak{u}(1)^*$$

According to the method of Thimm (see for instance [20]), if we consider the generators of the Casimirs of all $\mathfrak{u}(k)^*$ for $k = 1, \ldots, n$ and we pull them back to $\mathfrak{u}(n)^*$, we then obtain a set of functions which Poisson commute and define an integrable system. For particular choice of generators (as the ones described in [21], its restriction to an open subset of $\mathcal{O}$ gives the Gelfand-Cetlin system.

This system is defined not only when restricted to the coadjoint orbit with the Kirillov-Kostant-Souriau symplectic structure but on $\mathfrak{u}(n)^*$.

In [21] Guillemin and Sternberg compute the geometric quantization of the Gelfand-Ceitlin systems via Sniatycki's result that compares geometric quantization with Bohr-Sommerfeld orbits.

Sniatycki's result yields interesting results about representations of $U(n)$ as can be seen in [21]. As we mentioned before, the Bohr-Sommerfeld leaves of an integrable system which is fibrating (when restricted to each coadjoint orbit) can be computed once we obtain a whole set of action functions (when restricted to each Bohr-Sommerfeld leaf). The Bohr-Sommerfeld leaves are then identified as the leaves for which the moment map (when restricted to each Bohr-Sommerfeld leaf) has integral value. The theorem of Sniatycki for Gelfand-Ceitlin systems just gives information about irreducible representations of $U(n)$ (see Theorem 6.1 in [21]). Let us point our here that in all these computations singularities are avoided.

Magnetic flows on Homogeneous spaces

The geodesic flow can be interpreted as the inertial motion of a particle on a Riemannian manifold (Q, g_{ij}) with kinetic energy given by:

$$H(x,p) = \frac{1}{2}\sum_{i,j} g^{ij} p_i p_j$$

The symplectic structure associated to g comes from the cotangent symplectic structure with momenta defined via $p_j = g_{ij}\dot{x}_i$. The motion of a particle under additional magnetic field given by a closed 2-form on Q:

$$\Omega = \sum_{i,j} F_{ij}(x) dx_i \wedge dx_j$$

is described by the Hamilton equations associated to the symplectic structure $\omega + \rho^*(\Omega)$ where ρ is the standard projection from T^*Q to Q.

In the sequel, we follow the paper [7].

Let G be a compact Lie group and H a closed subgroup. Let $a \in \mathfrak{h}$ be H-adjoint invariant.

In particular, $H \in G_a$. Let $\mathcal{O}(a)$ stand for the adjoint orbit. We then have a submersion of homogeneous spaces $\sigma : G/H \longrightarrow G/G_a \cong \mathcal{O}(a)$ and $\omega = \sigma^*(\Omega_{KKS})$ gives a magnetic field on G/H.

We then have,

Theorem 5.4 (Bolsinov, Jovanovich) *The magnetic geodesic flows of normal metric ds_0^2 in G/H with respect to the magnetic form ω is completely integrable in the non-commutative sense.*

By using a Theorem of Mischenko-Fomenko which says that non-commutative integrability implies commutative integrability) we obtain integrability of the magnetic geodesic flows on G/H. Again, this can be viewed in the Poisson manifold $T^*(G)/H$.

6. Acknowledgments

The author is thankful to the organizers of this conference for having gathered people with complementary interests in the field which made the conference really enriching. The author is currently supported by a Juan de la Cierva contract reference number JCI-2005-1712-18 and partially supported by the DGICYT/FEDER, project number MTM2006-04353 (Geometría Hiperbólica y Geometría Simpléctica).

References

[1] V. I. Arnold, *Mathematical methods of classical mechanics*. Graduate Texts in Mathematics, **60**.

[2] A.Alekseev, A. Malkin, A, *Symplectic structure of the moduli space of flat connection[s] on a Riemann surface.* Comm. Math. Phys. 169 (1995), no. 1, 99–119.

[3] A. Alekseev and E. Meinrenken, *Poisson geometry and the Kashiwara-Vergne conjecture.* C. R. Math. Acad. Sci. Paris 335 (2002), no. 9, 723–728.

[4] A. Alekseev and E. Meinrenken, *Ginzburg-Weinstein via Gelfand-Zeitlin* to appear in Journal of Differential Geometry, math.DG/0506112.

[5] M, Atiyah and R. Bott, *The Yang-Mills equations over Riemann surfaces. Philos. Trans. Roy. Soc. London Ser.* A 308 (1983), no. 1505, 523–615.
[6] S. Bochner, *Compact groups of differentiable transformations.* Ann. of Math. (2) **46**, (1945). 372–381.
[7] Alexey V. Bolsinov, Bozidar Jovanovic, *Magnetic geodesic flow on coadjoint orbits* , J. Phys. A 39 (2006), no. 16, L247- L252.
[8] M. Chaperon, *Quelques outils de la thorie des actions diffrentiables* Third Schnepfenried geometry conference, Vol. 1 (Schnepfenried, 1982), 259–275, Astrisque, 107-108, Soc. Math. France, Paris, 1983.
[9] J. Conn, *Normal forms for smooth Poisson structures.* Ann. of Math. (2) 121 (1985), no. 3, 565–593.
[10] R. Cushman and L. Bates, *Global aspects of classical integrable systems.* Birkhäuser Verlag, Basel, 1997.
[11] C. Curras-Bosch and E. Miranda, *Symplectic linearization of singular Lagrangian foliations*, Differential Geometry and its applications, 18, (2), 195-205, 2003.
[12] T. Delzant *Hamiltoniens périodiques et images convexes de l'application moment.* Bull. Soc. Math. France 116 (1988), no. 3, 315–339.
[13] J. Dixmier, *Champs de vecteurs adjoints sur les groupes et algèbres de Lie semi-simples.* J. Reine Angew. Math. 309 (1979), 183–190.
[14] J.P. Dufour, P. Molino and A. Toulet, *Classification des systèmes intégrables en dimension* 2 *et invariants des modèles de Fomenko.* C. R. Acad. Sci. Paris Sér. I Math. 318 (1994), no. 10, 949–952.
[15] J.J. Duistermaat, *On global action-angle coordinates. Comm. Pure Appl. Math.* **33** (1980), no. 6, 687–706.
[16] L. H. Eliasson, *Normal forms for Hamiltonian systems with Poisson commuting integrals*, Ph.D. Thesis (1984).
[17] L.H. Eliasson, *Normal forms for Hamiltonian systems with Poisson commuting integrals—elliptic case*, Comment. Math. Helv. **65** (1990), no. 1, 4–35.
[18] Fomenko, *Topological classification of all integrable Hamiltonian systems of general types with two degrees of freedom*, MSRI Publ., Vol 22 (1991) 131-340.
[19] A. V. Bolsinov and A. T. Fomenko, *Integrable Hamiltonian systems: geometry, topology, classification*, Chapman & Hall/CRC, 2004.
[20] V. Guillemin and S. Sternberg, *Symplectic techniques in physics.* Cambridge University Press, Cambridge, 1990.
[21] V. Guillemin and S. Sternberg. The Gel'fand-Cetlin system and quantization of the complex flag manifolds, *J. Funct. Anal.*, 52(1):106–128, 1983.
[22] V. Ginzburg, *Momentum mappings and Poisson cohomology.* Internat. J. Math. **7** (1996), no. 3, 329–358.
[23] V. Guillemin, E. Lerman and S. Sternberg, *Symplectic fibrations and multiplicity diagrams.* Cambridge University Press, Cambridge, 1996.
[24] M. Hamilton and E. Miranda, *Geometric quantization of integrable systems with hyperbolic singularities*, Annales de l'Institut Fourier, to appear in n 59, 2009.
[25] F.Kirwan, *Cohomology of quotients in symplectic and algebraic geometry.* Mathematical Notes, 31. Princeton University Press, Princeton, NJ, 1984.
[26] J.Liouville, *Note sur l'intégration des équations differentielles de la dynamique, présenntée au bureau des longitudes le 29 juin 1853*, Journal de Mathématiques pures et appliquées **20** (1855), 137-138.
[27] C. Laurent-Gengoux, E. Miranda and P. Vanhaecke, *Action-angle variables on Poisson manifolds*, preprint 2008, arxiv 0808.0338.
[28] Jean-Pierre Marco and Eva Miranda, *Topological entropy of integrable systems with singularities*, work in progress.
[29] J. Marsden and T. Ratiu, *Introduction to Mechanics and Symmetry*, Springer, 1999.
[30] J. Mather, *Stability of* C^∞ *mappings. I. The division theorem.*, Ann. of Math. (2) 87 1968 89–104.
[31] J. Mather, *Stability of* C^∞ *mappings. II. Infinitesimal stability implies stability.* Ann. of Math. (2) 89 1969 254–291.
[32] J. Mather, *Stability of* C^∞ *mappings. III. Finitely determined mapgerms.* Inst. Hautes tudes Sci. Publ. Math. No. 35 1968 279–308.
[33] J. Milnor, *Morse theory. Based on lecture notes by M. Spivak and R. Wells.* Annals of Mathematics Studies, No. 51 Princeton University Press, Princeton, N.J. 1963 vi+153 pp.
[34] H. Mineur, *Sur les systèmes mécaniques admettant n intégrales premieres uniformes et l'extension à ces systèmes de la méthode de quantification de Sommerfeld.*, C. R. Acad. Sci., Paris **200** (1935), 1571–1573.
[35] ———, *Reduction des systèmes mecaniques à n degré de liberte admettant n intégrales premieres uniformes en involution aux système à variable separees.*, J. Math Pure Appl., IX Sér. **15** (1936), 385–389.
[36] ———, *Sur les systèmes mécaniques dans lesquels figurent des paramètres fonctions du temps. Étude des systèmes admettant n intégrales premieres uniformes en involution. Extension à ces systèmes des conditions de quantification de Bohr-Sommerfeld.*, Journal de l'Ecole Polytechnique, Série III, 143ème année (1937), 173–191 and 237–270.

Journal of Physics: Conference Series **175** (2009) 012011 doi:10.1088/1742-6596/175/1/012011

[37] E. Miranda, *On symplectic linearization of singular Lagrangian foliations*, Ph.D. Thesis, Universitat de Barcelona, 2003.
[38] E. Miranda and N.T.Zung, *A note on equivariant normal forms of Poisson structures*, Math. Research Letters, vol 13-6,1001-1012, 2006.
[39] E. Miranda and S. Vu Ngoc, *A singular Poincaré Lemma* , IMRN, n 1, 27-46, 2005.
[40] E. Miranda and N.T.Zung, *Equivariant normal form for nondegenerate singular orbits of integrable Hamiltonian systems*, Ann. Sci. Ecole Norm. Sup.,37 (2004), no. 6, 819–839 2004.
[41] E. Miranda, *Some rigidity results for Symplectic and Poisson group actions*, Proceedings of the XV International Workshop on Geometry and Physics, Publicaciones de la RSME, 2007.
[42] E. Miranda, Ph. Monnier and N.T.Zung, *Rigidity of Poisson group actions*, preprint 2009.
[43] J. Moser, *On the volume elements on a manifold.* Trans. Amer. Math. Soc. 120 1965 286–294.
[44] E. Noether, *Invariante Variationsprobleme.* Nachr. D. Knig. Gesellsch. D. Wiss. Zu Gttingen, Math-phys. Klasse 1918. 235257.
[45] R. Palais, *Equivalence of nearby differentiable actions of a compact group.* Bull. Amer. Math. Soc. 67 1961 362–364.
[46] J. Śniatycki, On Cohomology Groups Appearing in Geometric Quantization, *Differential Geometric Methods in Mathematical Physics* I (1975).
[47] S. Sternberg, *Minimal coupling and the symplectic mechanics of a classical particle in the presence of a Yang-Mills field.* Proc. Nat. Acad. Sci. U.S.A. 74 (1977), no. 12, 5253–5254.
[48] J. Vey, *Sur certains systèmes dynamiques séparables*, Amer. J. Math. **100** (1978), no. 3, 591–614.
[49] J. Vey, *Sur le lemme de Morse.* Invent. Math. **40** (1977), no. 1, 1–9.
[50] Y. Vorobjev, *Coupling tensors and Poisson geometry near a single symplectic leaf.* Lie algebroids and related topics in differential geometry (Warsaw, 2000), 249–274, Banach Center Publ., 54, Polish Acad. Sci., Warsaw, 2001.
[51] S. Vu Ngoc, *Sur le spectre des systèmes complètement intégrables semi-classiques avec singularités.* Thèse, Institut Fourier, 1998.
[52] S. Vu Ngoc, *Formes normales semi-classiques des systèmes complètement intégrables au voisinage d'un point critique de l'application moment.* Asymptot. Anal. **24** (2000), no. 3-4, 319–342.
[53] S. Vu Ngoc, *On semi-global invariants for focus-focus singularities.* Topology **42** (2003), no. 2, 365–380.
[54] A. Weinstein, *Lectures on symplectic manifolds.* Regional Conference Series in Mathematics, No.**29**. American Mathematical Society, Providence, R.I., 1977.
[55] A. Weinstein, *The local structure of Poisson manifolds.*, J. Differential Geom. 18 (1983), no. 3, 523–557.
[56] J. Williamson, *On the algebraic problem concerning the normal forms of linear dynamical systems,* Amer. J. Math. **58:1** (1936), 141-163.
[57] Nguyen Tien Zung, *Symplectic topology of integrable Hamiltonian systems. I: Arnold-Liouville with singularities.*, Compos. Math. **101** (1996), no. 2, 179–215.
[58] Nguyen Tien Zung, *A note on degenerate corank-one singularities of integrable Hamiltonian systems,* Comment. Math. Helv. **75** (2000), no. 2, 271–283.

Workshop on Higher Symmetries in Physics IOP Publishing
Journal of Physics: Conference Series **175** (2009) 012012 doi:10.1088/1742-6596/175/1/012012

Hamiltonian systems related to Invariant Metrics

Gabriela P. Ovando [1]

Depto de Matemática, ECEN-FCEIA, Pellegrini 250, 2000 Rosario. Argentina.

E-mail: gabriela@fceia.unr.edu.ar

Abstract. Via a non degenerate symmetric bilinear form we identify the coadjoint representation with a new representation which acts on the Lie algebra. By inducing the canonical symplectic structure of the coadjoint orbits to the orbits of the new representation, we study Hamiltonians on the orbits by determining the Hamiltonian vector fields for invariant functions and we find commuting conditions under the corresponding Lie-Poisson bracket.

1. Introduction

One of the more succesfully algebraic methods used in analytical Mechanics to investigate Hamiltonian systems is related to the well known Theorem of Adler-Kostant-Symes (AKS-Theorem) [1, 8, 18] (see for example [7]). By using an ad-invariant non-degenerate symmetric bilinear form it is possible to identify the coadjoint representation from the Lie group G into $Aut(\mathfrak{g}^*)$ with the adjoint representation from G into $Aut(\mathfrak{g})$ and so to induce a symplectic form on the orbits of the adjoint representation. The AKS-Theorem describes the solutions of the Hamiltonian systems corresponding to ad-invariant functions. Moreover it asserts that the ad-invariant functions are in involution with respect to the corresponding Lie-Poisson bracket. From this started point this scheme were extended by other authors to several situations and applications (see for instance [5, 12, 13, 16, 17]).

In this work we consider a representation which is equivalent to the coadjoint one, by choosing a left invariant pseudo Riemannian metric on the Lie group, that is, a non-degenerate symmetric bilinear form on the corresponding Lie algebra (not necessarly ad-invariant as in the original version of the AKS-scheme). This allows to induce a symplectic structure on the orbits and after that it is natural the study of the corresponding Hamiltonian systems. As in the original case one can derive the respective formulas for the Hamiltonian vector fields, Hamiltonian systems and also prove involution conditions for the invariant functions.

The advantage of working on the Lie algebra lies on the possibility of using the algebra deriving from the Lie bracket. A essential ingredient for the application of the original version of the AKS scheme is the existence of an ad-invariant metric. This is always possible for semisimple Lie algebras, since the Killing form is the natural candidate to be used but it becomes an obstacle for non semisimple Lie groups. In fact it is not longer true that any non semisimple Lie algebra could be equipped with such metric and the algebraic structure of Lie algebras admitting such a metric is rather strong (see for instance [4, 10, 11, 14]).

[1] Present address: Abteilung für Reine Mathematik, Mathematisches Institut der Universität Freiburg, Eckerstr. 1, D-79104 Freiburg, Germany.

In the final section we show some examples. We consider basically nilpotent Lie algebras in low dimensions and by fixing a certain decomposition we concentrate in finding invariant functions. For nilpotent Lie algebras Corwin and Greenleaf [3]) exposed a method to find generators of the ring of rational invariant functions. Notice that while the invariant functions are invariant by equivalent representations this could not be the case for the rational invariant ones.

We hope that the present approach helps to bring another insight to different geometrical problems by the use of this tool.

We will assume through this work that Lie groups and their corresponding Lie algebras are real and finite dimensional. The Lie algebra of a Lie group G will be identified with the left invariant vector fields of G and denoted with greek letters as usual.

2. Preliminaries

A Lie group G acts on its Lie algebra $\mathfrak{g}$ by the adjoint representation $Ad : G \to Aut(\mathfrak{g})$ given by $Ad(g)X = dI(g)_e X$ for all $g \in G$, $X \in \mathfrak{g}$ where $I(g) : G \to G$ is defined by $I(g)h = ghg^{-1}$ for all $g, h \in G$.

We use de notation $ad_X Y$ for $[X, Y]$, where X, Y are elements of the Lie algebra $\mathfrak{g}$.

Let $\mathfrak{g}^*$ be the dual space of $\mathfrak{g}$, i. e. the space of linear functionals on $\mathfrak{g}$. The coadjoint representation from G into $Aut(\mathfrak{g}^*)$ is defined by:

$$(Ad^*(g)\phi)(Y) = \phi(Ad(g^{-1})Y) \qquad \forall g \in G,\ Y \in \mathfrak{g},\ \phi \in \mathfrak{g}^*.$$

Under this action of G, any point $\phi \in \mathfrak{g}^*$ sweeps out an orbit denoted by $O_\phi = \{Ad^*(g^{-1})\phi,\ \text{for all } g \in G\}$. The tangent vectors to the orbit at a point $\phi \in \mathfrak{g}^*$ are the induced vector fields defined as usual by

$$X^*_\phi = \frac{d}{ds}_{|s=0} Ad^*(\exp sX)\phi = -\phi \circ ad_X, \quad X \in \mathfrak{g}, \quad \phi \in \mathfrak{g}^*,.$$

Each orbit can be equipped with a symplectic form ω as follows: let X, Y be vectors in $\mathfrak{g}$ and let X^*, Y^* be the induced vectors on $\mathfrak{g}^*$, then the 2-form ω given by

$$\omega_\phi(X^*, Y^*) = \phi([X, Y]) \qquad X, Y \in \mathfrak{g}, \quad \phi \in \mathfrak{g}^* \tag{1}$$

makes of every orbit a symplectic manifold, in fact ω is closed and non-degenerate (see for example [15]).

Whenever (M, ω) denotes a symplectic manifold and $f : M \to \mathbb{R}$ is a smooth function on M, then there is exactly one vector field X_f on M with the property:

$$v_p(f) = df_p(v) = \omega_p(v, X_f) \qquad \forall\, p \in M, \forall\, v \in T_pM.$$

The vector field X_f is called the *Hamiltonian* vector field associated to the function f, and f is said to be the *Hamiltonian* (or the Hamiltonian function). Moreover, the *Hamiltonian system* for the function f is

$$\frac{dx}{dt} = X_f(x(t))$$

for $x : \mathbb{R} \to M$ a curve on M.

Recall that a *metric* on a Lie group G is a map $\langle\, , \rangle$, which for every $g \in G$ $\langle\, , \rangle_g$ is a (not necessarily definite) metric on T_gG and such it is smooth in the usual sense, that is, if X, Y denote vector fields on G, the function $g \to \langle X, Y\rangle_g$ is smooth.

We are interested on left invariant metrics, that is, those such that left multiplication by elements of the group are isometries. In this situation the metric is determined by its value at at the tangent space of the identity element T_eG. Therefore one identifies a left invariant metric on G with a non-degenerate symmetric bilinear form on the Lie algebra $\mathfrak{g}$.

If $\langle\,,\,\rangle$ denotes such a bilinear form, to each $\phi \in \mathfrak{g}^*$ it corresponds exactly one $Z \in \mathfrak{g}$ such that $\phi = \ell_Z$, where $\ell_Z(Y) = \langle Z, Y\rangle$. The map $Z \to \ell_Z$ is a linear isomorphism which allows to transport the coadjoint representation to a representation τ on the Lie algebra, which is given by

$$\langle \tau(g)X, Y\rangle = \langle X, Ad(g^{-1})Y\rangle.$$

In particular, making use of the exponential map one has

$$\langle \tau(\exp X)Y, Z\rangle = \langle Y, Ad(\exp -X)Z\rangle = \langle Y, \sum_{j=0}^{\infty}(-ad_X)^j Z\rangle,$$

and so

$$\tau(\exp X)Y = \sum_{j=0}^{\infty}(-ad_X^t)^j Y \tag{2}$$

where $ad_X^t : \mathfrak{g} \to \mathfrak{g}$ is the the transpose of the adjoint map, thus it is the endomorphism defined by $\langle ad_X^t Y, Z\rangle = \langle Y, ad_X Z\rangle$. Notice that (2) holds in a neighborhood of $0 \in \mathfrak{g}$.

Remark 1 *In the special case when G is an n-step nilpotent simply connected Lie group then the exponential map* $\exp : \mathfrak{g} \to G$, *is a diffeomorphism from the Lie algebra $\mathfrak{g}$ onto G and so the representation τ is given by:*

$$\tau(\exp X)Y = \sum_{j=1}^{n}(-ad_X^t)^j Y.$$

If $U \in \mathfrak{g}$, the orbit of the action through this point is denoted by $G \cdot U$ and the isotropy by G_U:

$$G \cdot U = \{\tau(g)U,\ \forall g \in G\} \qquad G_U = \{g \in G \,/\, \tau(g)U = U\}.$$

A vector field at the orbit is induced by a vector field $Y \in \mathfrak{g}$ and at $U \in \mathfrak{g}$ it has the form

$$\tilde{Y}_U = \frac{d}{ds}_{|_{s=0}} \tau(\exp sY)U = \frac{d}{ds}_{|_{s=0}} \sum_{j=1}^{n}(-ad_{sY}^t)^j U = -ad_Y^t U.$$

The Lie algebra of the isotropy subgroup G_U is the set of elements

$$L(G_U) = \{Y \in \mathfrak{g} \text{ such that } 0 = \frac{d}{ds}_{|_{s=0}} \tau(\exp sY)U\} = \{Y \in \mathfrak{g} \,/\, ad_Y^t U = 0\}.$$

Every orbit of the representation τ has a natural symplectic structure which is induced by the canonical symplectic structure of the orbits of the coadjoint representation (see 1). Explicitely for $U \in \mathfrak{g}$ and $Y, Z \in \mathfrak{g}$ the symplectic form in M at the point U is given by

$$\omega_U(\tilde{Y}, \tilde{Z}) = \langle U, [Y, Z]\rangle.$$

Our aim is to find the Hamiltonian vector field X_H for H the restriction to a orbit of a function $f : \mathfrak{g} \to \mathbb{R}$.

Definition 1 *If $\langle\,,\,\rangle$ denotes a non-degenerate bilinear form on $\mathfrak{g}$ and $f:\mathfrak{g}\to\mathbb{R}$ is a function on $\mathfrak{g}$, the gradient of f at the point U, $\nabla f(U)$, is defined by*

$$\langle \nabla f(U), Y\rangle = df_U(Y) \qquad \textit{for all } U, Y \in \mathfrak{g}.$$

Let $H = f_{|_M}$ denote the restriction of a function $f:\mathfrak{g}\to\mathbb{R}$ to an orbit $M = G\cdot U$ of the representation τ. The question is to determine the Hamiltonian vector field associated to H. Let $U\in\mathfrak{g}$ and let $Y\in\mathfrak{g}$ which induces a vector field at the orbit $M = G\cdot U$, then

$$\begin{aligned} dH_U(\tilde{Y}_U) &= \langle \nabla f(U), \tilde{Y}_U\rangle = \langle \nabla f(U), -ad^t_Y U\rangle = \langle -[Y,\nabla f(U)], U\rangle \\ &= -\omega_U(\tilde{Y}, \tilde{\nabla} f). \end{aligned}$$

On the other hand

$$dH_U(\tilde{Y}_U) = \langle U, [Y, X_H]\rangle,$$

comparing this with the previous relation one gets that

$$X_H(U) = -\tilde{\nabla} f(U)(U) = ad^t_{(\nabla f(U))}(U).$$

Example. Let $\mathfrak{h}$ be a Lie algebra with a metric $\langle\,,\,\rangle$ and let $f:\mathfrak{h}\to\mathbb{R}$ be the function given by $f(V) = \langle Q, V\rangle$ for a fixed point $Q\in\mathfrak{h}$. Then the gradient of f is $\nabla f(V) = Q$. In fact,

$$\langle \nabla f(V), Y\rangle = df_V(Y) = \frac{d}{dt}_{|_{s=0}} f(V+sY) = \frac{d}{dt}_{|_{s=0}} \langle Q, V+sY\rangle = \langle Q, Y\rangle$$

Consider now the orbits by the representation τ with the symplectic form ω above defined. The Hamiltonian vector field associated to $H = f_{|_M}$ results

$$X_H(U) = ad^t_Q(U)$$

and the Hamiltonian system for H

$$\begin{cases} x' &= ad^t_Q(x) \\ x(0) &= P \quad \text{for any fixed element } P\in\mathfrak{h} \end{cases}$$

has the solution

$$x(t) = -\tau(\exp tQ)P.$$

In particular if the metric $\langle\,,\,\rangle$ is ad-invariant, then the solution x coincides with $x(t) = Ad(\exp tQ)P$ (compare with [7] chapter 2).

Definition 2 *A smooth function $f:\mathfrak{g}\to\mathbb{R}$ is said to be τ-invariant if*

$$f(\tau(g)U) = f(U) \qquad \textit{for every } g\in G, \textit{ and } U\in\mathfrak{g}.$$

Remark 2 *A τ-invariant function is trivial on the orbits $G\cdot U$ for all $U\in\mathfrak{g}$. However we shall make use of them later, in a context where they are not trivial.*

Remark 3 *A τ-invariant function is actually a coadjoint invariant function. In fact, by construction the τ representation $\tau: G\to End(\mathfrak{g})$ is equivalent to the coadjoint representation $Ad^*: G\to End(\mathfrak{g}^*)$, since there is a linear isomorphism $T:\mathfrak{g}\to\mathfrak{g}^*$ such that $\tau(g) = T^{-1}\circ Ad(g)^*\circ T$ (explicitly $T=\ell$). Now if $f:\mathfrak{g}\to\mathbb{R}$ is a τ-invariant function, then*

$$f(\tau(g)x) = f(T^{-1}\circ Ad(g)^*\circ Tx) = f(x) = f(T^{-1}\circ Tx), \tag{3}$$

now $\tilde{f} := f\circ T^{-1}:\mathfrak{g}^*\to\mathbb{R}$ is by the relation above a Ad^*-invariant function. Moreover the map $f\to\tilde{f}$ is a bijection.

Proposition 1 *Let f be a τ-invariant function, then the following relations hold:*
(i) $\nabla f(\tau(g)U) = Ad(g)\nabla f(U)$ *for every* $g \in G$ *and* $U \in \mathfrak{g}$*;*
(ii) $ad^t_{\nabla f(U)}U = 0$ *for all* $U \in \mathfrak{g}$.

Proof. (i) For the first assertion, differentiate $f(\tau(g)U) = f(U)$ with respect to U:

$$\frac{d}{ds}_{|s=0} f(\tau(g)(U+sV)) = \frac{d}{dt}_{|s=0} f(U+sV)$$

which gives

$$\begin{aligned} df_{\tau(g)U}(\tau(g)V) &= \langle \nabla f(\tau(g)U), \tau(g)V\rangle \\ &= \langle Ad(g^{-1})\nabla f(\tau(g)U), V\rangle \end{aligned}$$

On the other hand,

$$df_U(V) = \langle \nabla f(U), V\rangle \qquad \text{for all } V$$

and so we obtain,

$$Ad(g^{-1})\nabla(\tau(g)U) = \nabla f(U).$$

(ii) Let $U \in \mathfrak{g}$, since f is τ invariant $f(\tau(\exp sY)U) = f(U)$ for all $Y \in \mathfrak{g}$ and for s near 0. If we differentiate at $s = 0$ we obtain,

$$\begin{aligned} 0 &= \tfrac{d}{ds}_{|s=0} f(\tau(\exp sY)U) = df_U(\tilde{Y}_U) \\ &= \langle \nabla f(U), -ad^t_Y U\rangle \\ &= \langle -ad(Y)\nabla f(U), U\rangle \\ &= \langle ad(\nabla f(U))Y, U\rangle \\ &= \langle Y, ad^t_{(\nabla f(U))}U\rangle \qquad \forall Y \in \mathfrak{g} \end{aligned}$$

and that means

$$ad^t(\nabla f(U))(U) = 0 \qquad \text{for all } U \in \mathfrak{g}.$$

■

3. A involution condition

A symplectic structure on a differentiable manifold (M, ω) induces a Poisson bracket on $C^\infty M$. In fact, for smooth functions $H_1, H_2 : M \to \mathbb{R}$ the Poisson bracket $\{H_1, H_2\}$ of them is a new function on M defined as

$$\{H_1, H_2\}(m) = \omega_m(X_{H_1}, X_{H_2}).$$

Notice that $C^\infty M$ is an associative algebra and $\{\,,\,\}$ is a Lie bracket on $C^\infty M$ satisfying the Leibniz rule.

Two functions f, g are said to be *in involution* if $\{f, g\} = 0$.

Let $(M, \{\,,\,\})$ denote a differentiable manifold equipped with a Poisson bracket and let $f : M \to \mathbb{R}$ be a fixed function on M, f is called *completely integrable* if there exist n functions in $C^\infty M$ such that

- $\{f, f_i\} = 0 = \{f_i, f_j\}$ for all $i, j = 1, \ldots, n$;
- The differentials $df_1, \ldots, df_n$ are linearly independent on every point on a open set invariant under the flow of X_f.

Let $\langle\,,\,\rangle$ denote a metric on the Lie algebra $\mathfrak{g}$, that is, $\langle\,,\,\rangle$ is a non degenerate symmetric bilinear form on $\mathfrak{g}$. For a subset $\mathfrak{v} \subseteq \mathfrak{g}$ one defines the orthogonal subspace $\mathfrak{v}^\perp$ as

$$\mathfrak{v}^\perp = \{x \in \mathfrak{g} : \langle x, v\rangle = 0 \text{ for all } v \in \mathfrak{v}\}.$$

Note that $\mathfrak{v}^\perp$ is not necessarly complementary to $\mathfrak{v}$ unless the metric is definite, however it always holds $\dim \mathfrak{g} = \dim \mathfrak{v} + \dim \mathfrak{v}^\perp$.

Assume there are Lie subalgebras $\mathfrak{g}_+$ and $\mathfrak{g}_-$ in $\mathfrak{g}$ such that

$$\mathfrak{g} = \mathfrak{g}_+ \oplus \mathfrak{g}_-$$

as direct sum of vector subspaces, then $\mathfrak{g}$ can be decomposed as a direct sum of subspaces $\mathfrak{g} = \mathfrak{g}_+^\perp \oplus \mathfrak{g}_-^\perp$.

Proposition 2 *Suppose* $\mathfrak{g}$ *has a non degenerate, symmetric bilinear form* $\langle\, , \rangle$ *and that* $\mathfrak{g}_+$, $\mathfrak{g}_-$ *are Lie subalgebras of* $\mathfrak{g}$ *such that* $\mathfrak{g}$ *decomposes into*

$$\mathfrak{g} = \mathfrak{g}_+ \oplus \mathfrak{g}_-$$

as direct sum of linear subspaces.

i) The maps $\mathfrak{g}_+^\perp \to \mathfrak{g}_-^*$ *and* $\mathfrak{g}_-^\perp \to \mathfrak{g}_+^*$ *defined by* $X \to \ell_{X_{|\mathfrak{g}_-}}$ *and* $X \to \ell_{X_{|\mathfrak{g}_-}}$ *respectively are linear isomorphisms. So the coadjoint action of* G_- *on* $\mathfrak{g}_-^*$ *(respectively from* G_+ *on* $\mathfrak{g}_+^*$*) induces an action of* G_- *onto* $\mathfrak{g}_+^\perp$ *(and another one of* G_+ *onto* $\mathfrak{g}_-^\perp$*), denoted by* $\tilde{\tau}$.

ii) Let $\pi_{\mathfrak{g}_+^\perp}$ *be the projection of* $\mathfrak{g}$ *onto* $\mathfrak{g}_+^\perp$ *with respect to the splitting* $\mathfrak{g} = \mathfrak{g}_+^\perp \oplus \mathfrak{g}_-^\perp$. *If* $X \in \mathfrak{g}_+^\perp$, $g_- \in G_-$ *then*

$$\tilde{\tau}(g_-)X = \pi_{\mathfrak{g}_+^\perp}(\tau(g_-)X).$$

iii) The infinitesimal vector field induced by $Y \in \mathfrak{g}_-$ *at the point* $X \in \mathfrak{g}_+^\perp$ *is*

$$\tilde{Y}(X) = \pi_{\mathfrak{g}_+^\perp}(-ad_Y^t X).$$

Proof. i) is a consequence from linear algebra. Let $X \in \mathfrak{g}_+^\perp$, $V \in \mathfrak{g}_-$, $g_- \in G_-$, a direct computation gives

$$\begin{aligned} Ad^*(g_-)\ell_X(V) &= \langle X, Ad(g_-)^{-1}V\rangle \\ &= \langle \tau(g_-)X, V\rangle \\ &= \langle \pi_{\mathfrak{g}_+^\perp}(\tau(g_-)X), V\rangle \\ &= \ell_{\pi_{\mathfrak{g}_+^\perp}(\tau(g_-)X)}(V) \end{aligned}$$

which proves ii) and iii) follows from ii) by taking $g_- = \exp(tY)$ and differentiate at $t = 0$. ■

An orbit $G_- \cdot X \subseteq \mathfrak{g}_+^\perp$ of the $\tilde{\tau}$-action is the set

$$G_- \cdot X = \{\tilde{\tau}(g_-)X,\ g_- \in G_-\}$$

and the isotropy subgroup at X is the set

$$G_X = \{g_- \in G_- \ :\ \tilde{\tau}(g_-)X = X\}$$

whose Lie algebra is

$$L(G_X) = \{Y_- \in \mathfrak{g}_- \ :\ \langle ad_Y^t X, Z\rangle = 0 \text{ for all } Z \in \mathfrak{g}_-\}.$$

The orbit $G_- \cdot X \subseteq \mathfrak{g}_+^\perp$ can be equipped with a symplectic structure which is induced by the symplectic structure on the coadjoint orbits, namely

$$\omega_X(\tilde{Y}, \tilde{Z}) = \langle X, [Y, Z]\rangle \qquad \text{for } X \in \mathfrak{g}_+^\perp, \text{ and } Y, Z \in \mathfrak{g}_-.$$

Notation: Given a vector $Y \in \mathfrak{g}$ we denote by Y_- (resp. Y_+) the projection of Y onto $\mathfrak{g}_-$ with respect to the decomposition $\mathfrak{g} = \mathfrak{g}_- \oplus \mathfrak{g}_+$.

Theorem 1 *Let $\mathfrak{g}$, $\mathfrak{g}_+$, $\mathfrak{g}_-$ and $\langle\, ,\, \rangle$ be as in Proposition 2. Let $M := G_- \cdot X \subset \mathfrak{g}_+^\perp$ be a G_--orbit, ω the orbit symplectic structure and $f : \mathfrak{g} \to \mathbb{R}$ a smooth function. Then:*

(i) The Hamiltonian vector field of $H = f_{|_M}$ at $U \in M$ is

$$X_H(U) = -\tilde{\nabla} f(U)_-(U) = \pi_{\mathfrak{g}_+^\perp}(ad^t_{\nabla_- f(U)} U)$$

(ii) If $f : \mathfrak{g} \to \mathbb{R}$ is τ-invariant, then the Hamiltonian equation for $H = f_{|_M}$ is

$$\frac{du}{ds} = -ad^t_{\nabla f_+(u)} u = ad^t_{\nabla f_-(u)}(u)$$

(iii) Let $f_1, f_2 : \mathfrak{g} \to \mathbb{R}$ be τ-invariant functions and let $H_i = f_{i_{|_M}}$,i=1,2, the restrictions respectively. Then

$$\{H_1, H_2\} = 0$$

(iv) Assume the multiplication map from $G_+ \times G_- \to G$ defined by $(g_+, g_-) \to g_+ g_-$ is a diffeomorphism, and that $f : \mathfrak{g} \to \mathbb{R}$ is τ-invariant. Then the initial value problem

$$\begin{cases} \frac{du}{ds} &= -ad^t_{\nabla f_+(u)} u \\ u(0) &= U_0 \end{cases} \tag{4}$$

can be solved by factorization. In fact, if we write

$$\exp(t \nabla f(U_0)) = g_+(t) g_-(t) \in G_+ \times G_-$$

then

$$u(t) = \tau(\exp -t \nabla f_+(U_0)) U_0$$

is the solution of (4).

Proof.
Let $U \in \mathfrak{g}_+^\perp$, $Y \in \mathfrak{g}_-$, then

$$\begin{aligned} dH_U(\tilde{Y}(U)) &= \langle \nabla f(U), \tilde{Y}(U) \rangle = \langle \nabla f(U), \pi_{\mathfrak{g}_+^\perp}(-ad^t_Y(U)) \rangle \\ &= \langle \nabla f_-(U), -ad^t_Y(U) \rangle \\ &= \langle -ad(Y) \nabla f_-(U), U \rangle \end{aligned}$$

But on the other hand

$$dH_U(\tilde{Y}(U)) = \langle U, [Y, X_H] \rangle$$

which implies

$$X_H(U) = -\tilde{\nabla} f_-(U)(U) = \pi_{\mathfrak{g}_+^\perp}(ad^t_{\nabla f_-(U)}(U)).$$

If f is τ-invariant then by Proposition 1 we have

$$0 = ad^t_{\nabla f(U)} U = ad^t_{\nabla f_+(U)} U + ad^t_{\nabla f_-(U)} U$$

Now

$$X_H(U) = \pi_{\mathfrak{g}_+^\perp}(ad^t_{\nabla f_-(U)} U) = -\pi_{\mathfrak{g}_+^\perp}(ad^t_{\nabla f_+(U)} U)$$

and we compute for $Y \in \mathfrak{g}_+$

$$\langle ad^t_{\nabla f_+(U)} U, Y \rangle = \langle U, ad_{\nabla f_+(U)} Y \rangle = 0$$

that means that $ad^t_{\nabla f_+(U)}U \in \mathfrak{g}_+^\perp \ \forall U$, and so

$$X_H(U) = -\pi_{\mathfrak{g}_+^\perp}(ad^t_{\nabla f_+(U)}U) = -ad^t_{\nabla f_+(U)}U = ad^t_{\nabla f_-(U)}U$$

which proves (2).

To prove (3), let $U \in M \subset \mathfrak{g}_+^\perp$. Then we have

$$\begin{aligned}\{H_1, H_2\} &= \langle U, [\nabla f_{1-}(U), \nabla f_{2-}(U)]\rangle = \langle ad^t_{\nabla f_{1-}U}U, \nabla f_{2-}(U)\rangle \\ &= \langle -ad^t_{\nabla f_{1+}(U)}U, \nabla f_{2-}(U)\rangle = \langle U, -[\nabla f_{1+}(U), \nabla f_{2-}(U)]\rangle \\ &= \langle ad^t_{\nabla f_{2-}(U)}U, \nabla f_{1+}(U)\rangle \\ &= -\langle U, [\nabla f_{2+}(U), \nabla f_{1+}(U)]\rangle = 0\end{aligned}$$

It remains to prove (4). First note that $ad^t_{\nabla f(U_0)}U_0 = 0$ implies

$$\tau(\exp t\nabla f(U_0))U_0 \equiv U_0$$

Write $g(t) = \exp t\nabla f(U_0) = g_+(t)g_-(t)$. Since $\tau(g_+g_-)U_0 \equiv U_0$, then $\tau(g_+^{-1})U_0 = \tau(g_-)U_0$. But since

$$g'(t) = \frac{d}{ds}_{|s=0} \exp s\nabla f(U_0) \exp t\nabla f(U_0) = dR_g \nabla f(U_0)$$

one immediately obtains

$$dR_{g^{-1}}g' = \nabla f(U_0).$$

We compute directly to get

$$\begin{aligned}\nabla f(\tau(g_+^{-1})U_0) &= Ad(g_+^{-1})\nabla f(U_0) = dL_{g_+^{-1}}dR_{g_+}dR_{g^{-1}}g' \\ &= dL_{g_+^{-1}}dR_{g_+}dR_{g_+^{-1}}dR_{g_-^{-1}}(dR_{g_-}g'_+ + dL_{g_+}g'_-) \\ &= dL_{g_+^{-1}}(g'_+) + dR_{g_-^{-1}}(g'_-) \in \mathfrak{g}_+ \oplus \mathfrak{g}_-\end{aligned}$$

So $\nabla f_+(\tau(g_+^{-1})U_0) = dL_{g_+^{-1}}(g'_+)$. But

$$u' = -ad^t_{dL_{g_+^{-1}}(g'_+)}u = -ad^t_{\nabla f_+(u)}u$$

which says that u is a solution of (4). ■

4. Examples

The topic of having enough functions in involution is a question to solve when we try to apply the last theorem in the previous section. In the case of nilpotent Lie groups, Corwin and Greenleaf [3] developped a method to construct the generators of the ring of rational τ-invariant functions. Below we show some examples in low dimensions, essentially in nilpotent cases, where the theory of the previous section was applied.

In dimension four there are solvable Lie algebras admitting an ad-invariant metric, however no nilpotent Lie algebra can be equipped with an ad-invariant metric unless it is abelian (see for instance [11]). Solvable Lie algebras with ad-invariant metrics were applied in the study of quadratic Hamiltonian in $\mathbb{R}^n$ ([12] [13]).

We know that the cardinal of the set of invariant functions is invariant by equivalent representations. However the rational one is not invariant. In fact as relation (3) shows by moving from one representation to a equivalent one, one could loss the rationality of the function.

Example i) Consider the three step nilpotent Lie algebra $\mathfrak{g}$ generated by the elements e_1, e_2, e_3, e_4, the Lie bracket relations given by:

$$[e_4, e_1] = e_2 \qquad [e_4, e_2] = e_3$$

and the metric defined by

$$\langle e_2, e_2\rangle = 1 = \langle e_4, e_4\rangle = -\langle e_1, e_3\rangle.$$

Then for $x = \sum x_i e_i, y = \sum y_i e_i$, the τ-action is the following one

$$ad_x^t y = (y_1x_2 - x_1y_2)e_1 + (x_4y_1 - x_1y_4)e_2 + (x_4y_2 - y_4x_1)e_3.$$

Take the following decomposition of $\mathfrak{g}$ into a direct sum of subalgebras, $\mathfrak{g} = \mathfrak{g}_1 \oplus \mathfrak{g}_2$ where

$$\mathfrak{g}_1 = span\{e_4, e_2, e_3\} \qquad \mathfrak{g}_2 = span\{e_1\}.$$

It is not difficult to see that

$$\mathfrak{g}_1^\perp = span\{e_3\} \qquad \mathfrak{g}_2^\perp = span\{e_1, e_2, e_4\}.$$

As we said G_1 acts on $\mathfrak{g}_2^\perp$. The orbits at a point $y = (y_1, y_2, y_3, y_4)$ have dimension zero if $y_1 = 0$ and have dimension two if $y_1 \neq 0$. So for example for the element $p = (1, 0, 0, 0)$ the orbit at p is the set

$$M = \{(1, u, 0, v) \,/\, u, v \in \mathbb{R}\}.$$

By following the method developped by Corwin and Greenleaf we obtain two functions which generate the ring of rational τ-invariant functions. In fact, for a point $x = \sum x_i e_i$ they are

$$f_1(x) = x_1 \qquad f_2(x) = x_1x_3 - \frac{1}{2}x_2^2$$

The gradient of the function f_2 is $\nabla f_2(U) = -u_1e_1 - u_2e_2 - u_3e_3$ and so the Hamiltonian system for f_2 with initial value $u(0) = (u_1^o, u_2^o, u_3^o, u_4^0)$ is

$$\begin{cases} \frac{du}{ds} &= -u_1u_2e_4 \\ u(0) &= (u_1^o, u_2^o, u_3^o, u_4^0) \end{cases}$$

This system has the solution $u(t) = (u_1^o, u_2^o, u_3^o, u_4^0 + u_1^o u_2^o t)$.

For $x \in M$ the function f_2 take the value $f_2(x) = -2u^2$, which is not constant at the orbit. Moreover the set $N = \{y \in M \,:\, f_2(y) = c\}$ for a constant c, is empty if $c > 0$, is the set $\{(1, 0, 0, v)\}$ if $c = 0$ and is the set $\{(1, \pm\sqrt{\frac{-c}{2}}, 0, v)\}$ for $c < 0$ and that means N is never compact.

Example ii) In this example we work on a fixed Lie algebra. We show that it is possible to obtain a different number of generators of the ring of rational invariant functions by changing the metric.

Let $\mathfrak{g}$ be the two-step nilpotent Lie algebra generated by Z, X, Y, X^*, Y^*, Z^* with the relations

$$[X, Y] = Z \qquad [Z^*, X] = Y^* \qquad [Z^*, Y] = -X^*$$

and the metric given by

$$\langle X^*, X\rangle = 1 = \langle Y^*, Y\rangle = \langle Z, Z\rangle = \langle Z^*, Z^*\rangle.$$

In this situation the ring of rational τ-invariant functions is generated by

$$p_1(u) = y, \quad p_2(u) = x, \quad p_3(u) = z^*, \quad p_4(u) = xx^* + zz^* + yy^*$$

where $u = zZ + xX + yY + z^*Z^* + x^*X^* + y^*Y^*$. We observe that this Lie algebra admits an ad-invariant metric defined by

$$\langle X^*, X\rangle = 1 = \langle Y^*, Y\rangle = \langle Z^*, Z\rangle.$$

and in this case the ring of rational ad-invariant functions is generated by p_1, p_2, p_3, p_4 as above. Furthermore, if we take the metric which makes the basis Z, X, Y, X^*, Y^*, Z^* an orthonormal basis, then the ring of rational τ-invariant functions is generated by p_1, p_2, p_3 as above.

Acknowledgments
The author is grateful to V. Bangert for his support during the stay at the Universität zu Freiburg, where part of this work was written.

References

[1] Adler, M., On a trace for formal pseudodifferential operators and the symplectic structure for the KdV type equations, Invent. Math., **50**, 219-248, (1979).
[2] Arnold V. 1989 Mathematical Methods of Classical Mechanics (Graduate texts in mathematics 60); New York-Heidelberg-Berlin: Springer.
[3] L. Corwin, F. P. Greenleaf, Representations of nilpotent Lie groups and their applications. Part 1: Basic theory and examples, Cambridge studies in advanced mathematics 18, Cambridge University Press 1990.
[4] G. Favre, L.J. Santharoubane, *Symmetric, invariant, non-degenerate bilinear form on a Lie algebra*, J. Algebra **105** (1987) 451–464.
[5] L. Feher, A. Gabor Adler-Kostant-Symes systems as Lagrangian Gauge theories, Phys.Lett. A301 (2002) 58-64 (arXiv:math-ph/0202033).
[6] A. Fomenko 1988 Symplectic Geometry, Advanced Studies in Mathematics, vol 5, Gordon and Breach Science Publishers, 1988.
[7] Guest, M, *Harmonic Maps, Loop Groups and Integrable Systems.* (London Math. Soc. Student Texts; 38). New York: Cambridge University Press (1997).
[8] Kostant, B., The solution to a generalized Toda lattice and representation theory, Advances in Math., **39**, 195 - 338, (1979).
[9] P. Liberman, Ch. M. Marle 1987 Symplectic Geometry and Analytical Mechanics, D. Reidel Publishing Company.
[10] Medina, A., Revoy, Ph., Algèbres de Lie et produit scalaire invariant, Ann. scient. Éc. Norm. Sup., 4e série, **t. 18**, 391 - 404, (1985).
[11] G. Ovando, *Estructuras complejas y sistemas hamiltonianos en grupos de Lie solubles*, Tesis de doctorado, Fa.M.A.F. Universidad Nacional de Córdoba (2002).
[12] G. Ovando, *Small oscillations and the Heisenberg Lie algebra*, J. Phys. A: Math. Theor. **40**, 2407-2424 (2007).
[13] G. Ovando, *Small oscillations on* $\mathbb{R}^2$ *and Lie Theory* Rev. de la UMA, **47** (2), 115-123 (2006).
[14] G. Ovando, *Two-step nilpotent Lie algebras with ad-invariant metrics and a special kind of skew-symmetric maps*, J. Algebra and its Appl., **6** 6 (2007), 897-917.
[15] A. Perelomov 1990 Integrable Systems of classical Mechanics and Lie algebras, vol I, Birkäuser Verlag, Basel Boston Berlin.
[16] Ratiu, T., *Involution theorems*, Geometric methods in Math. Phys., Lect. Notes in Math., **775**, Procedings, Lowell, Massachusetts 1979, Springer Verlag, (1980).
[17] Ratiu, T., The motion of the free n-dimensional rigid body, Indiana Univ. Math. Journal, **29**, 609 - 629, (1980).
[18] Symes, W., Systems of Toda type, inverse spectral problems and representation theory, Invent. Math., **59**, 13 - 53, (1978).

Workshop on Higher Symmetries in Physics IOP Publishing
Journal of Physics: Conference Series **175** (2009) 012013 doi:10.1088/1742-6596/175/1/012013

Symmetry reduction and superintegrable Hamiltonian systems

M A Rodríguez[1], P Tempesta[1] and P Winternitz[2]

[1]Dept. Física Teórica II, Facultad de Físicas, Universidad Complutense de Madrid, 28040–Madrid, Spain

[2]Centre de recherches mathématiques et Departement de Mathématiques et Statistique, Université de Montréal, Montréal (QC) H3C 3J7 Canada

E-mail: rodrigue@fis.ucm.es, p.tempesta@fis.ucm.es, wintern@crm.umontreal.ca

Abstract. We construct complete sets of of invariant quantities that are integrals of motion for two Hamiltonian systems obtained through a reduction procedure, thus proving that these systems are maximally superintegrable. We also discuss the reduction method used in this article and its possible generalization to other maximally superintegrable systems.

1. Introduction

Superintegrable Hamiltonian systems, both in classical and in quantum mechanics, have attracted a lot of attention in the last decades, due to their physically relevant properties. The existence of closed periodic bounded orbits for the classical models, and of accidental degeneracy of the energy levels for the quantum ones (removed by the existence of complete sets of commuting observables), are among the most remarkable consequences of the rich symmetry structure possessed by these systems.

Recent investigations lead to the construction of new superintegrable potentials, in flat spaces and curved spaces. As is well known, a maximally superintegrable system in a N–dimensional configuration space has $2N-1$ integrals of motion, although only sets of N integrals (chosen in several different ways among the $2N-1$ ones) are in involution. Among the classical examples of this type of Hamiltonians are well known systems: the Coulomb potential and the harmonic oscillator, whose dynamics is described by Bertrand's theorem [2] (for a review see, for instance, [17])

Superintegrable systems are rare and not easy to construct. Nevertheless, the approach we describe here offers a systematic way to produce superintegrable systems.

In this contribution, we are primarily interested in the reduction of superintegrable systems by using an adaptation of the Marsden–Weinstein reduction scheme [12] and in the study of the properties of the corresponding reduced systems. We will not analyze here in full generality and in all details the setting of this method. Instead, we will focus on two particular, but very interesting, cases. Reduction methods have been used as a common tool to derive new dynamic Hamiltonians starting from free systems, defined in the original manifold (see, for instance, [3]). Instead, in our approach, we start with a dynamics that is not free, and study its reductions. The original potentials (Coulomb and harmonic oscillator) are already superintegrable, and their invariants are known.

The first system, i.e. the Coulomb potential plus centrifugal terms, was analyzed in [19] by using an interesting *ad hoc* approach. We will present here a different point of view, which can be easily generalized to other models. We apply our approach to this model and to that one obtained as a reduction of the classical anisotropic harmonic oscillator. We recall that these systems, are also exactly solvable, i.e. their quantum spectrum can be computed by purely algebraic means [18]. This paper is based on the work [15]. For a treatment of the quantum case, see [6].

2. A geometric approach to the construction of superintegrable systems

In this section, we briefly review the geometric settings motivating our results. We work on a $2N$–dimensional symplectic manifold (M, ω). Recall that a momentum map for the action of a group G on M, whose associated algebra is $\mathbf{g}$, is a map $J : M \to \mathbf{g}^*$ such that

$$d(\langle J, X\rangle) = i_X \omega, \tag{1}$$

for all $X \in \mathbf{g}$. The previous formula reduces to $J_X = i_X \theta$ if there exist a Liouville one–form, i.e $\omega = d\theta$. We also need the equivariance assumption

$$\langle \mathrm{Ad}_g^* \xi, X\rangle = \langle \xi, \mathrm{Ad}_{g^{-1}} X\rangle, \tag{2}$$

where Ad_g^* denotes the coadjoint action of G in $\mathbf{g}^*$.

Let us consider now an integrable system, whose configuration space is M. Then, there exist $2N$ functions in involution $I_1, ..., I_{2N}$:

$$\{I_i, I_j\} = 0, \quad i, j = 1, ..., 2N, \tag{3}$$

whose differentials dI_j are independent at each point of an open dense submanifold K of M. The reduction procedure can be sketched as follows. Due to the isomorphism between the algebra of smooth functions on M and that of the associated Hamiltonian vector fields, the vector fields $X_1, ..., X_{2N}$ corresponding to the set of integrals in involution are in involution as well: $[X_i, X_k] = 0$. Then, we can construct their symplectic action $G = R^{2N}$ on the manifold. Let J be the associated momentum map. The theorem of Marden and Weinstein ensures us that, if μ is a regular value of J, then $P_\mu := J^{-1}/G$ is a symplectic manifold, whose dimension is equal to $\dim M - N = N$.

The same construction can be used to generate new superintegrable system as a reduction of a given superintegrable one. In the following sections, we apply this approach, very general, to two physically interesting cases, like the Coulomb potential and the anisotropic oscillator. We point out that the integrals of the motion of the two reduced systems are obtained as a consequence of the reduction scheme. The corresponding Hamiltonian vector fields are still in involution, since the equivariance condition ensures that the isomorphism between algebras described above is preserved. Also, the functional independence of the reduced integrals is guaranteed, if we work on a sufficiently regular manifold. In our specific examples, the integrals $K_1, ..., K_N$ responsible for the reduction process are the components of the angular momentum, and the corresponding momentum mapping is $J = K_1 \times \cdots \times K_N$. Consequently, the group action is defined by the cartesian product of N copies of $SO(2)$. Finally, our reduced space is

$$P_\mu = J^{-1}(\mu)/T^N, \tag{4}$$

where T^N is the N–dimensional torus. As a general observation, the reduced Hamiltonians, as well as the integrals of motion, although keep memory of the original ones, are essentially different in their structure. In this sense, the procedure appears to be nontrivial, since it generates new dynamics from known ones, as will be transparent from the subsequent discussion.

3. The Coulomb potential in even dimension and its reduction

The Coulomb potential

$$H_6^{\rm C} = \frac{1}{2}\sum_{i=1}^{N}\hat{p}_i^2 - \frac{\gamma}{\hat{r}}, \quad \hat{r}^2 = \sum_{i=1}^{N}\hat{x}_i^2 \tag{5}$$

is a maximally superintegrable system, with first integrals which can be derived from the angular momentum and the Laplace–Runge–Lenz tensors:

$$L_{ij} = \hat{x}_i\hat{p}_j - \hat{x}_j\hat{p}_i, \quad A_i = \sum_{j=1}^{N}\hat{p}_j L_{ij} - \gamma\frac{\hat{x}_i}{\hat{r}}, \quad i,j = 1,\ldots,N \tag{6}$$

$$\{H_6^{\rm C}, L_{ij}\} = 0, \quad \{H_6^{\rm C}, A_i\} = 0 \tag{7}$$

Although the whole procedure can be established in arbitrary even dimensions, for the sake of simplicity we will study in this work the six dimensional case, since the reduced system lives in a physical three dimensional space.

Due to the rotational symmetry of the problem, we can introduce the set of multipolar coordinates

$$\begin{aligned} \hat{x}_1 &= x_1\cos x_1, \quad \hat{x}_2 = x_1\sin x_1 \\ \hat{x}_3 &= x_2\cos x_2, \quad \hat{x}_4 = x_2\sin x_2 \\ \hat{x}_5 &= x_3\cos x_3, \quad \hat{x}_6 = x_3\sin x_3 \end{aligned} \tag{8}$$

and write the Hamiltonian as

$$H_6^{\rm C} = \frac{1}{2}p^2 - \frac{\gamma}{r} + \frac{p_4^2}{2x_1^2} + \frac{p_5^2}{2x_2^2} + \frac{p_6^2}{2x_3^2}, \quad r = \sqrt{x_1^2 + x_2^2 + x_3^2}, \quad p = \sqrt{p_1^2 + p_2^2 + p_3^2} \tag{9}$$

Since the variables x_4, x_5 and x_6 are cyclic, the corresponding momenta are constants and then we obtain the reduced Hamiltonian:

$$H_3^{\rm C} = \frac{1}{2}p^2 - \frac{\gamma}{r} + \frac{k_1}{x_1^2} + \frac{k_2}{x_2^2} + \frac{k_3}{x_3^2} \tag{10}$$

This Hamiltonian lives in the reduced manifold P_μ (see Section 2), where the group G is the product of three copies of $SO(2)$.

The Hamiltonian $H_3^{\rm C}$ is maximally superintegrable, that is we can find a set of five functionally independent first integrals (although not all of them in involution). Four of them can be easily obtained, being quadratic in the momenta [11], [5]. However, the fifth one is not quadratic and was recently computed in [19]. The method used in that work was inspired by the case when one of the barriers (k_i) vanishes. In this situation (for instance $k_3 = 0$), the quantity ($\vec{L}$ is the usual angular momentum in the three dimensional space):

$$p_2L_1 - p_1L_2 - 2x_3\Big(-\frac{\gamma}{2r} + \frac{k_1}{x_1^2} + \frac{k_2}{x_2^2}\Big) \tag{11}$$

is an integral. The argument in [19] is that the fifth integral (which is not quadratic in the momenta) should reduce to this one in the limit $k_3 \to 0$. This observation allows to calculate the integral we need to prove maximal superintegrability (see [19] for the details)

We will follow here a different approach, based on the theory of the reduction of superintegrable systems described above. The set of first integrals of the first Hamiltonian $H_6^{\rm C}$ can be written in the new coordinates x_i. However, not all of them will be independent of x_4, x_5, x_6, so they do not yield first integrals of the reduced system. Our approach consists in finding which of these quantities will satisfy this constraint. Equivalently, we will compute first order integrals of the Hamiltonian H_6^C which Poisson commute with the generators of the rotation group L_{12}, L_{34} and L_{56} (which are the generators of the subgroup of the rotation group which is used to construct the symmetry reduction). In fact, if we call

$$\tilde{L} = L|_{\text{reduced}}, \quad \tilde{A} = A|_{\text{reduced}} \tag{12}$$

the following reduced operators are constants

$$\tilde{L}_{12} = \sqrt{2k_1}, \quad \tilde{L}_{34} = \sqrt{2k_2}, \quad \tilde{L}_{56} = \sqrt{2k_3} \tag{13}$$

In fact, as we have explained in Section 2, the reduced phase space is $J^{-1}(\sqrt{2k_1}, \sqrt{2k_2}, \sqrt{2k_3})/T^3$.

We will define a set of reduced quantities, which do not depend on x_4, x_5, x_6 and then, are first order integrals of the reduced system:

$$E = H \tag{14}$$
$$I_1 = \tilde{L}_{13}^2 + \tilde{L}_{14}^2 + \tilde{L}_{23}^2 + \tilde{L}_{34}^2 \tag{15}$$
$$I_2 = \tilde{L}_{15}^2 + \tilde{L}_{16}^2 + \tilde{L}_{25}^2 + \tilde{L}_{26}^2 \tag{16}$$
$$I_3 = \tilde{L}_{35}^2 + \tilde{L}_{36}^2 + \tilde{L}_{45}^2 + \tilde{L}_{46}^2 \tag{17}$$

$$\{H_3^C, I_i\} = 0, \quad i = 1, 2, 3 \tag{18}$$

We need a fifth quantity which is not quadratic. We can construct it by using the Laplace–Runge-Lenz tensor (remark that these quantities commute with L_{12}, L_{34} and L_{56}):

$$T_1 = \tilde{A}_1^2 + \tilde{A}_2^2 \tag{19}$$
$$T_2 = \tilde{A}_3^2 + \tilde{A}_4^2 \tag{20}$$
$$T_3 = \tilde{A}_5^2 + \tilde{A}_6^2 \tag{21}$$

$$\{H, T_i\} = 0, \quad i = 1, 2, 3 \tag{22}$$

The seven first integrals cannot be functionally independent. In fact we only need a fourth order first integral. For instance, we can choose

$$T = T_1 + T_2 + T_3 \tag{23}$$

which can be written as:

$$T = \left(\frac{1}{2r}(2p^2r^2 - (\vec{p}\vec{r})^2) + 2r\left(-\frac{\gamma}{r} + \frac{k_1}{x^2} + \frac{k_2}{y^2} + \frac{k_3}{z^2}\right)\right)^2 + \gamma\frac{(\vec{p}\vec{r})^2}{r} - \frac{(\vec{p}\vec{r})^4}{4r^2} \tag{24}$$

(see [19]). We have constructed a set of five functionally independent first integrals (including the Hamiltonian) showing the maximal superintegrability of the system. The generalization of the above discussion to any (even) dimension is straightforward.

4. The anisotropic harmonic oscillator in even dimension and its reduction

As in the previous section, our results can be easily extended to any even dimension, but we shall restrict this work to the four dimensional case and its reduction, the 2–dimensional case. Here we review the results contained in [15].

The anisotropic oscillator in the two–dimensional case, both in classical and quantum mechanics, was discussed by Jauch and Hill [10], [4], [9]. The system

$$H_N^{\mathrm{A}} = \frac{1}{2}\sum_{i=1}^{N} \hat{p}_i^2 + \frac{1}{2}\sum_{i=1}^{N} \omega_i^2 \hat{x}_i^2 \tag{25}$$

is also known to be superintegrable (in any dimension), if the ratios of the frequencies are rational. Let us consider a $N = 4$ dimensional space and assume

$$\frac{\omega_1}{n_1} = \frac{\omega_2}{n_1} = \frac{\omega_3}{n_2} = \frac{\omega_4}{n_2} = \omega \tag{26}$$

that is, consider the Hamiltonian

$$H_4^{\mathrm{A}} = \frac{1}{2}\sum_{i=1}^{4} \hat{p}_i^2 + \frac{1}{2\omega^2}\left(n_1^2 \sum_{i=1}^{2} \hat{x}_i^2 + n_2^2 \sum_{i=3}^{4} \hat{x}_i^2\right) \tag{27}$$

Following [10], we define a set of (complex) invariants, with coordinates $z_i, \bar{z}_i$, $i = 1, \ldots, 4$:

$$z_j = \hat{p}_j - in_j\omega\hat{x}_j, \qquad \bar{z}_j = \hat{p}_j + in_j\omega\hat{x}_j. \tag{28}$$

It is easily checked that the expressions

$$c_{jk} = z_j^{n_k} \bar{z}_k^{n_j} \tag{29}$$

provide integrals of motion. In particular, among these integrals we have the angular momenta

$$L_{ik} = \hat{x}_i\hat{p}_k - \hat{x}_k\hat{p}_i \tag{30}$$

(when $n_i = n_k$) and the tensor

$$T_{ik} = \hat{p}_i\hat{p}_k + n_i n_k \omega^2 \hat{x}_i \hat{x}_k \tag{31}$$

We will now study one of the several possible reductions of the anisotropic oscillator (27) and establish the superintegrability of the corresponding reduced system, computing, in an explicit way, three functionally independent constants of motion.

According to the scheme adopted for the Coulomb case, let us consider the following change of coordinates:

$$\begin{aligned} \hat{x}_1 &= x_1 \cos x_3, \quad \hat{x}_2 = x_1 \sin x_3 \\ \hat{x}_3 &= x_2 \cos x_4, \quad \hat{x}_4 = x_2 \sin x_4. \end{aligned} \tag{32}$$

The corresponding momenta read

$$\begin{aligned} \hat{p}_1 &= -p_3\tfrac{\sin x_3}{x_1} + p_1 \cos x_3, \quad \hat{p}_2 = p_3\tfrac{\cos x_3}{x_1} + p_1 \sin x_3, \\ \hat{p}_3 &= -p_4\tfrac{\sin x_4}{x_2} + p_2 \cos x_4, \quad \hat{p}_4 = p_4\tfrac{\cos x_4}{x_2} + p_2 \sin x_4. \end{aligned} \tag{33}$$

and the system in the new coordinates is:

$$H_4^{\mathrm{A}} = \frac{1}{2}(p_1^2 + p_2^2) + \frac{p_3^2}{2x_1^2} + \frac{p_4^2}{2x_2^2} + \frac{1}{2}\omega^2(n_1^2 x_1^2 + n_2^2 x_2^2) \tag{34}$$

Setting

$$p_3 = \sqrt{k_1}, \quad p_4 = \sqrt{k_2} \tag{35}$$

we get the reduced system:

$$H_2^A = \frac{1}{2}(p_1^2 + p_2^2) + \frac{k_1^2}{2x_1^2} + \frac{k_2^2}{2x_2^2} + \frac{1}{2}\omega^2\left(n_1^2 x_1^2 + n_2^2 x_2^2\right) \tag{36}$$

Notice the appearance of Rosochatius–type terms [16] in the reduced system.

As in he Coulomb case, this Hamiltonian is defined in a reduced manifold P_μ and the group used in this reduction is $SO(2) \times SO(2)$.

The integrals of the system (27), given in (29), will be used to construct a set of integrals for the reduced system. Those that will survive the reduction are the ones that are left invariant by the $SO(2) \times SO(2)$ rotations. The arguments are the same we used in the Coulomb potential, in fact, they must Poisson commute with

$$\begin{aligned} L_{12} =& \frac{\mathrm{i}}{2n_1\omega}(z_1\bar{z}_2 - z_2\bar{z}_1) = \hat{x}_1\hat{p}_2 - \hat{x}_2\hat{p}_1, \\ L_{34} =& \frac{\mathrm{i}}{2n_2\omega}(z_3\bar{z}_4 - z_4\bar{z}_3) = \hat{x}_3\hat{p}_4 - \hat{x}_4\hat{p}_3. \end{aligned} \tag{37}$$

The computation of these commuting quantities is not as straightforward as in the Coulomb potential, where we found the integrals by simple inspection. We will use z coordinates to solve our problem. The Poisson bracket can be written in terms of the z_i variables as

$$\{f(z_i, \bar{z}_i), g(z_i, \bar{z}_i)\} = -2\mathrm{i}\omega \sum_{k=1}^{N} \sum_{j=2k-1}^{2k} n_k \left(\frac{\partial f}{\partial z_j}\frac{\partial g}{\partial \bar{z}_j} - \frac{\partial f}{\partial \bar{z}_j}\frac{\partial g}{\partial z_j} \right) \tag{38}$$

hence, functions of $z_k, \bar{z}_k$ Poisson commuting with L_{12} and L_{34} must satisfy

$$\begin{aligned} z_2\partial_{z_1} f - z_1\partial_{z_2} f + \bar{z}_2\partial_{\bar{z}_1} f - \bar{z}_1\partial_{\bar{z}_2} f = 0, \\ z_4\partial_{z_3} f - z_3\partial_{z_4} f + \bar{z}_4\partial_{\bar{z}_3} f - \bar{z}_3\partial_{\bar{z}_4} f = 0. \end{aligned} \tag{39}$$

A basis for the corresponding $SO(2) \times SO(2)$ invariants is given by

$$\begin{aligned} \xi_1 = z_1^2 + z_2^2, \quad \bar{\xi}_1 = \bar{z}_1^2 + \bar{z}_2^2, \quad \eta_1 = z_1\bar{z}_1 + z_2\bar{z}_2, \\ \xi_3 = z_3^2 + z_4^2, \quad \bar{\xi}_3 = \bar{z}_3^2 + \bar{z}_4^2, \quad \eta_2 = z_3\bar{z}_3 + z_4\bar{z}_4, \end{aligned} \tag{40}$$

These rotationally invariants will be constants of motion of our system if they satisfy:

$$\{H_4^A, f(\xi_1, \bar{\xi}_1, \eta_1, \xi_3, \bar{\xi}_3, \eta_2)\} = 0. \tag{41}$$

The following expressions are solutions of equation (41)

$$\begin{aligned} E_1 = \tfrac{1}{2}(|z_1|^2 + |z_2|^2), \qquad & E_2 = \frac{1}{2}(|z_3|^2 + |z_4|^2), \\ Q_1 = (z_1^2 + z_2^2)^{n_2}(\bar{z}_3^2 + \bar{z}_4^2)^{n_1}, \quad & \bar{Q}_1 = (\bar{z}_1^2 + \bar{z}_2^2)^{n_2}(z_3^2 + z_4^2)^{n_1}, \\ I_1 = (z_1^2 + z_2^2)(\bar{z}_1^2 + \bar{z}_2^2), \qquad & I_2 = (z_3^2 + z_4^2)(\bar{z}_3^2 + \bar{z}_4^2). \end{aligned} \tag{42}$$

however, only five of these integrals are functionally independent.

We now perform the reduction for the integrals of motion using the change of variables (32), (33). The integrals (37) reduce to constants $L_{12} = \sqrt{k_1}$, $L_{34} = \sqrt{k_2}$. The first four integrals in (42) reduce to nontrivial integrals for the Hamiltonian (36), namely

$$\begin{aligned} E_1 &= \frac{1}{2}p_1^2 + \frac{k_1}{2x_1^2} + \frac{1}{2}n_1^2\omega^2 x_1^2, \\ E_2 &= \frac{1}{2}p_2^2 + \frac{k_2}{2x_2^2} + \frac{1}{2}n_2^2\omega^2 x_2^2 \\ Q_1 &= (p_1^2 + \frac{k_1}{x_1^2} - n_1^2\omega^2 x_1^2 - 2\mathrm{i}n_1\omega p_1 x_1)^{n_2}(p_2^2 + \frac{k_2}{x_2^2} - n_2^2\omega^2 x_2^2 + 2\mathrm{i}n_2\omega p_2 x_2)^{n_1}, \\ \bar{Q}_1 &= (p_1^2 + \frac{k_1}{x_1^2} - n_1^2\omega^2 x_1^2 + 2\mathrm{i}n_1\omega p_1 x_1)^{n_2}(p_2^2 + \frac{k_2}{x_2^2} - n_2^2\omega^2 x_2^2 - 2\mathrm{i}n_2\omega p_2 x_2)^{n_1}. \end{aligned} \tag{43}$$

The remaining two integrals in (42) give nothing new

$$I_1 = 4(E_1^2 - k_1 n_1^2 \omega^2), \quad I_2 = 4(E_2^2 - k_2 n_2^2 \omega^2). \tag{44}$$

Three functionally independent real integrals of motion of the system with Hamiltonian (27) can be chosen to be

$$\{E_1, E_2, Q = \frac{1}{2}(Q_1 + \bar{Q}_1)\}. \tag{45}$$

They are of order 2, 2 and $2(n_1 + n_2)$ in the momenta, respectively. Their existence is the proof of the maximal superintegrability of the considered system.

The integral of motion Q simplifies to give a second order one in two cases (that were known previously [7], [20]). They are

I) $n_1 = n_2 = 1$

$$\frac{4E_1E_2 - Q}{2\omega^2} = (p_1x_2 - p_2x_1)^2 + \frac{k_1x_2^2}{x_1^2} + \frac{k_2x_1^2}{x_2^2}. \tag{46}$$

II) $n_1 = 1, n_2 = 2, k_2 = 0$

$$\left(\frac{8E_1^2E_2 - Q}{8\omega^2} - k_1E_2\right)^{1/2} = p_1(x_2p_1 - x_1p_2) - \omega^2x_1^2x_2 + k_1\frac{x_2}{x_1^2} \tag{47}$$

The integrals (46) and (47) are responsible for the separation of variables in polar and parabolic coordinates, respectively. The integrals $\{E_1, E_2\}$ are responsible for the separation in cartesian coordinates.

5. Conclusions

From the previous examples it is clear that it would be very interesting to construct, in a systematic way, transformations mapping a superintegrable system into another system, that is also superintegrable, and defined in a reduced phase space. The role of higher order groups of transformations generated by the flow associated to integrals that are polynomials in the momenta remains to be fully investigated.

We also mention that an algebraic approach to superintegrable systems, independent of the geometry of the phase space, and based on a suitable nilpotency condition, is presently under investigation [8].

A quantum mechanical version of this reduction procedure is also to be understood.

Acknowledgments

The research of P.W. was partly supported by NSERC of Canada. The research of M.A.R. and P.T. was supported by MICINN (Spain) under grants no. FIS2008-00752, FIS2008-00752 and Universidad Complutense and Comunidad de Madrid under grant no. GR74/07-910556.

References

[1] Abraham R, Marsden J E 1978 *Foundations of Mechanics* (2nd ed., The Benjamin/Cummings Publishing Company, Reading).
[2] J. Bertrand, Comptes Rendus. Acad. Sci. Paris, **77**, 849 (1873).
[3] Cariñena J F, Clemente–Gallardo J and Marmo G 2007 *Int. J. Geom. Methods Mod. Phys.* **4** 1363
[4] Demkov Yu N 1963 *Sov. Phys. JETP* **17** 1349.
[5] Evans N W 1990 *Phys. Rev.* A **41**, 5666.
[6] Evans N W and Verrier P E 2008 *J. Math. Phys.* **49** 092902.
[7] Friš J, Mandrosov V, Smorodinsky Ya A, Uhlíř M and Winternitz P 1965 *Phys. Lett.* **16** 354.
[8] A. Ibort, G. Marmo, M. A. Rodríguez and P. Tempesta, *Nilpotent integrability and algebraic reductions in classical mechanics*, in preparation.

[9] Ilkaeva L A 1963 *Vestnik LGU* **22** 56 (in Russian).
[10] Jauch J and Hill E 1940 *Phys. Rev.* **57** 641.
[11] Makarov A A, Smorodinsky J A, Valiev Kh and Winternitz P 1967 *Nuovo Cimento* A **52** 1061.
[12] Marsden J E and A. Weinstein 1974 *Rep. Math. Phys.* **5** 121.
[13] Mischenko A S and A. T. Fomenko 1978 *Funct. Anal. Appl.* **12** 133
[14] Nekhoroshev N N 1972 *Trans. Moscow Math. Soc.* **26** 180
[15] Rodríguez M A, Tempesta P and Winternitz P 2008 *Phys. Rev.* E **78** 046608.
[16] Rosochatius E 1877 *Dissertation* (Gottingen, Gebr. Unger, Berlin)
[17] Tempesta P, Winternitz P, Harnad J, Miller W Jr, Pogosyan G, Rodríguez M A (eds) 2004 *Superintegrability in Classical and Quantum Systems*, (Montréal, CRM Proceedings and Lecture Notes, AMS vol. 37).
[18] Tempesta P, Turbiner A and Winternitz P 2001 *J. Math. Phys.* **42** 4248
[19] Verrier P E and Evans N W 2008 *J. Math. Phys.* **49** 022902.
[20] Winternitz P, Smorodinsky Ya, Uhlíř M and Friš J 1966 *Yad. Fiz.* **4** 625; 1967 *Sov. J. Nucl. Phys.* **4** 444.

Journal of Physics: Conference Series **175** (2009) 012014 doi:10.1088/1742-6596/175/1/012014

Quantum Hall effect on Riemann surfaces

Carlos Tejero Prieto
Departamento de Matemáticas and
Instituto Universitario de Física Fundamental y Matemáticas
Universidad de Salamanca
Spain

E-mail: carlost@usal.es

Abstract. We study the family of Landau Hamiltonians compatible with a magnetic field on a Riemann surface S by means of Fourier-Mukai and Nahm transforms. Starting from the geometric formulation of adiabatic charge transport on Riemann surfaces, we prove that Hall conductivity is proportional to the intersection product on the first homology group of S and therefore it is quantized. Finally, by using the theory of determinant bundles developed by Bismut, Gillet and Soul, we compute the adiabatic curvature of the spectral bundles defined by the holomorphic Landau levels. We prove that it is given by the polarization of the jacobian variety of the Riemann surface, plus a term depending on the relative analytic torsion.

1. Introduction

The classical Hall effect, giving rise to the so-called Hall conductivity, was observed in 1879 by E. Hall. Almost exactly a century later, in 1980, K. von Klitzing [1] discovered the integer quantum Hall effect which implies the quantization of Hall conductivity. Soon afterwards, in 1981, R. Laughlin [2] gave the first physical explanation based on a gauge argument on a cylinder. The remarkable exactness of this effect triggered a worldwide interest on it which eventually lead to the discovery of the the fractional quantum Hall effect by D. Tsui, H Stormer, A. Gossard in 1982 [3]. The same year, D. Thouless, M. Kohmoto, P. Nightingale and M. den Nijs [4] gave the first topological explanation for the quantization of the Hall conductivity by means of the perturbative Kubo's formula applied to a Flat torus configuration space. With the same geometry, J. Avron-R. Seiler [5], Q. Niu-D. J. Thouless [6], M. Kohmoto [7], and Q. Niu-D. J. Thouless-Y. S Wu [8] showed in 1985 that Kubo's formula is given by the Chern number of a line bundle over the torus of magnetic fluxes. The next conceptual step was carried out by J. Avron, R. Seiler and L. Yaffe in 1987 [9], when they interpreted Kubo's formula in the framework of adiabatic transport theory and proved its validity to second order. This approach allowed them to show that Hall conductivity was given by the adiabatic curvature of spectral projectors associated with a family of Schrödinger operators parametrized by the Aharanov-Bohm potentials. Later, in 1990, M. Klein, R. Seiler [10] proved the validity of Kubo's formula to any perturbation order. Finally, J. Avron, R. Seiler and P. G. Zograf [11] outlined in 1994 the computation of the adiabatic conductance for the first Landau level for strong magnetic fields.

In this paper we give a geometric description of the spectral bundles defined by the family of Landau Hamiltonians on a Riemann surface S by means of holomorphic spectral geometry techniques. This will allow us to determine the Hall conductivity for holomorphic Landau levels in terms of Chern classes of spectral bundles.

In order to achieve this goal we first use a particular instance of Nahm's transform to describe the spectral bundles associated with the holomorphic Landau levels. Then we show that this Nahm transform is equivalent to an integral functor associated with the jacobian of the Riemann surface S. This identification allows us to determine the topological invariants of the spectral bundles and hence to prove the quantization of the Hall conductivity of holomorphic Landau levels on a Riemann surface.

After this we go one step further and compute the adiabatic conductance for holomorphic Landau levels. We show that this is equivalent to determining the adiabatic curvature of the spectral bundles and in turn this is the same as computing the curvature of the determinant bundles of the spectral bundles with respect to the natural Hermitian metrics induced on them. These metrics are conformal to the Quillen metrics and we can compute the curvature of the latter by means of the techniques developed by Bismut, Gillet and Soulé [12, 13, 14]. As a final result we prove that the adiabatic conductance of the holomorphic Landau levels is determined by the polarization of the jacobian of the Riemann surface plus a term which depends on the relative analytic torsion.

The results expounded here are based in the papers [15, 16], the reader is referred to them for further details.

2. Geometric quantization of the Landau-Hall problem

Let (S, g) be an oriented Riemannian surface and $B \in \Omega^2(S)$ a covariantly constant magnetic field, thus $B = \hat{B}\Omega_2$, where $\hat{B} \in \mathbb{R}$ and Ω_2 is the Riemannian area element.

The Landau-Hall problem for particles of charge e and mass m is the Hamiltonian dynamical system

$$(T^*S, \omega_2, H),$$

where $\omega_2 = d\theta + e\,B$, and $H = \frac{\|\omega\|_g^2}{2m}$ is the Landau Hamiltonian.

It can be shown that the symplectic manifold (T^*S, ω_2) is quantizable if and only if (S, eB) is quantizable, that is, if and only if $[\frac{eB}{\hbar}] \in H^2(X, \mathbb{Z})$. Moreover, the Hamiltonian H is quantizable in the BKS scheme and if $\mathcal{L} = (L, \langle\,,\rangle, \nabla) \to S$ is a prequantization bundle with Hermitian metric $\langle\,,\rangle$ and unitary connection ∇, then the Schrödinger operator associated with H is

$$\hat{H} = \frac{\hbar^2}{2m}(\nabla^*\nabla + \frac{R}{6}),$$

where $\nabla^*\nabla$ is the Bochner Laplacian of the connection ∇ and R is the scalar curvature of the riemannian metric g.

The different Schrödinger operators associated with the magnetic field B are parametrized by the set of equivalence classes of prequantization line bundles of (S, eB). This set is given by $H^1(S, U(1))$, the space of Aharanov-Bohm potentials.

If we consider a prequantization bundle $\mathcal{L} = (L, \langle\,,\rangle, \nabla) \to S$ then any other prequantization bundle $\mathcal{L}' = (L', \langle\,,\rangle', \nabla')$ can be obtained by twisting $\mathcal{L}$ by a flat Hermitian line bundle L_0.

Since L_0 is flat, its is trivializable, hence

$$L' = L \otimes L_0 \simeq L,$$
$$\nabla' = \nabla - \frac{i}{\hbar}A \quad A \in \Omega^1(S) \text{ closed.}$$

3. Families of Landau operators

On a compact surface S there is an isomorphism

$$H^1(S, U(1)) \simeq H^1(S, \mathbb{R})/H^1(S, \mathbb{Z})$$

and $J(S) = H^1(S,\mathbb{R})/H^1(S,\mathbb{Z})$ is the Lazzeri model of the jacobian variety of S.

Any flat line bundle on the Riemann surface (S,g) is endowed with a natural holomorphic structure. Thus, $H^1(S,U(1))$ gets identified with the Picard group $\mathrm{Pic}^0(S)$ of degree zero holomorphic line bundles.

Therefore there are isomorphisms

$$J(S) \simeq H^1(S,U(1)) \simeq \mathrm{Pic}^0(S),$$

that correspond to the identifications

$$\left\{ \begin{array}{c} \text{Jacobian} \\ \text{variety of } S \end{array} \right\} \leadsto \left\{ \begin{array}{c} \text{Aharanov-Bohm} \\ \text{potentials} \end{array} \right\} \leadsto \left\{ \begin{array}{c} \text{Degree zero holomorphic} \\ \text{line bundles} \end{array} \right\}$$

To each presentation of the space of Aharanov-Bohm potentials either as $J(S)$ or $\mathrm{Pic}^0(S)$ we can associate a family of Schrödinger operators parametrized by these spaces.

3.1. Family of Landau operators parametrized by $J(S)$

In this case we keep the line bundle fixed and parametrize the connections.

Let $\mathcal{L} = (L, \langle\,,\rangle, \nabla) \to S$ be a prequantization bundle on (S, eB). We have seen that the Landau Hamiltonian H is quantizable and its Schrödinger operator, acting on $L^2(S,L)$, is:

$$\hat{H} = \frac{\hbar^2}{2m}(\nabla^*\nabla + \frac{R}{6}) \quad (\textit{Landau operator}).$$

The spaces of eigensections of $\hat{H}$ are called *Landau levels*. We have also seen that any other prequantization bundle is of the form $(L, \langle\,,\rangle, \nabla' = \nabla - \frac{i}{\hbar}A)$ where $A \in \Omega^1(S)$ is closed. We denote by

$$\hat{H}(A) = \frac{\hbar^2}{2m}(\nabla'^*\nabla' + \frac{R}{6}),$$

the corresponding Landau operator acting on $L^2(S,L)$.

We get this way a family of operators $\hat{H}(A)$ parametrized by $J(S)$, whose elements act on the Hilbert space $L^2(S,L)$. We consider the trivial Hilbertian bundle

$$\mathcal{H} = J(S) \times L^2(S,L) \to J(S).$$

$\mathcal{H}$ is endowed with the flat connection $\tilde{\nabla}$ associated with the trivialization. The family of operators $\hat{H}(A)$ can be thought as a section $\hat{H} \in \Gamma(J(S), \mathrm{End}\,(\mathcal{H}))$.

3.2. Family of operators defined by a relative operator

From a geometrical point of view, a C^∞ family of differential operators parametrized by a differentiable manifold T is represented by a relative differential operator.

Definition 1 *Let $\pi \colon X \to T$ be a submersion and let $E \to X$, $F \to X$ be two vector bundles. A k-th order relative differential operator is a $C^\infty(T)$-linear map*

$$D \colon \Gamma(X,E) \to \Gamma(X,F),$$

that factorizes through the k-th order relative jet extension

$$j^k_{X/T} \colon \Gamma(X,E) \to \Gamma(X, J^k_{X/T}(E)),$$

by means of a $C^\infty(X)$-linear mapping $\overline{D} : \Gamma(X, J^k_{X/T}(E)) \to \Gamma(X, F)$. That is, there is a commutative diagram

$$\begin{array}{ccc} \Gamma(X,E) & \xrightarrow{\quad D \quad} & \Gamma(X,F) \\ & {\scriptstyle j^k_{X/T}} \searrow \qquad \nearrow {\scriptstyle \overline{D}} & \\ & \Gamma(X, J^k_{X/T}(E)) & \end{array}$$

For any $t \in T$, set $X_t = \pi^{-1}(t)$ and let $F_t \to X_t$, $E_t \to X_t$ be the restrictions of the vector bundles E, F to the fiber X_t.

A relative differential operator $D\colon \Gamma(X, E) \to \Gamma(X, F)$, induces in a natural way differential operators

$$D_t : \Gamma(X_t, E_t) \to \Gamma(X_t, F_t)$$

in this way we obtain the family of differential operators defined by D and parametrized by T.

3.3. Family of Landau operators parametrized by $\mathrm{Pic}^0(S)$

Now we deform the line bundle.

It is well known that $\mathrm{Pic}^0(S)$ parametrizes the gauge equivalence classes of flat Hermitian line bundles on S.

Let $\mathcal{P}_S \to S \times \mathrm{Pic}^0(S)$ be a Poincaré line bundle. $\mathcal{P}_S$ can be endowed with a unitary connection $\nabla_{\mathcal{P}}$, such that the restriction of $(\mathcal{P}_S, \nabla_{\mathcal{P}})$ to every fiber $S_\xi \equiv S \times \{\xi\}$ of the natural projection $\pi_{\mathrm{Pic}^0(S)}\colon S \times \mathrm{Pic}^0(S) \to \mathrm{Pic}^0(S)$ is a flat bundle in the gauge equivalence class defined by $\xi \in \mathrm{Pic}^0(S)$.

Let $\mathcal{L} = (L, \langle\, , \rangle, \nabla) \to S$ be a prequantization bundle for (S, eB) and let $\pi_S : S \times Pic^0(S) \to S$ be the natural projection.

We endow $\pi_S^* L \otimes \mathcal{P}_S \to S \times \mathrm{Pic}^0(S)$ with the product connection $\check{\nabla} = \pi_S^*\nabla \otimes 1 + 1 \otimes \nabla_{\mathcal{P}}$, that can be considered as a relative differential operator for the projection $\pi_{\mathrm{Pic}^0(S)} : S \times \mathrm{Pic}^0(S) \to \mathrm{Pic}^0(S)$.

Definition 2 *The family of Landau operators parametrized by* $\mathrm{Pic}^0(S)$ *is the family defined by the relative Schrödinger operator*

$$\hat{H} = \frac{\hbar^2}{2m}(\check{\nabla}^*\check{\nabla} + \frac{R}{6}), \quad (\textit{Universal Landau operator})$$

defined on the line bundle $\pi_S^* L \otimes \mathcal{P}_S \to S \times \mathrm{Pic}^0(S)$.

Comparing it with the family parametrized by $J(S)$, in this case we get a Hilbertian bundle

$$\widehat{\mathcal{H}} \to \mathrm{Pic}^0(S),$$

whose fiber at $[L_0] \in \mathrm{Pic}^0(X)$ is $L^2(S, L \otimes L_0)$. Moreover, $\widehat{\mathcal{H}}$ is endowed with a flat connection $\check{\nabla}$ introduced by Bismut [12].

3.4. Equivalence of the families of Landau operators parametrized by $J(S)$ and $\mathrm{Pic}^0(S)$

It can be proved that the Abel-Jacobi isomorphism $J(S) \simeq \mathrm{Pic}^0(S)$ establishes an isomorphism between the families of Landau operators parametrized by $J(S)$ and $\mathrm{Pic}^0(S)$.

We call any of them the family of Landau operators on the Riemann surface S corresponding to the magnetic field B. Bearing in mind the isomorphism induced by the Abel-Jacobi map we can use these families in an equivalent way.

4. Adiabatic charge transport on Riemann surfaces

By means of an extension of Kato's Adiabatic Theorem, Avron, Seiler and Jaffe [9] interpreted Hall conductivity as the adiabatic curvature of a family of spectral projectors parametrized by the Aharanov-Bohm potentials.

A time dependent quantum system is described by a selfadjoint operator $\hat{H}(t)$ defined on a Hilbert space. Let τ be a parameter defining the temporal scale of the system. The dynamics of the system is determined by the time dependent Schrödinger equation

$$i\hbar\frac{\partial\psi_\tau(t)}{\partial t} = \hat{H}(t/\tau)\psi_\tau(t).$$

The adiabatic limit of the system corresponds to the limit $\tau \to \infty$.

To formulate the Adiabatic Theorem one has to make several hypothesis on the uniparametric family of operators $\hat{H}(\sigma)$ with $\sigma = \frac{t}{\tau}$:

A1. The family $\hat{H}(\sigma)$ is smooth.

A2. The spectrum of $\hat{H}(\sigma)$ has a gap whose size is uniformly bound. Therefore, we can define a projector $P(\sigma)$ onto the part of the spectrum separated by the gap.

A3. The rank of $P(\sigma)$ is finite.

Let $\psi_\tau(0) \in \operatorname{Im} P(0)$, the Adiabatic Theorem states that the evolved state $\psi_\tau(\sigma)$, where $\sigma \sim 1$, is in $\operatorname{Im} P(\sigma)$ up to correction terms.

Theorem 1 *If $\hat{H}(\sigma)$ satisfies conditions A1-A3 and $U_\tau(\sigma)$ is the evolution operator of the Schrödinger equation, then one has*

$$U_\tau(\sigma)P(0)U_\tau(\sigma)^{-1} = P(\sigma) + O(\tau^{-1})$$

The magnitude of the error term depends on τ and the size of the spectral gap.

4.1. Charge transport on a Riemann surface S

Let $\{\alpha_1, \dots, \alpha_{2p}\}$ be a symplectic basis of $H_1(S, \mathbb{Z})$ and let $\{\omega_1, \dots, \omega_{2p}\}$ be the dual basis of $H^1(S, \mathbb{Z})$ formed by harmonic 1-forms, that is

$$\int_{\alpha_i} \omega_j = \delta_{ij}.$$

We say that ω_j is the Aharanov-Bohm potential associated with α_j.

If $A(t)$ is a curve in $J(S) = H^1(S, \mathbb{R})/H^1(S, \mathbb{Z})$ then $D = \frac{\partial A}{\partial t} \in \mathfrak{X}(J(S))$ is a vector field that by Faraday's law gets identified with an electromotive force.

Let ψ be a solution to the Schrödinger equation $i\hbar\frac{\partial\psi}{\partial t} = \hat{H}(A(t))\psi$. The intrinsic expression of the current 1-form associated with the state of the system described by ψ and induced by the variation of the Aharanov-Bohm potentials given by $A(t)$, is

$$I(\psi) = \frac{e}{\hbar}d\langle\psi, \hat{H}\psi\rangle + i\frac{e^2}{\hbar}i_D\langle\tilde{\nabla}\psi, \tilde{\nabla}\psi\rangle$$

where $\tilde{\nabla}$ is the connection on the Hilbertian bundle $\mathcal{H} \to J(S)$.

The 2-form $C = i\frac{e^2}{\hbar}\langle\tilde{\nabla}\psi, \tilde{\nabla}\psi\rangle$ relating currents with electromotive forces is called the conductance 2-form. The current induced along $\alpha_j \in H_1(S, \mathbb{Z})$ by the variation $A(t)$ is given by:

$$I(\alpha_j, \psi) = i_{D_j} I(\psi)$$

where D_j is the vector field on $J(S)$ associated with ω_j.

The charge transport along the homology class α_j due to the variation of the Aharanov-Bohm potentials $A(t)$ is given by

$$Q(\alpha_j, \psi) = \int_0^\tau I(\alpha_j, \psi)dt$$

Finally, the charge transport along α_j while the Aharanov-Bohm potential associated with α_k experiments a unity increment, that is $D = D_k$, is denoted $Q(\alpha_k, \alpha_j)$.

Journal of Physics: Conference Series **175** (2009) 012014 doi:10.1088/1742-6596/175/1/012014

4.2. Adiabatic charge transport

The idea behind adiabatic transport is to replace ψ by its adiabatic evolution and estimate the difference by means of the Adiabatic Theorem.

Let P be the spectral projector associated with ψ and define the spectral bundle $\widehat{P} = \mathrm{Im}\, P \to J(S)$. This endowed with the connection $\widehat{\nabla} = P \circ \tilde{\nabla}$ with curvature

$$\Omega^{\widehat{\nabla}} = P \circ (\tilde{\nabla} P \wedge \tilde{\nabla} P) \circ P,$$

where $\tilde{\nabla} P$ denotes the covariant derivative of $P \in \mathrm{End}\,(\mathcal{H})$.

Definition 3 *The ordinary 2-form*

$$\Omega^P = \mathrm{Tr}\left(P \circ (\tilde{\nabla} P \wedge \tilde{\nabla} P) \circ P\right),$$

*is called the **adiabatic curvature** associated with the spectral bundle $\widehat{P} \to J(S)$.*

We get this way the adiabatic currents $I_{Ad}(\psi)$ and the adiabatic charge transports $Q_{Ad}(\alpha_k, \alpha_j)$. One checks that the adiabatic conductance 2-form is proportional to the adiabatic curvature $C_{Ad} = i\frac{e^2}{\hbar}\Omega^P$.

The average of the adiabatic quantities are given by

$$\langle I_{Ad} \rangle(\alpha_j, \psi) = \frac{e^2}{h}\frac{i}{2\pi}\int_{\alpha_j} i_D \Omega^P,$$

$$\langle Q_{Ad} \rangle(\alpha_k, \alpha_j) = e\frac{i}{2\pi}\int_{\alpha_k \star \alpha_j} \Omega^P,$$

where $\alpha_k \star \alpha_j$ is the Pontryagin product of the homology classes $\alpha_k,\ \alpha_j \in H_1(J(S), \mathbb{Z})$.

Theorem 2 *Let us suppose that the family of Landau operators on a Riemann surface S satisfies conditions (A1-A3) of the Adiabatic Theorem, then one has*

(i) $C = C_{Ad} + O(1/\tau)$.

(ii) $\langle Q \rangle(\alpha_k, \alpha_j) = \langle Q_{Ad} \rangle(\alpha_k, \alpha_j) + O(\tau^{-\infty})$ *as* $\tau \to \infty$.

The formula for the adiabatic transport

$$\langle Q_{Ad} \rangle(\alpha_k, \alpha_j) = e\frac{i}{2\pi}\int_{\alpha_k \star \alpha_j} \Omega^P,$$

is the geometric formulation of Kubo's formula.

Therefore, on a surface S of genus p there are $p(2p-1)$ different transport coefficients, and any of them gives rise to a mean Hall conductivity

$$\langle \sigma_H \rangle(\alpha_k, \alpha_j) = \frac{e^2}{h}\frac{i}{2\pi}\int_{\alpha_k \star \alpha_j} \Omega^P.$$

This is the most common form of Kubo's formula.

Since $\frac{i}{2\pi}\Omega^P = \frac{i}{2\pi}\mathrm{Tr}\,(\Omega^{\widehat{\nabla}})$ represents the first Chern class $c_1(\widehat{P}) \in H^2(J(S), \mathbb{Z})$ of the spectral bundle $\widehat{P} \to J(S)$, one has that $\langle Q_{Ad} \rangle$ and $\langle \sigma_H \rangle$ depend only on the cohomology class $c_1(\widehat{P})$.

As a consequence, Theorem 2 implies that the charge transport, as well as the Hall conductivity are quantized up to infinitesimal terms in the adiabatic parameter.

5. Nahm transforms on Riemann surfaces

In order to be able to describe the spectral bundles associated with the Landau levels and to calculate their topological invariants, we introduce the Nahm transform associated to the jacobian variety $J(S)$ of the Riemann surface S. We assume the identification $J(S) \simeq \mathrm{Pic}^0(S)$.

We have seen that $J(S)$ parametrizes the gauge equivalence classes of flat line bundles and the restriction of the Poincaré bundle $\mathcal{P}_S \to S \times J(S)$ endowed with its unitary connection $\nabla_{\mathcal{P}}$ to any fiber $S_\xi = S \times \{\xi\}$ is a flat line bundle $(\mathcal{P}_\xi, \nabla_{\mathcal{P},\xi}) \to S$ in the equivalence class defined by $\xi \in J(S)$.

Let $E \to S$ be a Hermitian vector bundle with a unitary connection ∇. We endow $E \otimes \mathcal{P}_\xi$ with the product connection $\nabla_\xi = \nabla \otimes 1 + 1 \otimes \nabla_{\mathcal{P},\xi}$.

Let us consider the spinor bundle $\mathbb{S} = \mathbb{S}^+ \oplus \mathbb{S}^-$ of S as a spin^c manifold, with $\mathbb{S}^+ = \Lambda^{0,0}T^*S$, $\mathbb{S}^- = \Lambda^{0,1}T^*S$ and let $\nabla_{\mathbb{S}}$ be the spinorial connection of $\mathbb{S}$.

We get a family of coupled Dirac operators:

$$D_\xi \colon \Omega^0(S, \mathbb{S}^+ \otimes E \otimes \mathcal{P}_\xi) \to \Omega^0(S, \mathbb{S}^- \otimes E \otimes \mathcal{P}_\xi).$$

By the Atiyah-Singer theorem for families, the difference bundle

$$\{\mathrm{Ker} D_\xi\} - \{\mathrm{Coker} D_\xi\},$$

defines an object $\mathrm{Ind}(D)$ in the K-theory of $J(S)$, the index of D.

If either $\{\mathrm{Ker} D_\xi\}$ or $\{\mathrm{Coker} D_\xi\}$ has constant rank, then $\mathrm{Ker} D$ and $\mathrm{Coker} D$ are vector bundles over $J(S)$ and moreover

$$\mathrm{Ind}(D) = [\mathrm{Ker} D] - [\mathrm{Coker} D] \in K(J(S)).$$

5.1. The transformed bundle

Definition 4 *A Hermitian vector bundle with unitary connection* $(E, \nabla) \to S$ *is an* IT*-pair (Index Theorem) if either* $\mathrm{Coker}\, D = 0$ *or* $\mathrm{Ker}\, D = 0$. *It is an*

- IT_0*-pair if* $\mathrm{Coker}\, D = 0$.
- IT_1*-pair if* $\mathrm{Ker}\, D = 0$.

The transformed bundle of an IT_k*-pair is the finite rank vector bundle*

$$\widehat{E} = (-1)^k \mathrm{Ind}(D) \to J(S).$$

$\mathcal{P}_S \to S \times J(S)$ is a holomorphic line bundle and the connection $\nabla_{\mathcal{P}}$ is compatible with the holomorphic structure.

The curvature F^∇ of (E, ∇) has $(1,1)$ type, therefore E is a holomorphic vector bundle. The Dirac operator D_ξ coincides with the Dolbeault-Dirac operator $D_\xi = \sqrt{2}(\bar{\partial}^*_{E\otimes\mathcal{P}_\xi} + \bar{\partial}_{E\otimes\mathcal{P}_\xi})$ of $E \otimes \mathcal{P}_\xi$.

5.2. The transformed Hermitian metric

Let $\pi_S : S \times J(S) \to S$ be the natural projection and let $\mathcal{H}^\infty_\pm$ be the spaces of C^∞-sections of the vector bundles $\pi_S^*(\mathbb{S}^\pm \otimes E) \otimes \mathcal{P}_S \to S \times J(S)$.

$\mathcal{H}^\infty_\pm$ can be thought of as infinite dimensional vector bundles $\mathcal{H}^\infty_\pm \to J(S)$ whose fibers $\mathcal{H}^\infty_{\pm,\xi}$ at $\xi \in J(S)$ are the spaces of smooth sections $\Gamma(S, \mathbb{S}^\pm \otimes E \otimes \mathcal{P}_\xi)$. On each fiber we define the L^2 Hermitian metric

$$\langle s_1, s_2 \rangle_\xi = \int_S \langle s_1, s_2 \rangle\, \omega, \qquad s_1, s_2 \in \mathcal{H}^\infty_{\pm,\xi}.$$

Journal of Physics: Conference Series **175** (2009) 012014 doi:10.1088/1742-6596/175/1/012014

We introduce the Hilbertian bundles $\mathcal{H}_\pm$ whose fibers $\mathcal{H}_{\pm,\xi}$ are the L^2-completion of $\mathcal{H}^\infty_{\pm,\xi}$ with respect to the previous metric.

If (E,∇) is an IT_i-pair, by the Regularity Theorem for elliptic operators $\widehat{E}$ is a subbundle of $\mathcal{H}^\infty_\pm$, hence we get an induced Hermitian metric on $\widehat{E}$.

5.3. The transformed connection
Following Bismut, we define a connection $\widetilde{\nabla}$ on $\mathcal{H}^\infty_\pm$ as follows

$$\widetilde{\nabla}_D h = \nabla^1_{D^H} h, \qquad \text{for every } \; D \in \mathfrak{X}(J(S)), \; h \in H^\infty_\pm,$$

D^H is the natural lift of the vector field D on $J(S)$ to $S \times J(S)$ and ∇^1 is the product connection of $\pi_S^*(\mathbb{S}^\pm \otimes E) \otimes \mathcal{P}_S$. One checks that $\widetilde{\nabla}$ is a flat connection.

We endow $\widehat{E}$ with the connection

$$\widehat{\nabla} = P \circ \widetilde{\nabla},$$

where P is the orthogonal projection onto $\widehat{E}$

Definition 5 *Let (E,∇) be an* IT*-pair. The pair $(\widehat{E},\widehat{\nabla})$ is called the Nahm transform of (E,∇).*

5.4. The curvature of the transformed connection
Let $(\widehat{E},\widehat{\nabla})$ be the Nahm transform of an IT_i-pair. Taking into account that $\tilde{\nabla}$ is flat, one proves that the curvature of $\widehat{\nabla}$ is

$$F^{\widehat{\nabla}} = P \circ (\widetilde{\nabla} P \wedge \widetilde{\nabla} P) \circ P.$$

6. Integral functors and Fourier-Mukai transforms

Now we introduce an integral functor on the Riemann surface S related to the Fourier-Mukai transform of its jacobian variety $J(S)$. This is the holomorphic analogue of the C^∞ Nahm transform just described.

The use of this integral functor has the following advantages:

(i) Computation of the topological invariants of the spectral bundles.

(ii) Establish geometrical properties of these bundles.

We will see in Section 6.4 that this integral functor is compatible, in a precise sense, with the Nahm transform.

If X is a complex projective variety, $D(X)$ denotes its bounded derived category of complexes of coherent $\mathcal{O}_X$-modules.

Let X, Y be two projective varieties and let $\mathcal{P} \to X \times Y$ be a vector bundle. We have the natural projections

$$\begin{array}{ccccc} & & X \times Y & & \\ & \overset{\pi_X}{\swarrow} & & \overset{\pi_Y}{\searrow} & \\ X & & & & Y \end{array}$$

We define an integral functor with kernel $\mathcal{P}$ as

$$\begin{aligned} \Phi^{\mathcal{P}}_{X\to Y} \colon D(X) &\to D(Y) \\ \mathcal{E}^\bullet &\mapsto \Phi^{\mathcal{P}}_{X\to Y}(\mathcal{E}^\bullet) = \mathbf{R}\pi_{Y*}(\pi_X^*(\mathcal{E}^\bullet) \overset{\mathbf{L}}{\otimes} \mathcal{P}). \end{aligned}$$

If $\Phi^{\mathcal{P}}_{X\to Y} \colon D(X) \to D(Y)$ is an equivalence of categories, we say that it is a *Fourier-Mukai transform*.

6.1. WIT (Weak Index Theorem) and IT (Index Theorem) conditions.

We write $\Phi = \Phi^{\mathcal{P}}_{X\to Y}$ and $\Phi^i(\mathcal{E}^\bullet) = \mathcal{H}^i(\Phi(\mathcal{E}^\bullet))$ is the i-th cohomology sheaf of the complex $\Phi(\mathcal{E}^\bullet)$.

We denote by $\mathcal{P}_y$ the restriction of $\mathcal{P}$ to $X_y = X \times \{y\}$.

Definition 6 *A sheaf E on X is Φ-WIT$_i$ if $\Phi^j(E) = 0$ for $j \neq i$ and we denote by $\widehat{E}$ the unique non vanishing sheaf $\Phi^i(E)$.*

A sheaf E on X is Φ-IT$_i$ if $H^j(X, E \otimes \mathcal{P}_y) = 0$ for every $j \neq i$ and every $y \in Y$.

By the base change theorem of algebraic geometry one has:

Theorem 3 *E is Φ-IT$_i$ $\iff$ it is Φ-WIT$_i$ and $\widehat{E} = \Phi^i(E)$ is a vector bundle.*

6.2. The integral functor associated to the jacobian

The first example of a Fourier-Mukai transform was given by Mukai. Let X be an abelian variety, $\widehat{X}$ its dual abelian variety and let $\mathcal{P} \to X \times \widehat{X}$ be the Poincaré bundle then the Mukai transform is $\mathcal{S} \equiv \Phi^{\mathcal{P}}_{X\to\widehat{X}}$.

Let S be a Riemann surface and let $\mathcal{P}_S \to S \times J(S)$ be a Poincaré bundle. We define the integral functor $\Phi_J : D(S) \to D(J(S))$ as $\Phi_J = \Phi^{\mathcal{P}_S}_{S\to J(S)}$. Since $J(S)$ is an autodual abelian variety, its Mukai transform defines a functor

$$\mathcal{S} : D(J(S)) \to D(J(S))$$

There is a functorial isomorphism

$$\Phi_J \simeq \mathcal{S} \circ \alpha_*$$

where $\alpha : S \to J(S)$ is the Abel-Jacobi immersion.

6.3. Change of topological invariants under Φ_J

Later, in Section 7, we will see that the spectral bundles associated to the family of Landau operators can be expressed as $\widehat{P} = \Phi_J(L)$ for a suitable line bundle $L \to S$. Since we have seen that the Hall conductivity is determined by the first Chern class of $\widehat{P}$, a key problem is the computation of the topological invariants of the transformed sheaves $\Phi_J(L)$.

Proposition 1 *If E is a coherent sheaf on S with Chern character $\mathrm{ch}(E) = (r, d)$, then $\mathrm{ch}(\Phi_J(E)) \in H^\bullet(J(S), \mathbb{Q})$ is given by*

$$\mathrm{ch}(\Phi_J(E)) = (d + r(1-p), -r[\Theta], 0, \dots, 0)$$

where p is the genus of S and $[\Theta]$ is the cohomology class defined by the theta divisor $\Theta \subset J(S)$.

There exists a line bundle $\mathcal{L}_\Theta \to J(S)$ such that $c_1(\mathcal{L}_\Theta) = [\Theta]$ and the Appel-Humbert theorem identifies $c_1(\mathcal{L}_\Theta)$ with the cup product

$$c_1(\mathcal{L}_\Theta)(\omega_1, \omega_2) = \textstyle\int_S \omega_1 \wedge \omega_2.$$

6.4. Compatibility between Φ_J and the Nahm transform

Let $E \to S$ be a Hermitian bundle with unitary connection ∇.

We have seen that the spin^c Dirac operator D_ξ gets identified with the Dolbeault-Dirac operator of $E \otimes \mathcal{P}_\xi$.

Hodge theory and the Dolbeault isomorphism give isomorphisms

$$\mathrm{Ker}\, D_\xi \simeq H^0(S_\xi, E \otimes \mathcal{P}_\xi) \tag{1}$$
$$\mathrm{Coker}\, D_\xi \simeq H^1(S_\xi, E \otimes \mathcal{P}_\xi). \tag{2}$$

Thus, E is IT_i with respect to $\Phi_J \Leftrightarrow (E, \nabla)$ is an IT_i-pair with respect to the Nahm transform.

There is a natural isomorphism of C^∞ vector bundles induced by Hodge theory

$$\phi_E \colon \widehat{E} \xrightarrow{\sim} \Phi^i_J(E).$$

Theorem 4 *The connection $\widehat{\nabla}$ is compatible with the holomorphic structure of $\Phi^i_J(E)$ and $\phi_E \colon \widehat{E} \xrightarrow{\sim} \Phi^i_J(E)$ is a holomorphic isomorphism.*

7. Spectral bundles of the family of Landau operators

Let (S, g) be a compact surface of genus p with constant curvature R, let $\mathcal{L} = (L, \langle\,,\rangle, \nabla) \to S$ be a prequantization bundle and let $\hat{H} = \frac{\hbar^2}{2m}(\nabla^*\nabla + \frac{R}{6})$ be the associated Landau operator.

In order to explicitly describe the spectral bundles we must recall the relationship between the spectral geometry of $\hat{H}$ and the holomorphic structure of $\mathcal{L}$. The results we need are contained in [15]. what we need is the following result.

Theorem 5 *Suppose that $|\operatorname{gr}(L)| > \operatorname{gr}(K_S)$, where K_S is the canonical line bundle of S.*

(i) If $p = 1$, the spectrum of $\hat{H}$ is the set

$$\operatorname{Spec}(\hat{H}) = \{E_q = \hbar|\omega|(q + \tfrac{1}{2}),\ \forall\, q \in \mathbb{Z},\ q \geq 0\},$$

where $\omega = \frac{e\hat{B}}{m}$.

(ii) If $p > 1$ and $k(L) = \frac{\deg(L)}{\deg(K_S)}$, $R = -\frac{2}{r^2}$, then the set

$$\operatorname{Spec}^d(\hat{H}) = \{E_q = \frac{\hbar^2}{2mr^2}\left(|k(L)|(2q+1) - q(q+1) - \frac{1}{3}\right), \\ \forall\, q \in \mathbb{Z},\ 0 \leq q < |k(L)| - 1\}.$$

is contained in the spectrum of $\hat{H}$ and are the lowest eigenvalues.

In both cases the space of eigensections of eigenvalue E_q gets identified either with $H^0(S, K_S^{-q} \otimes L)$ if $\deg L > 0$ or $H^0(S, K_S^{-q} \otimes L^{-1})$ if $\deg L < 0$.

Definition 7 *The subset $\operatorname{Spec}_{hol}(\hat{H})$ of $\operatorname{Spec}(\hat{H})$ defined in the previous theorem is called the holomorphic spectrum of $\hat{H}$. The eigensections with eigenvalue $E_q \in \operatorname{Spec}_{hol}(\hat{H})$ form the q-th holomorphic Landau level.*

According to Theorem 5, $\operatorname{Spec}_{hol}(\hat{H})$ does not depend on the chosen prequantization bundle. Thus, $\operatorname{Spec}_{hol}(\hat{H})$ is constant for the family of Landau operators parametrized by $J(S) \simeq \operatorname{Pic}^0(S)$ and we denote it σ_{hol}.

In particular the eigensections of the Landau operator $\hat{H}(L_0)$ corresponding to the flat line bundle L_0 are given by

- $H^0(S, K_S^{-q} \otimes L \otimes L_0)$ if $\deg L > 0$, and by
- $H^0(S, K_S^{-q} \otimes L^{-1} \otimes L_0)$ if $\deg L < 0$.

We assume that $\operatorname{gr} L > 0$, since if it were not the case it would be enough to replace L by L^{-1} in order to obtain the corresponding results. Hence, for any integer $q \geq 0$, with $\operatorname{gr}(K_S^{-q} \otimes L) > \operatorname{gr} K_S$, there exists an eigenvalue $E_q \in \sigma_{hol}$ corresponding to a holomorphic Landau level.

Lemma 1 *The family P_q of spectral projections associated with an eigenvalue $E_q \in \sigma_{hol}$ fulfills the hypothesis (A1-A3) of the adiabatic Theorem.*

Therefore, the q-th holomorphic Landau level defines a spectral bundle $\widehat{P}_q \to J(S)$ and we endow it with the induced connection $\widehat{\nabla}_q = P_q \circ \tilde{\nabla}$. Let ∇_q be the connection on $K_S^{-q} \otimes L$ obtained by twisting the connection of L with the connection of K_S^{-q} induced by the Levi-Civita connection.

Theorem 6 *Let $\widehat{P}_q \to J(S)$ be the spectral bundle defined by the q-th holomorphic Landau level, we have:*

(i) The pair $(\widehat{P}_q, \widehat{\nabla}_q)$ is the Nahm transform of $(K_S^{-q} \otimes L, \nabla_q)$.

(ii) There exists a holomorphic isomorphism $\widehat{P}_q \xrightarrow{\sim} \Phi_J^0(K_S^{-q} \otimes L)$.

7.1. Determination of adiabatic transport and Hall conductivity

The equivalence of the Nahm transform with the integral functor associated to the jacobian, allows us to study the spectral bundles by means of the machinery of integral functors. In particular by Proposition 1 we have the following result.

Corollary 1 *The first Chern class of the spectral bundle $\widehat{P}_q \to J(S)$ is $c_1(\widehat{P}_q) = c_1(\mathcal{L}_\Theta)$, where $\mathcal{L}_\Theta$ is the principal polarization of $J(S)$ defined by the theta divisor Θ.*

Now we can determine the adiabatic transport and Hall conductivity.

Theorem 7 *The mean adiabatic transport $\langle Q_{Ad}(\widehat{P}_q)\rangle$ and the mean Hall conductivity $\langle \sigma_H(\widehat{P}_q)\rangle$ of the q-th holomorphic Landau level are*

$$\langle Q_{Ad}(\widehat{P}_q)\rangle(\alpha, \beta) = -e\,(\alpha, \beta),$$

$$\langle \sigma_H(\widehat{P}_q)\rangle(\alpha, \beta) = -\tfrac{e^2}{h}(\alpha, \beta),$$

where (,) denotes the intersection product on $H_1(S, \mathbb{Z})$.

7.2. Stability of spectral bundles

The vector bundle $\widehat{L}_d = \Phi_J^0(L_d)$, where $L_d \to S$ is a line bundle of degree d, with $d > \deg K_S$, is called in the literature the d-th Picard sheaf on $J(S)$. These sheaves have been studied by several authors, among them we may cite Ein and Lazarsfeld [17]. Using their results we can prove the following:

Theorem 8 *The spectral bundles $\widehat{P}_q \to J(S)$ associated with the holomorphic Landau levels are holomorphic vector bundles with respect to the polarization of the jacobian $J(S)$ defined by the theta divisor Θ.*

The stability of the spectral bundles $\widehat{P}_q \to J(S)$ is a remarkable fact, since in the description of the fractional quantum Hall effect proposed by Varnhagen [18] there also appear stable bundles. These facts seem to suggest a connection between the stability of spectral bundles and the interpretation of quantum Hall effect.

8. Analytic torsion and Quillen metrics on holomorphic determinant bundles

Although we have determined the Hall conductivity as the integral of the adiabatic conductance, our final aim is the computation of the adiabatic conductance that controls the fluctuations of the Hall conductivity and is proportional to the adiabatic curvature.

The adiabatic curvature Ω^{P_q} of the spectral bundle $(\widehat{P}_q, \widehat{\nabla}_q) \to J(S)$ is the trace of its curvature, that is $\Omega^{P_q} = \operatorname{Tr}\Omega^{\widehat{\nabla}_q}$. Equivalently, Ω^{P_q} is the curvature of the connection $\det\widehat{\nabla}_q$ induced on the determinant bundle $\det\widehat{P}_q \to J(S)$, thus

$$\Omega^{P_q} = \Omega^{\det\widehat{\nabla}_q}.$$

On the other hand, the connection $\widehat{\nabla}_q$ on the holomorphic vector bundle $\widehat{P}_q \to J(S)$ gets identified with the Chern connection of the Hermitian metric $\langle\ ,\ \rangle_{L^2}$ induced by the L^2 metric of the Hilbertian bundle $\mathcal{H}_+ \to J(S)$.

Hence, the computation of the adiabatic curvature is reduced to finding the curvature of the Chern connection of the determinant bundle $\det\widehat{P}_q \to J(S)$ for the Hermitian metric $\langle\ ,\ \rangle_{L^2}$ induced by the L^2 metric.

However, our strategy is to compute first the curvature of the Chern connection on the determinant bundle with respect to the Quillen metric, since:

(i) The metric $\langle\ ,\ \rangle_{L^2}$ is conformal to the Quillen metric.
(ii) It is the natural C^∞ metric defined on the determinant bundles in terms of the analytic torsion, or equivalently, in terms of the zeta-regularized determinants of certain elliptic operators.

In order to calculate the curvature of Quillen metric we use the techniques developed by Bismut, Gillet and Soulé [12, 13, 14].

Let $\pi : X \to T$ be a holomorphic submersion and let $E \to X$ be a holomorphic vector bundle. The direct image $\mathbf{R}\pi_* E$ admits a determinant that is a holomorphic line bundle $\det\mathbf{R}\pi_*E \to T$ whose dual $\lambda^{\mathrm{KM}}(E) = (\det\mathbf{R}\pi_*E)^{-1} \to T$ is called the Knudsen-Mumford determinant.

For every $t \in T$, let $X_t = \pi^{-1}(t)$ be the fiber over t and let $H^i(X_t, E_t)$ be the cohomology of the restrictionE_t of E to X_t. For every fiber $\lambda^{\mathrm{KM}}(E)_t$ of the Knudsen-Mumford determinant there is a canonical identification

$$\lambda^{\mathrm{KM}}(E)_t = \bigotimes_{i\geq 0} \det H^i(X_t, E_t)^{(-1)^{i+1}}.$$

Let $g^{X/T}$ be a C^∞ relative Kähler metric on $\pi : X \to T$, H_E a Hermitian metric on $E \to X$ and let $D = \sqrt{2}(\overline{\partial}_E + \overline{\partial}_E^*)$ be the Dolbeault-Dirac operator of E.

One defines a family of vector spaces $\lambda(E) \to T$ whose fibers are

$$\lambda(E)_t = (\det \mathrm{Ker} D_t)^{-1} \otimes \det \mathrm{Coker} D_t.$$

$\lambda(E) \to T$ admits a natural structure of holomorphic line bundle, called the Bismut-Gillet-Soulé determinant.

The L^2 metric induces a Hermitian metric $\langle\ ,\ \rangle_{L^2}$ on $\lambda(E)$ that in general, due to the changes of dimension in $\mathrm{Ker}D_t$, it is not C^∞. However, if $\det' D^*D$ is the function on the parameter space T whose value at $t \in T$ is the regularized determinant of $\det' D^*D(t) = \det' D_t^* D_t$ of D_t then Quillen proved the following result.

Theorem 9 (Quillen) *The Quillen metric, defined on the determinant bundle $\lambda(E)$ as*

$$\langle\ ,\ \rangle_Q = \det' D^*D \langle\ ,\ \rangle_{L^2},$$

is a C^∞ Hermitian metric.

The value of the function $T(E, H_E, g^{X/T}) = (\det' D^*D)^{\frac{1}{2}}$ at $t \in T$ coincides with the Ray-Singer analytic torsion $T(E_t, H_{E_t}, g^{X_t})$ of the vector bundle $E_t \to X_t$. We say that $T(E) = T(E, H_E, g^{X/T})$ is the relative analytic torsion of $E \to X$.

For every $t \in T$ there is a canonical isomorphism between the fibers

$$\lambda(E)_t \simeq \lambda^{\mathrm{KM}}(E)_t,$$

of the Bismut-Gillet-Soulé and Knudsen-Mumford determinants.

One says that $\pi : X \to T$ is locally Kähler if there exists an open cover $\mathfrak{U}$ of T such that for every $U \in \mathfrak{U}$ there exists a Kähler metric on $\pi^{-1}(U)$.

Let $R^{X/T}$, Ω^E be the curvatures of $(\Lambda^{1,0}T^*(X/T), g^{X/T})$ and (E, H_E), respectively. We have the following key result.

Theorem 10 (Bismut-Gillet-Soulé) *Suppose that $\pi : X \to T$ is locally Kähler. Then*

(i) The canonical identification of the fibers $\lambda(E)_t \simeq \lambda^{\mathrm{KM}}(E)_t$ defines a holomorphic isomorphism

$$\lambda(E) \simeq \lambda^{\mathrm{KM}}(E).$$

(ii) The curvature of the Quillen metric $\lambda(E) \simeq \lambda^{\mathrm{KM}}(E)$ is the degree 2 component of the differential form on T

$$2\pi i \int_{X/T} \mathrm{Td}(\frac{iR^{X/T}}{2\pi}) \wedge \mathrm{Tr}(\exp(\frac{i\Omega^E}{2\pi})),$$

where $\int_{X/T}$ denotes integration along the fibers.

9. Adiabatic conductance of spectral bundles

The natural projection $\pi_{J(S)} : S \times J(S) \to J(S)$ is locally Kähler. Therefore, we can apply the results of Bismut, Gillet and Soulé.

Theorem 11 *The adiabatic curvature Ω^{P_q} of $\widehat{P}_q \to J(S)$ is*

$$\frac{i\Omega^{P_q}}{2\pi} = \mathrm{Im}\, H_\Theta + \frac{i}{\pi}\bar{\partial}\partial \ln(T(\pi_S^*(K_S^{-q} \otimes L) \otimes \mathcal{P}_S)),$$

where H_Θ is the principal polarization of $J(S)$ and $T(\pi_S^*(K_S^{-q} \otimes L) \otimes \mathcal{P}_S)$ is the relative analytic torsion.

Corollary 2 *The adiabatic conductance $C_{Ad}(\widehat{P}_q) = i\frac{e^2}{\hbar}\Omega^{P_q}$ of the q-th holomorphic Landau level is*

$$C_{Ad}(\widehat{P}_q) = 2\pi\frac{e^2}{\hbar}\{\mathrm{Im}\, H_\Theta + \frac{i}{\pi}\bar{\partial}\partial \ln(T(\pi_S^*(K_S^{-q} \otimes L) \otimes \mathcal{P}_S))\}.$$

References

[1] Klitzing K V, Dorda G and Pepper M 1980, New method for high accuracy determination of the fine structure constant based on the quantized Hall resistance, *Phys. Rev. Lett.* **45** 494-7.
[2] Laughlin R B 1981, Quantized Hall conductivity in two dimensions, *Phys. Rev. B* **23** 5652-4.
[3] Tsui D,
[4] Thouless D J, Kohmoto M, Nightingale M P and den Nijs M 1982, Quantized Hall conductance in a two-dimensional periodic potential, *Phys. Rev. Lett.* **49** 405-8.
[5] Avron J E and Seiler R 1985, Quantization of the Hall Conductance for general, multiparticle Schrödinger Hamiltonians, *Phys. Rev. Lett.* **54** 259-62.
[6] Niu Q and Thouless D J 1984, Quantised adiabatic charge transport in the presence of substrate disorder and many-body interaction, *J. Phys. A: Math. Gen.* **17** 2453-62.
[7] Kohmoto M 1985, Topological invariant and the quantization of the Hall conductance, *Ann. Physics* **160** 343-54.

[8] Niu Q, Thouless D J and Wu Y-S 1985, Quantized Hall conductance as a topological invariant, *Phys. Rev. B* **31** 3372-7.
[9] Avron J E, Seiler R and Yaffe L G 1987, Adiabatic theorems and applications to the quantum Hall effect, Commun. Math. Phys. **110** 33-49.
[10] Klein M and Seiler R 1990, Power-law corrections to the Kubo formula vanish in quantum Hall systems, *Commun. Math. Phys.* **128** 141-60.
[11] Avron J E, Seiler R and Zograf P G 1994, Adiabatic quantum transport: quantization and fluctuations, *Phys. Rev. Lett.* **73** 3255-66.
[12] Bismut, J.M., Gillet, H.,Soulé, C.: Analytic torsion and holomorphic determinant bundles I. Bott-Chern forms and analytic torsion. Commun. Math. Phys. **115**, 49-78 (1988).
[13] Bismut, J.M., Gillet, H.,Soulé, C.: Analytic torsion and holomorphic determinant bundles II. Direct images and Bott-Chern forms. Commun. Math. Phys. **115**, 79-126 (1988).
[14] Bismut, J.M., Gillet, H.,Soulé, C.: Analytic torsion and holomorphic determinant bundles III. Quillen metrics and holomorphic determinants. Commun. Math. Phys. **115**, 301-351 (1988).
[15] Tejero Prieto C 2006, Holomorphic spectral geometry of magnetic Schrödinger operators on Riemann surfaces. *Diff. Geom. and Appl.* **24** 288-310.
[16] Tejero Prieto C 2006, Fourier-Mukai Transform and adiabatic curvature of spectral bundles for Landau Hamiltonians on Riemann surfaces *Commun. Math. Phys.* **265** 373-96.
[17] Ein, L., Lazarsfeld, R.: Stability and restrictions of Picard bundles, with an application to the normal bundles of elliptic curves. In: Ellingsrud, G., Peskine, C., Sacchiero, G., Strømmeand S. A. (eds.) Complex Projective Geometry (Trieste, 1989 / Bergen, 1989). Cambridge: London Math. Soc. Lecture Note Ser. **179**, Cambridge University Press, 1992, pp. 149-156.
[18] Varnhagen, R.: Topology and fractional quantum Hall effect. Nucl. Phys. B **443**, 501-515 (1995).

Tensor product decompositions for $\mathfrak{su}(3)$ of an irreducible representation with itself and with its conjugate

Maria S M Wesslén

Department of Mathematics, University of Toronto, 40 St. George Street, Toronto, Ontario, M5S 2E4, Canada

E-mail: maria.wesslen@utoronto.ca

Abstract. In this article a method to obtain the decomposition, including multiplicities, of an $\mathfrak{su}(3)$ irreducible representation with itself and with its conjugate, is provided. The decomposition of such tensor products plays an important role in physics. The objective is to provide a simple method for performing these decompositions in addition to showing how this method can be used to derive several interesting facts about the decompositions, particularly in regards to multiplicities.

1. Introduction

The decomposition of tensor products for $\mathfrak{su}(3)$ are important in many parts of physics, for example in elementary particle physics [1], nuclear physics [2] and in the analysis of quantum interferometers [3].

The Clebsch-Gordan series for the tensor product of two $\mathfrak{su}(2)$ representations is well documented and has been used extensively (see for example [4]). The literature on the decompositions of two irreducible representations (irreps) of $\mathfrak{su}(3)$ is not as familiar, in part due to multiplicities in the decompositions, arising from the presence of weight multiplicities in $\mathfrak{su}(3)$ representations, but also because closed form expressions are often cumbersome to use.

A lot of attention has been devoted to the $\mathfrak{su}(3)$ decomposition problem; see for instance work by Biedenharn et. al. (e.g. [5]) as well as King (e.g. [6]), but also several others [7]. In addition, a range of computer programs have been developed to compute various types of tensor product decompositions [8]. Some of these methods are used mainly on individual tensor products, making it difficult to see the general structure, or are in other ways not suitable for this particular problem. The method presented in this article takes full advantage of the geometric features of the weight diagram of an irrep and its conjugate, thus making it less complicated and more transparent than the corresponding restrictions of more general results. The methods given in Ref. [9], [10] and [11] are similar to the one discussed here, but either do not provide multiplicities or are only valid for certain tensor products for $\mathfrak{su}(3)$, not including those of interest here.

In a previous paper [12] a simple geometric method for decomposing tensor products for $\mathfrak{su}(3)$ was given together with several applications. Some of the same techniques also appear in Ref. [13]. Here, this geometric method is discussed in the context of the tensor product of an

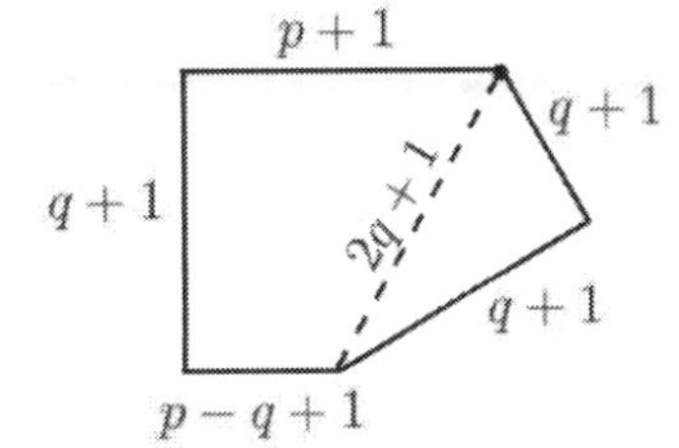

Figure 1. The boundary of the tensor diagram of $(p, q) \otimes (p, q)$ when $p \geq q$. The values refer to the number of vertices along each line, and this number will be called the *length* of the line.

irreducible representation in $\mathfrak{su}(3)$ with itself, or with its contragredient. These tensor products are especially important because they occur when a particle interacts with an identical particle or antiparticle as well as in the action of $\mathfrak{su}(3)$ on the density matrix of a system for which states carry representations of $\mathfrak{su}(3)$. An application to the problem of tomographic reconstructions of density matrices of $\mathfrak{su}(3)$ systems can be found in Ref. [14].

Although many of the aspects of these tensor products can be deduced from the study of the general case in Ref. [12], they involve, in addition to important applications, several interesting characteristics that justifies studying them further. Contrary to the general case, the symmetry of the highest weights simplifies the analysis, thus allowing for more compact results as well as for additional conclusions to be drawn.

Before describing the method, some notation needs to be established. Let $\mathfrak{g} = \mathfrak{su}(3)$ and let (p, q) be an irrep with highest weight $p\lambda_1 + q\lambda_2$, where $\lambda_i, i = 1, 2$ are the fundamental weights and $p, q \in \mathbb{Z}_{\geq 0}$. In this article the tensor products $(p, q) \otimes (p, q)$ and $(p, q) \otimes (q, p)$ will be considered.

We are interested in the decomposition of these tensor products into direct sums of irreps. These, as any other tensor product decompositions, can be described by tensor diagrams [12]. A *tensor diagram* is a collection of vertices in a triangular lattice (a weight lattice) with each vertex representing the irrep with the corresponding highest weight. Each vertex is assigned a multiplicity according to the number of copies of the relevant irrep that appear in the direct sum decomposition of the tensor product. The irreps not appearing in the direct sum can be thought of as having multiplicity zero, or they can be thought of as not being part of the tensor diagram. Moreover, $\alpha_i, i = 1, 2$ are the simple roots of $\mathfrak{su}(3)$ and the $(\alpha_1 + \alpha_2)$-diagonals refer to the line segments in the tensor diagram, in the direction of $\alpha_1 + \alpha_2$ (parallel to the dashed line in Figure 1). The *centre diagonal* is the $(\alpha_1 + \alpha_2)$-diagonal with the highest vertex as an endpoint.

2. The Tensor Product $(p, q) \otimes (p, q)$

We are now in a position to describe the decomposition method for $(p, q) \otimes (p, q)$, which involves determining its tensor diagram. W.l.o.g. assume $p \geq q$; if $p < q$ it is possible to interchange the coordinates by reflecting all diagrams in the centre diagonal.

(i) The boundary of the tensor diagram of $(p, q) \otimes (p, q)$ is given in Figure 1. All vertices on or inside this boundary have positive multiplicity, and all vertices outside it have multiplicity zero.

(ii) Now find the multiplicities along each $(\alpha_1 + \alpha_2)$-diagonal as follows: The top vertex of each diagonal has multiplicity one. Moving downwards along the diagonal, the multiplicities then increase by one at a time, until they plateau at some upper bound (described below). Near the end, the multiplicities start decreasing by one at a time, so that they are back at one again

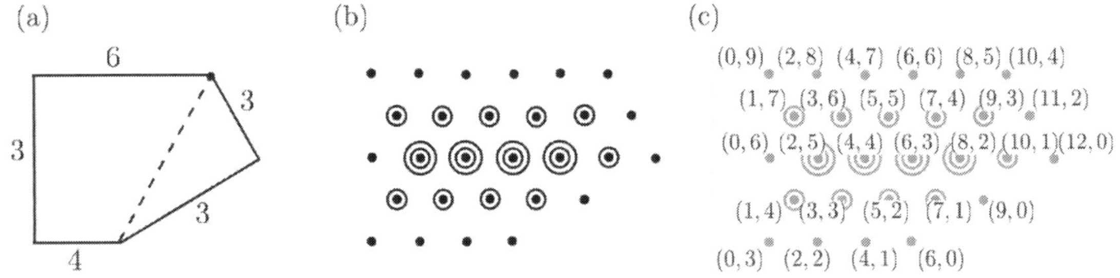

Figure 2. The tensor decomposition of $(5,2) \otimes (5,2)$. Part (a) shows the boundary adapted from Figure 1, part (b) shows the tensor diagram and part (c) shows the tensor diagram with the irreps labelled by their highest weights.

for the final vertex.

(iii) The upper bounds along the diagonals behave in a similar way; the leftmost diagonal has maximum one, the next one two and so on until it reaches the upper bound for the entire diagram, which is $q+1$. Near the end, the upper bounds decrease by one at a time, so that they are back at one again for the final diagonal. The resulting diagram is the tensor diagram, and it determines the direct sum decomposition.

Remarks: 1. This can be proved using methods similar to those used in proofs in Ref. [12].

2. In the special case $(p,0) \otimes (p,0)$ i.e., where $q = 0$, this tensor diagram reduces to a horizontal straight line, of length $p+1$.

In order to illustrate the procedure, let us consider the tensor product $(5,2) \otimes (5,2)$. The boundary of the tensor diagram for this example is given in part (a) of Figure 2. Using the rules for multiplicities in (ii) and (iii) above gives the tensor diagram in part (b) of Figure 2. The highest highest weight is given by $(2p, 2q)$ which in this example is (10,4). Part (c) of Figure 2 shows the tensor diagram with the vertices labelled. In conclusion, the tensor product decomposition is:

$$\begin{aligned}(5,2) \otimes (5,2) = (10,4) \oplus (8,5) \oplus (6,6) \oplus 2(9,3) \oplus (11,2) \oplus 2(7,4) \oplus \\ (4,7) \oplus 2(5,5) \oplus 3(8,2) \oplus (2,8) \oplus 3(6,3) \oplus 2(3,6) \oplus 2(10,1) \oplus \\ 3(4,4) \oplus (12,0) \oplus 2(5,2) \oplus 3(2,5) \oplus 2(7,1) \oplus 2(1,7) \oplus 2(3,3) \oplus \\ (9,0) \oplus (0,9) \oplus (4,1) \oplus 2(1,4) \oplus (6,0) \oplus (0,6) \oplus (2,2) \oplus (0,3)\end{aligned}$$

In a notation commonly found in physics where representations are labelled by their dimension, we find:

$$\begin{aligned}81 \otimes 81 = 440 \oplus 405 \oplus 343 \oplus 2 \cdot 280 \oplus 270 \oplus 2 \cdot 260 \oplus \underline{260} \oplus 2 \cdot 216 \oplus \\ 3 \cdot 162 \oplus \underline{162} \oplus 3 \cdot 154 \oplus 2 \cdot \underline{154} \oplus 2 \cdot 143 \oplus 3 \cdot 125 \oplus 91 \oplus 2 \cdot 81 \oplus \\ 3 \cdot \underline{81} \oplus 2 \cdot 80 \oplus 2 \cdot \underline{80} \oplus 2 \cdot 64 \oplus 55 \oplus \underline{55} \oplus 35 \oplus 2 \cdot \underline{35} \oplus 28 \oplus \underline{28} \oplus 27 \oplus \underline{10}\end{aligned}$$

Further examples are provided in Table A1.

We now return to the general discussion. From the decomposition method above it is clear that the upper bound for the multiplicities in the tensor diagram of $(p,q) \otimes (p,q)$ is $q+1$. Calculations similar to those in Ref. [12] show that this upper bound is always attained, i.e., there is always at least one vertex with this multiplicity. It also follows that the vertices with maximum multiplicity are arranged in a horizontal line of length $p - q + 1$. This is illustrated in Figure 3.

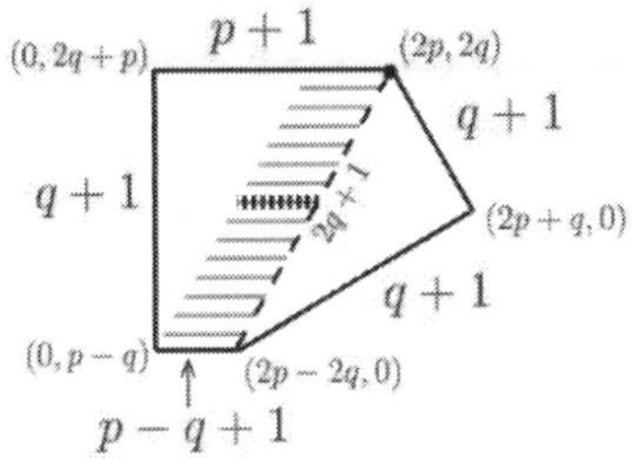

Figure 3. Tensor diagram of $(p, q) \otimes (p, q)$. The lined region is where the diagonals are longest. The multiplicities are stable along each of these indicated lines, and the thicker dashed line is the line where multiplicities are maximal.

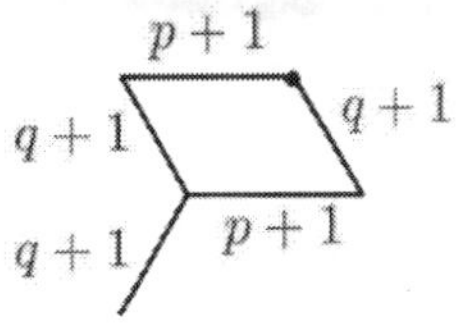

Figure 4. Boundary of region where the vertices in tensor diagram have the same multiplicities as in the weight diagram. The highest highest weight is indicated.

Note also that the length of the centre diagonal is $2q + 1$, so the maximum multiplicity must occur q vertices below the highest vertex (which is at $(2p, 2q)$), i.e., at $(2p - q, q)$. The lengths and directions of the lines of maximum multiplicity are already known, so the complete list of irreps with maximum multiplicity can be found. It is:

$$(2p - q, q), (2p - q - 2, q + 1), ..., (q, p) \quad (p - q + 1 \text{ vertices})$$

With a similar argument, using the decomposition method it can be shown that the irrep with lowest highest weight is $(0, p - q)$. It is also clear from the decomposition method that the vertices of multiplicity one are exactly those on the boundary of the convex hull of the tensor diagram. In particular the irrep with lowest highest weight has multiplicity one.

Note that although all vertices on the boundary have multiplicity one, this does not imply that the first/last vertex on any given line has multiplicity one, as these vertices may not lie on the boundary of the convex hull. For example, there are many horizontal lines where the first vertex has multiplicity two. This is the case in the example shown in Figure 2. See also Ref. [12].

Figure 3 shows the tensor diagram of $(p, q) \otimes (p, q)$. The thick dashed line indicates the line where the multiplicities are maximal. This falls within the lined region, which is where the diagonals are longest. In this region, the multiplicities are stable along each of the horizontal lines. The lowest highest weight is in agreement with the one stated above.

Let $\lambda = p\lambda_1 + q\lambda_2$ i.e., λ is the highest weight in (p, q). Consider the weight diagram of (p, q) shifted by λ. The purpose of this is to ensure that the highest vertex of the weight diagram is shifted to coincide with the highest vertex of the tensor diagram, in order to place the tensor diagram in the context of shifted weight diagram. It is known that the tensor diagram of $(p, q) \otimes (p, q)$ is contained in this shifted weight diagram of (p, q).

Near the upper right corner, the vertices in the tensor diagram have the same multiplicities as those in the shifted weight diagram of (p, q). The region where the multiplicities coincide, is illustrated in Figure 4. Inside or on the boundary of this region, the multiplicities are the same for both diagrams, and outside this region the multiplicities are strictly less for the tensor diagram. This can be proved using the techniques developed in Ref. [12].

Note that a result similar to this was proved in Ref. [12]. However, in the general case only part of this region was given, so that outside, the tensor multiplicities were less than or equal to the weight multiplicities. In the case of $(p, q) \otimes (p, q)$ we are able to give the exact region.

It is interesting to see the tensor diagram of $(p, q) \otimes (p, q)$ in the context of the weight diagram of (p, q) shifted by λ, and in particular how the regions where the multiplicities are maximal are

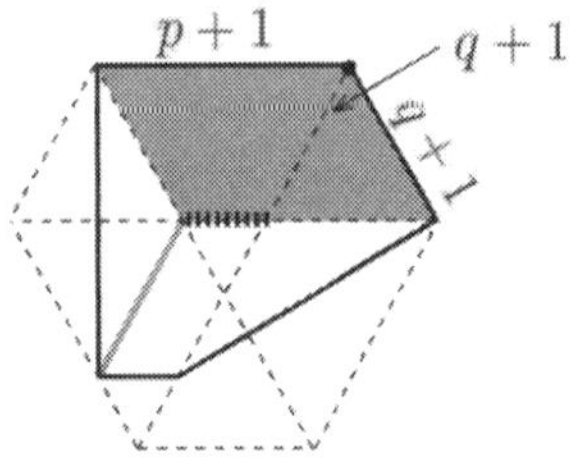

Figure 5. Tensor diagram of $(p,q) \otimes (p,q)$ in the context of the shifted weight diagram of (p,q). The thin dashed lines show the boundary of the shifted weight diagram of (p,q), and the lines forming the triangle where the multiplicity is largest. The thick dashed line indicates the line where the multiplicities are maximal in the tensor diagram. The shaded region shows where the multiplicities of the tensor and weight diagrams coincide.

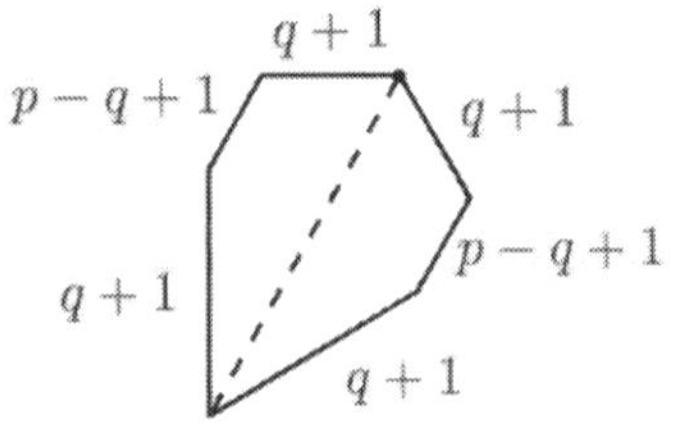

Figure 6. The boundary of the tensor diagram of $(p,q) \otimes (q,p)$ when $p \geq q$. The dashed line is the centre diagonal, and it has length $2q+1$.

related (see Figure 5). As expected, the tensor diagram is contained within the shifted weight diagram. The shaded region shows where the multiplicities of the tensor and weight diagrams coincide, as discussed above. Although the tensor diagram is larger than this region, the tensor multiplicities are strictly less than the weight multiplicities outside the shaded area.

The maximum multiplicity is $q+1$ in both the weight and tensor diagrams. The centre triangle formed by the thin dashed lines show the region where the multiplicities are maximal in the weight diagram. Figure 5 shows that the line of maximum multiplicities in the tensor diagram is one side of this triangle.

3. The Tensor Product $(p,q) \otimes (q,p)$

Now consider the decomposition of the tensor product $(p,q) \otimes (q,p)$. To ease notation assume w.l.o.g that $p \geq q$ (otherwise interchange the two irreps). Many of the same properties that were considered for $(p,q) \otimes (p,q)$ in Section 2 will be considered here but now for $(p,q) \otimes (q,p)$. First the decomposition method for $(p,q) \otimes (p,q)$ is adjusted to fit $(p,q) \otimes (q,p)$.

i) The boundary of the tensor diagram of $(p,q) \otimes (q,p)$ is given in Figure 6. All vertices on or inside this boundary have positive multiplicity, and all vertices outside it have multiplicity zero.

ii) and iii) The multiplicities along each $(\alpha_1+\alpha_2)$-diagonal satisfy the same properties as in the method in Section 2.

Remarks: 1. This tensor diagram is symmetric about the centre $(\alpha_1+\alpha_2)$-line. This is to be expected because interchanging the first and second coordinates in $(p,q) \otimes (q,p)$ would result in the same tensor product.

2. In the special case $(p,0) \otimes (0,p)$ i.e., where $q=0$, this tensor diagram reduces to a straight line in the $(\alpha_1+\alpha_2)$-direction, of length $p+1$. See Ref. [14] for an application of decompositions of these particular tensor products.

As an example, consider the tensor product $(5,2) \otimes (2,5)$. (Cf. the example of $(5,2) \otimes (5,2)$ in Section 2.) Following the procedure with this example, Figure 6 becomes part (a) of Figure 7. Using the rules for multiplicities in (ii) and (iii) above gives the tensor diagram in part (b) of Figure 7. The highest highest weight is given by $(p+q, p+q)$ which in this example is (7,7). Part (c) of Figure 7 shows the tensor diagram with the vertices labelled. In conclusion, the

Journal of Physics: Conference Series **175** (2009) 012015 doi:10.1088/1742-6596/175/1/012015

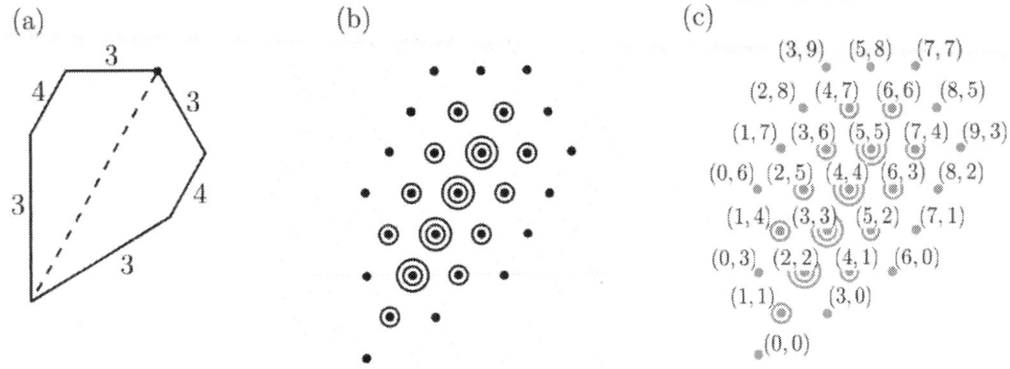

Figure 7. The tensor decomposition of $(5,2)\otimes(2,5)$. Part (a) shows the boundary adapted from Figure 6, part (b) shows the tensor diagram and part (c) shows the tensor diagram with the irreps labelled by their highest weights.

tensor product decomposition is:

$$\begin{aligned}(5,2)\otimes(2,5) = (7,7)\oplus(8,5)\oplus(5,8)\oplus 2(6,6)\oplus(9,3)\oplus(3,9)\oplus\\ 2(7,4)\oplus 2(4,7)\oplus 3(5,5)\oplus(8,2)\oplus(2,8)\oplus 2(6,3)\oplus 2(3,6)\oplus\\ 3(4,4)\oplus 2(5,2)\oplus 2(2,5)\oplus(7,1)\oplus(1,7)\oplus(3,3)\oplus 2(4,1)\oplus\\ 2(1,4)\oplus(6,0)\oplus(0,6)\oplus 3(2,2)\oplus(3,0)\oplus(0,3)\oplus 2(1,1)\oplus(0,0)\end{aligned}$$

In dimension notation this becomes:

$$\begin{aligned}81\otimes\underline{81} = 512\oplus 405\oplus\underline{405}\oplus 2\cdot 343\oplus 280\oplus\underline{280}\oplus 2\cdot 260\oplus 2\cdot\underline{260}\oplus\\ 3\cdot 216\oplus 162\oplus\underline{162}\oplus 2\cdot 154\oplus 2\cdot\underline{154}\oplus 3\cdot 125\oplus 2\cdot 81\oplus 2\cdot\underline{81}\oplus 80\oplus\\ \underline{80}\oplus 3\cdot 64\oplus 2\cdot 35\oplus 2\cdot\underline{35}\oplus 28\oplus\underline{28}\oplus 3\cdot 27\oplus 10\oplus\underline{10}\oplus 2\cdot 8\oplus 1\end{aligned}$$

More examples can be found in Table A1.

We can now discuss multiplicities. The upper bound for the multiplicities in the tensor diagram of $(p,q)\otimes(q,p)$ is $q+1$ and, as was the case for $(p,q)\otimes(p,q)$, this upper bound is always attained. The vertices with maximum multiplicity are again arranged in a straight line of length $p-q+1$, but for $(p,q)\otimes(q,p)$ the line goes along the centre diagonal. The length of the centre diagonal is $2q+1$, so the top vertex with maximum multiplicity must be q vertices below the highest vertex (which is at $(p+q,p+q)$), i.e., at (p,p). The subsequent $p-q$ vertices along the centre diagonal have maximum multiplicity as well; hence the irreps with maximum multiplicity are exactly:

$$(p,p),(p-1,p-1),\ldots,(p-q,p-q)\quad(p-q+1\text{ vertices})$$

From the decomposition method it also follows that the vertices of multiplicity one are exactly those on the convex hull of the tensor diagram. Also, the irrep with lowest highest weight is $(0,0)$ and it has multiplicity one, as it is on the boundary.

As in the case of $(p,q)\otimes(p,q)$, even though the vertices on the boundary have multiplicity one, this does not imply that the first/last vertex in any given line has multiplicity one.

Now let $\lambda = p\lambda_1+q\lambda_2$ and $\lambda' = q\lambda_1+p\lambda_2$ i.e., λ and λ' are the highest weights in (p,q) and (q,p), respectively. Shifting the weight diagram of (p,q) by λ and the weight diagram of

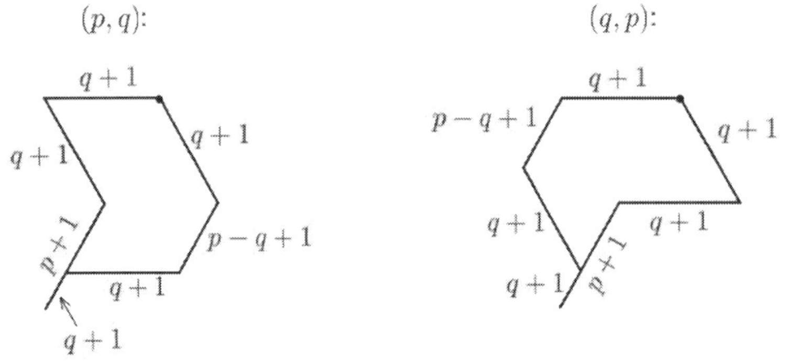

Figure 8. Boundary of region where the vertices have the same multiplicities in the tensor diagram as in the shifted weight diagrams ((p, q) and (q, p) respectively). The highest highest weight is indicated by a dot. See also Figure 9.

(q, p) by λ' brings the highest vertex of each weight diagram to the highest vertex of the tensor diagram so that part of the diagrams coincide.

In a region near the upper right corner, the vertices in the tensor diagram have the same multiplicities as those in the shifted weight diagram of (p, q) and/or (q, p). This region is shown in Figure 8.

As in the case of $(p, q) \otimes (p, q)$, it is interesting to see the tensor diagram of $(p, q) \otimes (q, p)$ in the context of the shifted weight diagram (p, q) although here the shifted weight diagram of (q, p) is also relevant (see Figure 9). The given highest and lowest vertices refer to the tensor diagram (and coincide with those given above).

The shaded region shows where the multiplicities of the tensor and shifted weight diagrams coincide, as discussed above. Outside this region the tensor multiplicities are strictly less than the weight multiplicities.

The thick dashed line shows where irreps in the tensor diagram has maximum multiplicity. As can be seen in the figure, this line is one side of the triangle where the weight multiplicities are maximal.

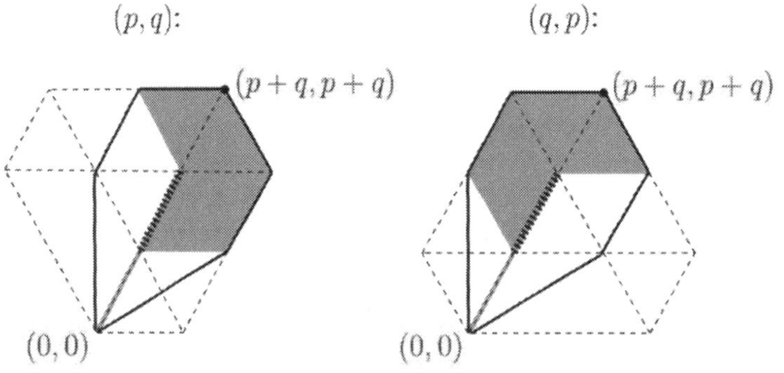

Figure 9. The tensor diagram of $(p, q) \otimes (q, p)$ in the context of the shifted weight diagrams of (p, q) and (q, p). The dashed lines show the boundary of the shifted weight diagram, and the lines forming the triangle where the weight multiplicities are largest. The thick dashed line indicates the line where the multiplicities are maximal in the tensor diagram. Note that when $p = q$ the two diagrams coincide.

Remark: When $p = q$, $(p, q) \otimes (q, p)$ coincides with $(p, q) \otimes (p, q)$, and in particular the lines of maximum multiplicity reduce to a single point in both cases; some examples are provided in Table A1.

4. Conclusions

The $\mathfrak{su}(3)$ tensor products $(p, q) \otimes (q, p)$ and $(p, q) \otimes (p, q)$ were considered from a geometric perspective, and a method for decomposing them into direct sums was given. This geometric approach makes it particularly easy to rapidly obtain multiplicities of irreps. In particular, irreps of maximum multiplicity were discussed.

The work in this article can be applied to many different physical problems and there are several directions in which further work should be considered.

One possible application is the missing label problem, where geometrical constructions such as those presented here could be used to find additional arguments that justify the choice of labeling operators. The information regarding multiplicities given in this article provide new insights into the problem of deciding whether the appearing degeneracies can be solved by means of the given Casimir operators, or whether the full number of missing label operators are needed.

Also, tensor products for $\mathfrak{su}(3)$ are related to affine $\hat{\mathfrak{su}}(3)_k$ fusions [15] and when k is large the fusion coefficients coincide with a triple tensor product decomposition [16]. It is therefore natural to ask whether a similar method can be developed for triple tensor products.

Other potential further developments include constructing a similar procedure for the analysis of the symmetric and antisymmetric tensor products of multiplets, as well as the decomposition of tensor products into symmetric and antisymmetric components. These have applications in the problem of conflicting symmetries in atomic spectroscopy [17], where they are studied in the context of vanishing matrix elements.

The procedure could perhaps also be extended to find decompositions in the asymptotic limit [18], which are important in the study of the non-compact rigid rotor algebra [19].

Acknowledgments

The author would like to thank Joe Repka (University of Toronto) for his continued help and encouragement, as well as Hubert de Guise (Lakehead University) and Rutwig Campoamor-Stursberg (Universidad Complutense de Madrid) for their valuable suggestions, especially with regards to the physical background and applications.

This work was supported by the University of Toronto, Canada.

Appendix A. Table

Table A1. Some small examples of decompositions of tensor products of the form $(p,q)\otimes(p,q)$ and/or $(p,q)\otimes(q,p)$. Note that the tensor diagram of e.g. $(1,2)\otimes(1,2)$ is the diagram of $(2,1)\otimes(2,1)$ reflected in the centre diagonal.

Decomposition in highest weight notation	Decomposition in dimension notation	Tensor diagram
$(1,0)\otimes(1,0)=(2,0)\oplus(0,1)$	$3\otimes 3=6\oplus\underline{3}$	
$(1,0)\otimes(0,1)=(1,1)\oplus(0,0)$	$3\otimes\underline{3}=8\oplus\ 1$	
$(1,1)\otimes(1,1)=(2,2)\oplus$ $(3,0)\oplus(0,3)\oplus 2(1,1)\oplus(0,0)$	$8\otimes 8=27\oplus 10\oplus$ $\underline{10}\oplus 2\cdot 8\oplus 1$	
$(2,0)\otimes(2,0)=$ $(4,0)\oplus(2,1)\oplus(0,2)$	$6\otimes 6=15\oplus 15\oplus\underline{6}$	
$(2,0)\otimes(0,2)=$ $(2,2)\oplus(1,1)\oplus(0,0)$	$6\otimes\underline{6}=27\oplus 8\oplus 1$	
$(2,1)\otimes(2,1)=(4,2)\oplus$ $(2,3)\oplus 2(3,1)\oplus(5,0)\oplus$ $2(1,2)\oplus(0,4)\oplus(2,0)\oplus(0,1)$	$15\otimes 15=60\oplus$ $\underline{42}\oplus 2\cdot 24\oplus 21\oplus$ $2\cdot\underline{15}\oplus\underline{15}\oplus 6\oplus\underline{3}$	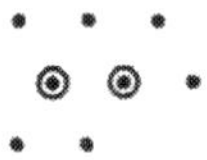
$(2,1)\otimes(1,2)=(3,3)\oplus$ $(4,1)\oplus(1,4)\oplus 2(2,2)\oplus$ $(3,0)\oplus(0,3)\oplus 2(1,1)\oplus(0,0)$	$15\otimes\underline{15}=64\oplus$ $35\oplus\underline{35}\oplus 2\cdot 27\oplus$ $10\oplus\underline{10}\oplus 2\cdot 8\oplus 1$	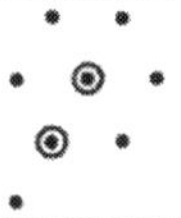
$(2,2)\otimes(2,2)=(4,4)\oplus(5,2)\oplus$ $(2,5)\oplus 2(3,3)\oplus 2(4,1)\oplus$ $2(1,4)\oplus(6,0)\oplus(0,6)\oplus 3(2,2)\oplus$ $(3,0)\oplus(0,3)\oplus 2(1,1)\oplus(0,0)$	$27\otimes 27=125\oplus 81\oplus$ $\underline{81}\oplus 2\cdot 64\oplus 2\cdot 35\oplus$ $2\cdot\underline{35}\oplus 28\oplus\underline{28}\oplus 3\cdot 27\oplus$ $10\oplus\underline{10}\oplus 2\cdot 8\oplus 1$	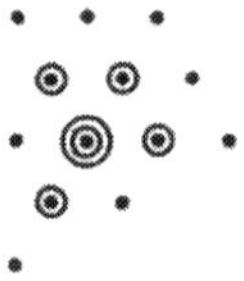
$(3,0)\otimes(3,0)=(6,0)\oplus$ $(4,1)\oplus(2,2)\oplus(0,3)$	$10\otimes 10=28\oplus$ $35\oplus 27\oplus\underline{10}$	
$(3,0)\otimes(0,3)=(3,3)\oplus$ $(2,2)\oplus(1,1)\oplus(0,0)$	$10\otimes\underline{10}=64\oplus$ $27\oplus 8\oplus 1$	
$(3,1)\otimes(3,1)=(6,2)\oplus$ $(4,3)\oplus(2,4)\oplus 2(5,1)\oplus$ $2(3,2)\oplus(7,0)\oplus 2(1,3)\oplus$ $(0,5)\oplus(4,0)\oplus(2,1)\oplus(0,2)$	$24\otimes 24=105\oplus 90\oplus$ $\underline{60}\oplus 2\cdot 48\oplus 2\cdot 42\oplus 36\oplus$ $2\cdot\underline{24}\oplus\underline{21}\oplus 15\oplus 15\oplus\underline{6}$	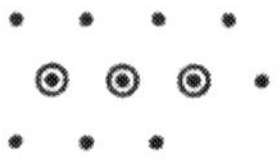
$(3,1)\otimes(1,3)=(4,4)\oplus$ $(5,2)\oplus(2,5)\oplus 2(3,3)\oplus$ $(4,1)\oplus(1,4)\oplus 2(2,2)\oplus$ $(3,0)\oplus(0,3)\oplus 2(1,1)\oplus(0,0)$	$24\otimes 24=125\oplus 81\oplus$ $\underline{81}\oplus 2\cdot 64\oplus 35\oplus\underline{35}\oplus$ $2\cdot 27\oplus 10\oplus\underline{10}\oplus 2\cdot 8\oplus 1$	

References

[1] Gell-Mann M and Ne'eman Y 1964 *The Eightfold Way* (New York: Benjamin)
Ne'eman Y 1961 *Nucl. Phys.* **26** 222-9
[2] Elliott J P 1958 *Proc. R. Soc. Lond.* A **245** 128-45
[3] Reck M, Zeilinger A, Bernstein H J and Bertani P 1994 *Phys. Rev. Lett.* **73** 58-61
Sanders B C, de Guise H, Rowe D J and Mann A 1999 *J. Phys.* A **32** 7791-801
[4] Wigner E P 1959 *Group Theory* (New York: Academic)
[5] Baird G E and Biedenharn L C 1963 *J. Math. Phys.* **4** 1449-66
Baird G E and Biedenharn L C 1964 *J. Math. Phys.* **5** 1723-30
Baird G E and Biedenharn L C 1964 *J. Math. Phys.* **5** 1730-47
Baird G E and Biedenharn L C 1965 *J. Math. Phys.* **6** 1847-54
Biedenharn L C 1963 *J. Math. Phys.* **4** 436-45
[6] King R C 1988 *Invariant Theory and Tableaux* ed D Stanton (New York: Springer-Verlag) pp 226-61
King R C 1990 *Topics in Algebra: Part 2 Commutative Rings and Algebraic Groups* ed S Balcerzyk *et al* (Warsaw: PWN-Polish Scientific Publishers) pp 327-44
[7] Cornwell J F 1984 *Group Theory in Physics Volume II* (London: Academic Press)
Fuchs J and Schweigert C 1997 *Symmetries, Lie Algebras and Representations* (Cambridge: Cambridge University Press)
Itzykson C and Nauenberg M 1966 *Rev. Mod. Phys.* **38** 95-120
Rowe D J and Repka J 1997 *J. Math. Phys.* **38** 4363-88
Speiser D 1962 *Group Theoretical Concepts and Methods in Elementary Particle Physics, Lectures of the Istanbul Summer School of Theoretical Physics* ed F Gürsey (New York: Gordon and Breach) pp 201-76
de Swart J J 1963 *Rev. Mod. Phys.* **35** 916-39
[8] Moody R V, Patera J and Rand D 1996 "SimpLie^{TM}" software, (Montréal: Centre de Recherches Mathématiques)
[9] Bhatia R, 2001 *Am. Math. Month.* **108** 289-318
[10] Fulton W, 2000 *Bull. Am. Math. Soc.* **37** 209-49
[11] Preziosi B, Simoni A and Vitale B 1964 *Il Nuovo Cimento* **34** 1101-13
[12] Wesslén M S M 2008 *J. Math. Phys.* **49** 073506
[13] Wesslén M S M 2009 *A Diagrammatic Description of Tensor Product Decompositions for SU(3)* PhD Thesis (Toronto: University of Toronto)
[14] Klimov A B and de Guise H 2008 *J. Phys.* A **41** 025303
[15] Felder G and Varchenko A 1996 *Lett. Math. Phys.* **37** 167-72
Rasmussen J and Walton M A 2002 *J. Phys.* A **35** 6939-52
[16] Bégin L, Mathieu P and Walton M A 1992 *Mod. Phys. Lett.* A **7** 3255-65
[17] McLellan A G 1960 *Proc. Phys. Soc.* **76** 419-22
Oss S 1996 *Adv. Chem. Phys.* **93** 455-649
[18] de Guise H and Rowe D J 1995 *J. Math. Phys.* **36** 6991-7008
de Guise H and Rowe D J 1998 *J. Math. Phys.* **39** 1087-106
[19] Rowe D J, LeBlanc R and Repka J 1989 *J. Phys.* A **22** L309-16

AUTHOR INDEX

Ballesteros, A 88
Barbero G, J Fernando 114
Blasco, A 88
Boya, Luis J 122
Calzada, J A 129
Campoamor-Stursberg, R 145
Cariñena, José F 155
De Azcárraga, J A 1
De Traubenberg, M Rausch 40
Del Olmo, M A 129
Figueroa-O'Farrill, José Miguel 25
Guha, Partha 155
Herranz, F J 88
Izquierdo, J M 1
Miranda, Eva 178
Musso, F 88
Negro, J 129
Ovando, Gabriela P 194
Prieto, Carlos Tejero 212
Ragnisco, O 88
Rañada, Manuel F 155
Rodríguez, M A 204
Tempesta, P 204
Vergel, Daniel Gómez 170
Villaseñor, Eduardo J S 170
Wesslén, Maria S M 226
Winternitz, P 204

Institute of Physics
Dirac House, Temple Back
Bristol BS1 6BE UK